Elementary Mathematical Models

Order Aplenty and a Glimpse of Chaos

Inspired by the work of M. C. Escher, the cover design exhibits order aplenty, and provides a glimpse of chaos. Intertwined with the fish are images of a bifurcation diagram that arises in the discussion of chaos in Chapter 14. And just as a study of fish populations is transformed into an introduction to chaos, so the fish in the cover design are transformed into bifurcation diagrams as they swim across the page.

© 1997 by
The Mathematical Association of America (Incorporated)

Library of Congress Catalog Card Number 97-74331
ISBN 0-88385-707-3

Printed in the United States of America

Current Printing (last digit):
10 9 8 7 6 5 4 3

Elementary Mathematical Models

Order Aplenty and a Glimpse of Chaos

Dan Kalman
American University

Published and distributed by
The Mathematical Association of America

CLASSROOM RESOURCE MATERIALS

Classroom Resource Materials is intended to provide supplementary classroom material for students—laboratory exercises, projects, historical information, textbooks with unusual approaches for presenting mathematical ideas, career information, etc.

MAA Service Center
P. O. Box 91112
Washington, DC 20090-1112
1-800-331-1622   fax: 1-301-206-9789

Preface

College professors could learn something from comedian Don Novello. In his Father Guido Sarducci personna, Novello described a revolutionary approach to higher education: the five minute university. This is an accelerated course of study in which the students take five minutes to learn what the average college graduate remembers five years after graduation. For example, the five minute university course in economics is *Supply and Demand*. That is not the title of the course, it *is* the course. Similarly, the five minute university course in English literature is *To Be, or Not To Be*.

There is insight in Novello's satire, and a pointed question for educators. What is it that we really want our students to take away from their college courses? After the cramming and exams are done, after the students leave school and go about the business of living, what is it that we want them to retain?

They Don't Teach College Algebra at the Five Minute University. As a case in point, consider a typical course in college algebra. The goal of this course is to train students in techniques of algebraic manipulation. There is a heavy dose of algebraic operations that students will seldom see again, if ever, even in more advanced mathematics courses: the simplification of complicated rational functions, compound radical expressions with several different radices, and logarithmic equations with variable bases, to name a few. Students are expected to master the rules that apply to each kind of symbolic form, the assumed precedence for the order of operations, and the proper use of parentheses, as well as a variety of standard forms, and algorithms for obtaining these forms. How much of all this does the average student remember five years after graduation? My guess is zero. That is why the five minute university doesn't have a college algebra course.

Prompted by this insight, I set out to create a new kind of course. I ended up writing the book you now hold. Its development was influenced by both the immediate needs of students to hone mathematical skills for other courses, as well as longer term goals that might warrant a mention at the five minute university.

In this Preface I will describe the students for whom the book is intended, and what I hope they will get out of the course. There will also be a discussion of the philosophy followed in developing the course, followed by teaching suggestions. These will pertain both to methods of presentation, as well as to the pace and coverage appropriate for a 14 week semester course. To simplify the exposition, *EMM* will be used as an abbreviation for *Elementary Mathematical Models*.

Students Served.   The EMM course seeks to serve the same students as traditional college algebra and liberal arts mathematics courses. These are students who have studied at least a year of algebra in high school, but who are not headed for calculus. They may need to take a mathematics course for a general education requirement. They may also need familiarity with the main ideas of college algebra for quantitative general education courses in areas outside of mathematics, such as science, economics, and business. For most of these students this will be the only mathematics course completed in college.

Development Philosophy.   The development of EMM reflects two important concerns. First, I wanted the course to possess an internal integrity. My aim was to make the material intrinsically interesting and worthwhile, to present a coherent story throughout the semester, and to convey something of the utility, power, and method of mathematics. Among other things, this demands genuine and understandable applications.

In formulating these goals, I was strongly influenced by the ground-breaking statistics text by Freedman, Pisani, and Purves[1]. The authors' account[2] of the development of this book is a model of clarity and insight. It describes a succession of attempts to refine an introductory statistics course. Ultimately, the authors were forced to focus on a single fundamental question: What are the main ideas the field of statistics has to offer the intelligent outsider? Their answer to this question became the foundation for a new course in statistics. In the same way, I have tried to capture some of the main ideas that arise in the applications of mathematics, and to present these ideas in an interesting and intelligible context. I am pleased to acknowledge the influence of Freedman, Pisani, and Purves on my thinking.

Course development was also guided by a second fundamental concern, to emphasize the topics that students are most likely to meet in mathematical applications in other disciplines. I am thinking here of the most elementary applications, formulated in terms of arithmetic and simple algebraic operations: linear, quadratic, polynomial, and rational functions; square roots; exponential and logarithmic functions. These functions are the building blocks for the simple models that appear in first courses in the physical, life, and social sciences. Many students in my course have learned about some or all of these functions in prior classes. The challenge is to make the material fresh and interesting for these students, and at the same time, accessible as a first exposure. I tried to do this by creeping up on each mathematical topic in the context of an application. The applications are all analyzed using a common methodology, involving difference equations. This point will be discussed in greater detail presently. It suffices here to indicate that difference equation methods are applicable to a number of problems that (I hope) have obvious significance and relevance. The mathematical topics I want to cover come up in a natural way as difference equations are used to study the applications. The algebraic emphasis is

---

[1] David Freedman, Robert Pisani, and Roger Purves. *Statistics.* W. W. Norton, New York, 1978.
[2] See the first 6 pages of the introduction to the instructors manual for the text.

restricted to what is really required for working with these simple models. More arcane aspects of algebra, like those described earlier, have been banished from my course.

That is not to say that I have eliminated all algebra. Far from it. Algebra is an important and powerful tool, and students should appreciate and remember that fact. This can even be a part of the five minute university curriculum. *Algebra is an important and powerful tool.* But I want the students to make that observation themselves, from seeing algebra in action. The algebra that does appear in the book is presented in a way that makes it clear *why* algebra is needed, and what it contributes to formulating and analyzing models.

I believe EMM is a hybrid of liberal arts mathematics courses and traditional college algebra courses. The goals of instruction, particularly as regards the five minute university ideas, are like those of a liberal arts mathematics course. The content emphases are more like those of a college algebra course. I have set out to infuse quadratic equations and logarithms with the same kind of enthusiasm and vitality that authors like Jacobs[3] have brought to their works. Mine is a terminal mathematics course that doesn't foredoom the student to terminating the study of mathematics. He or she will be well prepared to go on to precalculus and calculus. It is an entry level mathematics course that students don't have to slog through motivated only by vague promises that the material will be needed in future studies. They can see as they take the course what this material is good for and how it can be used. Hersh and Davis[4] have described mathematics as the part of the humanities that has the qualities of science. In an analogous way, my course is the entry level college algebra course that has the qualities of liberal arts mathematics.

Based on the concerns discussed above, my development of EMM was guided by the following principles:

- Introduce each new mathematical operation in the context of a believable problem;
- Weave all of the topics into an integrated whole;
- Provide as a theme a methodology that can be used in a large number of problem contexts;
- Emphasize conceptual understanding of how the mathematics contributes to solving problems over technical mastery of each mathematical topic for its own sake.

The second and third principles are reflected in the choice of topics for the course. A brief discussion of the organization of topics will be presented next.

Course Outline. In overview, EMM concerns discrete and continuous models of growth. Each application starts with a discrete model built around a sequence $\{a_n\}$. There is also a simple hypothesis describing the way successive terms depend on preceding terms. One example of such a hypothesis is that each new term a_{n+1} can be obtained by adding a fixed constant to the preceding term, a_n. The algebraic expression of this hypothesis,

$$a_{n+1} = a_n + \text{a constant}$$

is a *difference equation*. Throughout the course, a succession of simple hypotheses are considered: the terms of the sequence increase by a constant amount; the terms increase

---

[3] Harold R. Jacobs. *Mathematics, A Human Endeavor, Second Edition.* W. H. Freeman, New York, 1982.
[4] Reuben Hersh and Phillip J. Davis. *The Mathematical Experience.* Birkhauser, Boston, MA, 1980.

by varying amounts which themselves increase by a constant amount; the terms increase by a constant percentage; and so on. Each of these hypotheses finds expression as a particular kind of difference equation. Each new class of difference equation is studied using a common methodology. In outline, the unifying methodology is

- Formulate a family of difference equations;
- Develop *solutions* to the difference equations;
- Study the behavior of the solutions.

While the models are initially formulated in terms of discrete variables, the solutions of the difference equations can generally be interpreted in the context of continuous variables.

The course starts with arithmetic growth and builds up through successively more complicated models: quadratic growth, geometric growth, mixtures of arithmetic and geometric growth, and finally logistic growth. EMM climaxes with a study of the chaotic behavior that can occur in logistic models. Interspersed among the discussions of the various growth models are units on the families of functions that appear as solutions to difference equations. Thus, the study of arithmetic growth leads into a unit on linear functions; quadratic growth models provide a setting for studying the properties of quadratic functions, and so on.

Unifying Themes. Throughout the course, several themes are touched on repeatedly. Obviously, the formulation of a model in terms of a difference equation, and the development of a solution to that difference equation has a constant presence. The use of systematic investigation and patterns to identify the solutions of the difference equations is another common theme. Numerical and graphical methods are used systematically in all the topics. The idea of expressing the solution to the difference equation as a function, and of evaluating or inverting that function to answer questions about the model is also stressed repeatedly. The progression of topics demonstrates the incremental nature of modeling, as earlier models are refined or modified to obtain later, more realistic models. Finally, the most fundamental theme of EMM is the applicability of the mathematical topics. Students recognize this from the start, because the entire course evolves out of the investigation of real problems. By the end of the course, they are studying sophisticated models with unexpected and nontrivial behaviors. The power of algebra to answer important questions about these models is always on display.

Teaching EMM

Scope and Sequence. There is more in the book than I have been able to cover in a semester. Here are a few suggestions for a subset that will make up a reasonable course. The core material in the course consists of chapters 1–6, 9, 10, 12. This will cover most of the same functions as the standard college algebra course, though in less algebraic detail. When I teach the course, I cover this core material, plus chapters 7, 13, and 14. That is everything in the book except chapters 8 and 11. If that pace turns out to be too fast, you can eliminate the sections of chapter 13 that involve harvesting, and part or all of chapter 7. The result will still be a coherent course with a definite story to tell, and a climactic finish.

You may prefer to cover some things that I leave out. Chapter 8 uses what has already been developed about quadratic functions to present the ideas behind least squares linear regression. I put it in the book as a significant application of quadratic functions, and one which connects to an important tool used throughout the book: fitting a line to data. Chapter 11 gives greater depth about logarithmic functions, particularly the ideas behind logarithmic scales and log-linear regression. If you want to cover this material, you might elect to skip chaos (chapter 14) and possibly logistic growth (chapter 13), as well.

I go quickly through the first two chapters. The first chapter presents an overview of the course that will not make a great deal of sense at the start. It is provided to give the students a feel for what is coming. It might be productive to have the students reread it after finishing chapter 6. Chapter 2 introduces difference equations in a general setting. Some students have difficulty with the subscript notation and the idea of difference versus functional equations. My approach is to go over it quickly, indicating that the students will have plenty of opportunity for reviewing the same ideas in later chapters, and then provide time in chapters 3 and 5 to again work on the troublesome ideas. An alternative is to slow down in chapter 2, and then go faster through chapters 3 and 5.

Even if you don't do chapter 14, take a look at the first section. It provides a retrospective review of most of the difference equation models used in the course. You may wish to assign some or all of this material as review at the end of the course.

A word of warning. You will not be able to present in class everything in each chapter that you cover. I wrote this book so that I could say on paper just about everything I wanted to present to the students. I let them know early that they needed to read the material before class, and that I would not lecture on every topic. Instead, I used a combination of students working in groups, lab activities, and presenting solutions to problems to augment what the students learned by studying outside class. If you are not comfortable with that approach, you may have to reduce the number of chapters you cover.

The schedule below shows how I organize my semester course.

Week	Topics
1	Chapters 1 and 2
2	Chapter 3
3	Chapter 4
4	Chapter 5
5	Exam, review
6	Chapter 6
7	Chapter 7
8	Chapter 9
9	Chapter 10
10	Exam, review
11	Chapter 12
12	Chapter 13
13	Chapter 14
14	Review

Teaching Methods. The methods used to teach EMM are independent of the course content. I think one could successfully present the material in a very traditional format. My own preference is to use a combination of teaching methods. My students listen to some lectures, and complete fairly traditional homework assignments. But they also work in groups; they do reading and writing assignments; they develop models based on data they find in magazines or newspapers; they investigate aspects of the models in a computer laboratory. I have tried to develop the material for EMM in a way that lends itself to a variety of approaches.

Nevertheless, I feel that the material works best when there is a strong emphasis on reading and writing. There are reading comprehension questions throughout the text, and I work very hard at the beginning of the course to get the students to actually *read* their math books. This is a new experience for many students, and it may require some coaching. As teachers, we are well aware that reading a math book with comprehension requires different strategies and skills than for other kinds of literature. Students generally do not know that this is the case.

Technology. EMM lends itself to numerical exploration and experimentation. Graphing calculators which support difference equations (for example the Texas Instruments TI-82) can be used to observe numerical and graphical behavior of the growth models studied in the course. This is particularly appropriate in the later chapters of the course when the effects of changing parameters is studied.

Of course, with appropriate software, a computer can also be used. In fact, there is a computer component that can be used with EMM, developed using the Mathwright software. Mathwright is an authoring tool that permits math teachers to create highly structured mathematical environments with which students interact using a point and click methodology. The main developer of Mathwright, James White, developed a series of laboratories for EMM at the California State University at Monterey Bay. These modules are available on the world wide web as part of an internet library sponsored by the NSF. The internet library provides free of charge the software necessary to use the modules. The library can be accessed at http://ike.engr.washington.edu/mathwright, along with instructions for downloading the software and modules. The modules for EMM are included under the heading of *Collections*, and referred to as finite mathematics laboratories.

Classroom Experience

The EMM course has been taught for several semesters at American University. Generally speaking the student reaction has been favorable. The students seem to appreciate the fact that the applications are so tightly integrated into the development of the ideas. They also have generally found the course to be demanding. But on the whole, they find the material within reach.

In my courses, the students have diverse backgrounds. Those whose preparation is weak find the text quite challenging. Better prepared students sometimes complain that the prose is repetitive or long-winded. In most cases, I have erred on the side of too little reliance on the power of algebra to summarize and justify conclusions, and students with a superior understanding of symbolic methods might sometimes perceive the exposition as inefficient. Overall, though, my students have generally liked the text.

For almost all of my students, EMM (or some alternative mathematics course) is a requirement. I see my share of unmotivated students, as well as students who really seem to make an honest effort but find the material too difficult. I cannot honestly claim to have objective proof that my students are happier, more motivated, more interested, or more successful than students in a traditional college algebra course. But I do feel that EMM provides a better opportunity for contributing to the general education of these students. It remains to be seen what, if anything, the five minute university version of EMM will contain.

Acknowledgments

In acknowledging the help and support which made this volume possible, I should start with my sincere thanks to Linda, my partner for over 25 years, and to our children Jennifer and Chris. I am grateful to them for their love and affection, for encouraging me in my work, and for their patience when that work consumed too much of my time.

This book was developed in conjunction with a course at American University. The original design of that course was supported by an American University Curriculum Development Grant. Professors Steve Casey and Virginia Stallings, colleagues at American, taught from preliminary versions of the text, and provided valuable comments and suggestions. Angela Hare was my closest collaborator in the development of the course and text, and she contributed in a variety of significant roles: teaching from the text; carefully reading and reacting to each chapter as it was written; developing Mathwright modules for use in the computer laboratory; joining me in presenting material about the course in workshops for colleagues; and just sharing her ideas about teaching and evaluation for the new course. James White was supportive of the project from the beginning. In addition to the thanks I owe him for developing the Mathwright software, he has my gratitude for general collegial support, and for conducting the first classroom test of the course away from American University.

Several people at the MAA are also deserving of thanks. Don Albers, Director of Publications, was enthusiastic about the project from the very beginning, and encouraged me to persist in the effort. I think he knew before I did that I was writing a book. Andy Sterrett and the Editorial Board for Classroom Resource Materials encouraged me throughout the review process, and suggested several valuable improvements. June and Richard Kraus did a thorough and painstaking job of copy editing. The production of the book was overseen by Elaine Pedreira and Beverly Ruedi.

Finally, I would like to thank the students of American University who took part in my classes as this book was developed. They made significant contributions to improving earlier drafts, were patient with my missteps, and generous with their support and praise.

To all of the members of the MAA staff at the headquarters office in Washington, DC. Your efforts and dedication make a tremendous contribution to the MAA and the larger mathematics community.

Contents

Preface   v

1   Overview   **1**

An Example 1

Three Methods: Numerical, Graphical, Theoretical 2

Mathematical Models 5

Chaos 7

Summary 8

Exercises 8

Solutions to Selected Exercises 9

2   Sequences and Difference Equations   **11**

Sequences 11

Difference Equations 14

The Numerical Approach 18

The Graphical Approach 19

The Theoretical Approach 21

Back to Reality 22

Sample Difference Equations 24

Summary 28

Exercises 28

Solutions to Selected Exercises 31

3   Arithmetic Growth   **37**

The Arithmetic Growth Difference Equation 37

Examples 39

Numerical, Graphical, and Theoretical Properties 40

Proportional Reasoning and Continuous Models 44

Beyond Arithmetic Growth: Oil Reserves 46

Summary   47
Exercises   47
Solutions to Selected Exercises   50

4   Linear Graphs, Functions, and Equations   **55**
Linear Functions and Equations   55
Continuous Models   57
Algebra and Solving Linear Equations   59
Graphs of Linear Equations   61
Fitting a Line to Data   68
Formulating Linear Models   69
Proportional Reasoning   70
Summary   72
Exercises   72
Solutions to Selected Exercises   77

5   Quadratic Growth Models   **79**
A New School Population Example   79
Quadratic Growth Difference Equations   82
Applications of Quadratic Growth   91
Properties of Quadratic Growth Models 101
Whole Number Sums and Σ Notation 104
Summary 107
Exercises 108
Solutions to Selected Exercises 111

6   Quadratic Graphs, Functions, and Equations   **117**
Quadratic Functions and Equations 119
Algebraic Considerations 120
Graphs 122
Solving Equations 126
An Example 133
Summary 139
Exercises 139
Solutions to Selected Exercises 145

7   Polynomial and Rational Functions   **149**
Algebraic Aspects of Polynomials 150
Graphing Polynomials 151
Solving Polynomial Equations 153
Applications of Polynomials 156
Graphs of Rational Functions 157
Summary 161
Exercises 162
Solutions to Selected Exercises 163

8 Fitting a Line to Data **165**

Defining "Best" 165

An Error Function 167

Finding the Best Slope 170

Three Dimensional Parabolas 172

Summary 176

Exercises 176

Solutions to Selected Exercises 178

9 Geometric Growth **181**

The Geometric Growth Assumption 182

The Geometric Growth Difference Equation 187

Functional Equation 191

Exponential Functions 193

Additional Applications 196

Summary 199

Exercises 200

Solutions to Selected Exercises 206

10 Exponential Functions **213**

Graphs 214

Algebraic Properties 217

Solving Equations 220

The number e 226

Summary 228

Exercises 229

Solutions to Selected Exercises 231

11 More On Logarithms **233**

Rules for Logarithms 234

Logarithmic Scales 236

The pH Scale 241

Transforming Data 243

Summary 248

Exercises 248

Solutions to Selected Exercises 252

12 Geometric Sums and Mixed Models **255**

Geometric Sums 256

A Medicine Dosage Model 259

A Mixed Model for Repeated Loan Payments 267

Geometric Sum Models 271

A Mixed Model for Oil Consumption and Reserves 274

Summary 279

Exercises 279
Solutions to Selected Exercises 282

13 Logistic Growth   287
A Linear Growth Factor Model 288
General Features of Logistic Models 293
A Logistic Population Model with Harvesting 303
Summary 311
Exercises 312
Solutions to Selected Exercises 315

14 Chaos in Logistic Models   319
An Overview of Chaos 319
Chaos in the Logistic Model 325
Order in Chaos: A New Kind of Graph 332
Summary 338
Exercises 339
Solutions to Selected Exercises 339

Index   341

1

Overview

Real problems rarely work out as neatly as math problems. In elementary mathematics courses, the problems are usually stated in terms of mathematical language, and there is usually a short mathematical procedure that leads to a clear answer. But real problems are not at all like that. They are about complicated phenomena or situations: the AIDS epidemic, global warming, political representation, pollution. There are no short procedures to solve problems of this type, and the answers are not clear-cut. Nevertheless, mathematics is an indispensable tool in studying real problems. This book is an introduction to some of the methods that are used in applying mathematics to real problems.

This chapter will provide an overview of the course as a whole. It will introduce several ideas and terms that may not be completely clear at the start. The point is to give you an idea of what is to come later in the course. You may find it useful to review this chapter after studying several of those that follow. At that point, the general ideas presented here will make more sense, and may provide a useful perspective on the other chapters.

An Example

Before going any further, let us consider an example. The table below shows the levels of carbon dioxide concentration in the atmosphere for several years, as compiled at the Mauna Loa Observatory in Hawaii.[1] The level of carbon dioxide concentration is one of the key variables in the study of global warming. The table gives the concentration in *parts per million*. That is, the entry of 319.9 for 1965 means that the amount of carbon dioxide in a typical sample from the atmosphere would be equal to 319.9 one millionths of the sample. Put another way, if the air sample was divided into one million equal

---

[1] *Earth Algebra,* preliminary edition, by Christopher Schaufele and Nancy Zumhoff. Harper Collins, 1993, page 34.

Year	Carbon Dioxide Concentration
1965	319.9
1970	325.3
1980	338.5
1988	351.3

TABLE 1.1
Carbon Dioxide Concentration in Parts per Million

parts, and if all of the carbon dioxide was crowded into as few of these parts as possible, it would fill up 319 of the parts, and nine tenths of one additional part.

This is just one of countless tables of data that are compiled, analyzed and reported. They appear in government reports, magazines, newspapers, textbooks, and research publications. They are used to understand questions like these:

- How fast is the carbon dioxide level growing?

- When did it first reach 300 parts per million?

- When is it expected to reach 400 parts per million?

- What will it be 10 years from now?

Three Methods: Numerical, Graphical, Theoretical

In this course, you will study mathematical methods that are especially useful in this kind of analysis. Three important kinds of methods will be stressed: numerical, graphical, and theoretical. Numerical methods involve the actual data that appear in the table. A simple example of this kind of method might be to compute the differences between successive entries. From 1965 to 1970 the carbon dioxide concentration increased by 5.4 parts per million. In the next ten years the increase was 13.2 parts per million. Working directly with the numbers and trying to understand relationships in this way is what we mean by numerical methods. In general, a numerical method uses direct operation on numbers as its main tool.

Graphical methods use pictorial or spatial representations to help us understand and communicate relationships. It is common to use bar graphs or line graphs to depict data of the type shown in the table. These are both shown in Fig. 1.1. These figures allow us to use our understanding of spatial relationships to understand the data. For example, we can compare the sizes of bars in the bar graph, or see how steeply the lines slope upward in the line graph. Learning some of the ways that data can be graphically portrayed, and how to interpret those portrayals visually is what is referred to as graphical methods.

The third category of methods is what we call theoretical methods, and what other authors sometimes call analytic or symbolic methods. Just as numerical methods make use of the mathematics of computation, and graphical methods make use of the mathematics of graphs, so theoretical methods employ mathematical tools for expressing and manipulating

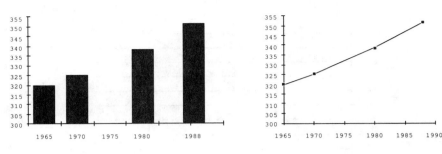

FIGURE 1.1
Bar and Line Graphs for Carbon Dioxide Data

relationships or patterns. These tools typically involve variables and algebra, as well as an understanding of how different objects in the mathematical universe behave. In the example of the carbon dioxide data, important variables include the carbon dioxide level and the year, and these can be related by an equation

$$\text{Carbon Dioxide Level} = 1.366 \cdot \text{Year} - 2365 \tag{1}$$

Here, as in many applications, the equation does not exactly agree with the data. If the year is 1970, the equation says

$$\text{Carbon Dioxide Level} = 1.366 \cdot 1970 - 2365$$

$$= 326.02$$

This is close to, but not exactly equal to, the carbon dioxide level for of 325.3 given for 1970 in Table 1.1. Take a minute now to repeat this computation for two other years shown in the table. Put the year in the right side of the equation, use a calculator to compute the carbon dioxide level, and then compare the result to the value given in the table. You will see that in each case the equation gives a result that is close to, but not exactly equal to the value in the table.

Finding an equation like this, and using it to answer questions about the carbon dioxide level, illustrate the theoretical method. Where did the equation come from? That is part of what you will learn in the course: when an equation can be expected to closely approximate data, how to find the equation, and how to use the equation to answer questions.

One of the questions raised above asks when the carbon dioxide level will reach 400 parts per million. This provides a good opportunity to compare graphical, numerical, and theoretical methods. First, to get a rough idea of the situation, use a graphical method. Enlarge the line graph shown in Fig. 1.1 so that the vertical axis goes up to 400 parts per million, as illustrated in Fig. 1.2. In the larger graph, the data points appear to fall nearly on a straight line, and that line has been extended far enough to cross the 400 mark on the vertical axis. Based on the graph, it appears that the carbon dioxide level will reach 400 in about 2023. With modern computer graphing tools, it is very easy to get accurate extended graphs in this way. The result may be only approximately correct, but

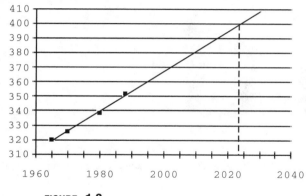

FIGURE 1.2
Extended Line Graph for Carbon Dioxide Data

it provides an overall picture of the original data and the carbon dioxide level considered in the question.

Next we will use a numerical method. This time, rather than using just the original data, we will use Eq. (1). From the graphical method, we expect the carbon dioxide level to reach 400 sometime around 2023. In the equation, with the year equal to 2023, we get

$$\text{Carbon Dioxide Level} = 1.366 \cdot 2023 - 2365$$
$$= 398.418$$

That is a little too low. What about in the year 2025?

$$\text{Carbon Dioxide Level} = 1.366 \cdot 2025 - 2365$$
$$= 401.15$$

That is a little too high. Continuing in this way, you might next try the year 2024, and find 399.784. This shows that a level of 400 will be reached sometime between 2024 and 2025. For even greater accuracy, try making the year 2024.5 (half way from 2024 to 2025) or 2024.75. This type of calculator trial and error is typical of numerical methods.

Finally, a theoretical method for this problem would be to use algebra. In the Eq. (1), put 400 in place of the carbon dioxide level, and leave the year as an unknown. The resulting equation can be solved to find the correct year. The steps look like this:

$$\text{Carbon Dioxide Level} = 1.366 \cdot \text{Year} - 2365$$
$$400 = 1.366 \cdot \text{Year} - 2365$$
$$400 + 2365 = 1.366 \cdot \text{Year}$$
$$2765 = 1.366 \cdot \text{Year}$$
$$2765/1.366 = \text{Year}$$
$$2024.16 = \text{Year}$$

Assuming that the equation for carbon dioxide level and year is correct, this pinpoints the time when a level of 400 is reached.

This comparison of the three methods leads naturally to three observations that will be met repeatedly during the course. First, the three methods complement each other and contribute to understanding and insight about our problem. Each is important, and each helps us think about the problem in a special way.

The second observation is that the theoretical method has greater power and generality than the other two methods. In the example, the process of finding the solution to the problem numerically quickly becomes tedious. The graphical method is quicker, but has limited accuracy. The theoretical method is quickly applied and yields an answer with perfect accuracy. More important, though, if we decide to ask when the carbon dioxide reaches a level different from 400, the results are quickly modified. If you review the steps used to solve the equation, you will see that the final answer was given by the computation $(400 + 2365)/1.366$. Now suppose we want to know when the carbon dioxide level will reach 380, instead of 400. The theoretical method we used earlier needs only a slight modification to replace the 400 with 380, and we will find that the answer is $(380 + 2365)/1.366$. This illustrates how the results of the theoretical method can be easily extended to related problems. In fact, by analyzing the steps used to solve a problem, it is possible to formulate procedures or rules that can be applied to a large number of closely related problems. In this way, theoretical methods give a kind of leverage for analyzing problems. We will see examples of this leverage over and over. It is really what makes mathematics an important field of study in its own right.

Finally, the third observation has to do with the entire approach to studying real problems. All three approaches to finding when the carbon dioxide level reaches 400 suffer from a common limitation. The graphical method depends on assuming that a straight line is an accurate portrayal of the data. The other two methods assume that the equation is a valid means of predicting the future. In fact, we know that these assumptions are not exactly correct. The data points do not fall right on a straight line, nor do they agree exactly with the equation. However, we simplify the problem by ignoring the differences between the real data and the equation. This simplification allows us to proceed and make predictions.

Mathematical Models

The simplification mentioned above is characteristic of the applications of mathematics. There are always simplification and idealization of the actual situation. For this reason, it has become a common expression to talk about the mathematics as providing a *model* for the real problem. In this context, the word *model* carries two meanings: ideal (as in *model citizen*) as well as simplified or reduced in scale (as in a *model train*). Often, the development and use of mathematical models is cyclical in nature. The very first model leads to predictions about a problem that can be tested by collecting data. The tests lead to refinements or improvements in the model, and then to new predictions. By successively refining the model, highly accurate mathematical descriptions can be found. These kinds of descriptions can be so accurate that they actually take the place of building and testing physical models. They are used in applications of mathematics to problems in all walks of life. A significant goal of this course is to instill in you an understanding and appreciation of this method of applying mathematics.

Although the problems discussed in the course will come from a wide variety of subjects, the mathematical tools will be limited in scope. In fact, we will spend most of our time looking at models that involve change over time, and change in a very restricted sense. Looking at regularly spaced data values (every hour, every year, every decade, etc.) we will try to find patterns that follow simple mathematical descriptions. In the carbon dioxide example, the use of a straight-line model is equivalent to assuming that the increase in carbon dioxide level is exactly the same every year. Of course, that assumption is not exactly correct. It is the kind of simplification that models always involve. We will study this and other simple patterns of change, and see what kinds of models emerge. One common aspect to all of these models is the way past data values are used to compute future values. Patterns of this type are described by mathematical equations called *difference equations*. The difference equation for the carbon dioxide study says, in essence, the carbon dioxide level next year will be this year's level plus 1.366. It can be expressed as an equation in the form

$$\text{Carbon Dioxide Level} = \text{Level From Previous Year} + 1.366$$

This simple statement completely defines a pattern of future growth for the carbon dioxide level. If the level is 319.9 in 1965, then in 1966 it must be $319.9 + 1.366 = 321.266$. Reasoning in the same way, we see that it must be $321.266 + 1.366 = 322.632$ in 1967, $322.632 + 1.366 = 323.998$ in 1968, and so on. Continuing in this way, it is possible to find the carbon dioxide level for any year we choose. It is tedious to continue the process for very long by hand. To find the carbon dioxide level for 1999, for example, it would require computing the level year by year from 1966 on, for 33 years! Fortunately, computers and calculators can be used to greatly streamline the process. But the repetitive nature of the process is characteristic of computations based on difference equations.

In addition to difference equations, we will also use *functional equations* in our study of models. The equation

$$\text{Carbon Dioxide Level} = 1.366 \cdot \text{Year} - 2365$$

is an example of a functional equation. To be more complete, the equation gives the carbon dioxide level *as a function of* year. This means that, as soon as a number is specified for the year, the equation can be used immediately to compute the carbon dioxide level. To find the carbon dioxide level for 1999, simply substitute 1999 for the year in the equation:

$$\text{Carbon Dioxide Level} = 1.366 \cdot 1999 - 2365$$

Notice that both the difference equation and the functional equation can be used to predict future levels of carbon dioxide, but in quite different ways. The difference equation is used *recursively,* meaning that the equation must be used repeatedly, year after year, to work up from 1966 to 1999. The functional equation is direct. It allows you to calculate the 1999 level immediately.

There is another distinction between these two kinds of equations. In very many applications, the way something changes over time reveals a pattern that is easily described by a difference equation. It is typically much harder to describe these patterns directly with functional equations. So, while the functional equations are more convenient to use, the difference equations are easier to formulate. In the applications we consider in this course, both difference equations and functional equations will play a part. As in the carbon dioxide example, it is often possible to find both a difference equation and a functional equation for the same set of data. In many different contexts we will follow a common method: start with a difference equation that is easy to formulate and then figure out a functional equation that is easy to use.

This methodology has been found to be remarkably effective at producing models that are useful in real problems. During the course we will see applications to population growth, economics, medicine, and the physical sciences. These problems also come up in studies of the spread of diseases, the survival of endangered species, and how the effects of pollution dissipate.

Chaos

In all of these application areas, the predictability of future data is possible because of the inherent order in the phenomena under study. Early success with this type of prediction in the physical sciences led to great optimism about the chances for effective prediction in a wide host of problem areas. Many mathematicians believed in an unspoken assumption: simple mathematical models would always lead to the ability to make predictions. It is an attractive idea. If two systems have mathematical models with very similar looking equations, then our ability to make predictions in the two systems should also be very similar.

Studies in recent years have shown that this assumption is not generally true. In some systems with very simple equations, future prediction is essentially impossible. To be more accurate, for certain cases, the mathematical equations themselves prevent meaningful future prediction. These equations can be very simple in nature and, yet, inherently unpredictable. The emerging body of theory regarding inherently unpredictable equations is called *chaos*. This is one topic on the forefront of applied mathematics today. It has been discussed in a popular book,[2] and appeared in a significant way in the plot of the novel *Jurassic Park,* (although the novel badly distorts the kinds of conclusions that are reached by studying chaos). Chaos is also closely connected with the intricate geometric patterns, called fractals, that are being produced by computers these days. Indeed, chaos is catching the imagination of a broad audience, and not just specialists in mathematics.

We shall catch a glimpse of chaos near the end of this course. As already mentioned, most of the course will focus on models using difference equations. We will begin with the simplest models, and then progress to more and more complicated ones. At first, the difference equations we study will lead to models in which meaningful predictions

---

[2] James Gleick, *Chaos: Making a New Science,* Viking, New York, 1987. This book is written for a general audience. It became a bestseller shortly after its publication.

of future evolution are possible. But, as we shall see near the end of the course, one cannot go very far in the development of ever more complicated difference equations before chaos arises. In particular, we will see how studying population growth leads in a very natural way to a family of difference equations for which chaos can occur. When it does, we lose the ability to make accurate predictions of a model's future. This will serve as an introduction to some of the main ideas connected with chaos.

The name *chaos* seems to imply the complete absence of order, but that is not an accurate description of the subject. Indeed, the kind of regularity that permits future prediction is just one kind of order. The study of chaos has revealed other kinds of order that emerge for unpredictable systems. These kinds of order can be useful in categorizing and understanding chaotic phenomena, even when precise predictions are not possible. In the simple case where we encounter chaos, we will also encounter this type of order.

Summary

This is a course about studying real problems through the use of mathematical models. All mathematical models involve simplifying assumptions. For the models we study, these assumptions will take the form of difference equations. A difference equation describes a pattern governing how something changes over time. More specifically, a difference equation relates a future data value to the immediately preceding data values. By applying the pattern over and over, we can extend predictions far into the future. Models can also be described by functional equations. These also describe patterns of change, but in a way that permits predictions without applying the pattern repeatedly. In many cases, we will formulate a difference equation based on an understanding of how a real phenomenon evolves, and then try to figure out appropriate functional equations. In all of these activities we will use three kinds of methods: numerical, graphical, and theoretical. We will also see some situations that defy meaningful prediction. These are called chaotic systems. Even in chaotic systems there are patterns and order. This order can be used for purposes other than prediction of future data.

Exercises

Reading Comprehension

1. Explain what is meant by graphical, numerical, and theoretical methods. Give an example of each type of method.

2. Explain what is meant by a difference equation. What is meant by a functional equation? How are the two the same? Different?

3. Explain the idea of a mathematical model.

4. What are the areas of application of mathematical models that were mentioned in the reading?

5. What is meant by chaos?

Problems in Context

1. Use the graphical method to estimate the year when the carbon dioxide level will reach 325 parts per million, according to Eq. (1). It is recommended that you use a graphing calculator or a computer graphing tool for this step, although the graph shown in Fig. 1.2 can also be used. Repeat this exercise for 350 parts per million and 360 parts per million.

2. Use the numerical method to improve your answers to the preceding problem.

3. Use the theoretical method to find the years when the carbon dioxide level will reach 325, 350, and 360 parts per million. Compare your results with the answers to the preceding problem. [Note: **an example of the use of theoretical method was given in the reading, with a carbon dioxide level of** 400. **If you do not understand the steps that were used in that example, discuss it with another student in the class, the instructor, or a tutor.**]

Group Activities

1. Look in a local newspaper or a publication such as USA Today, or Newsweek, for bar or line graphs. For each example, discuss whether the graphs give the appearance of a pattern that could be used for future predictions. Also discuss who might be interested in the future predictions, and how the predictions might be used. Write a one-page summary of the group's discussions.

2. The level of carbon dioxide is connected with concerns about global warming. Discuss the limitations of the mathematical model used for the carbon dioxide level data in the reading. If the model is used to predict when the carbon dioxide level reaches 400 parts per million, how accurate do you think that answer is? What additional kinds of evidence would you want to see before using predictions from the model to decide on national and international policies? Write a one-paragraph summary of the group's conclusions.

Solutions to Selected Exercises

Problems in Context

1. In Fig. 1.2, the diagonal line shows how carbon dioxide level and year are related. Find a point on this line with a carbon dioxide level of 325. (This can be done by making a mark on the vertical axis indicating a carbon dioxide level of 325, and placing a ruler on the figure so that it extends horizontally from your mark to the diagonal line.) For the point you have found on the diagonal line, find the year by looking straight down to the horizontal axis. The result should be about 1970, as shown in Fig. 1.3.

2. Put 1970 into Eq. (1). This gives the carbon dioxide level as $1.366 \cdot 1970 - 2365 = 326.02$. This is higher than the desired level of 325, so try a smaller number for the year, say 1969: $1.366 \cdot 1969 - 2365 = 324.654$. That is too low. Try 1969.5:

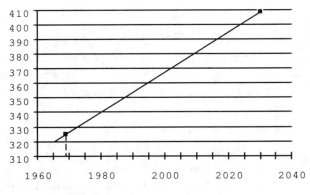

FIGURE 1.3
Graphical Method with Carbon Dioxide Level of 325

$1.366 \cdot 1969.5 - 2365 = 325.337$. That is an improvement over the answer found using the graphical method.

3. For 325, the equation becomes

$$325 = 1.366 \cdot \text{Year} - 2365$$

$$325 + 2365 = 1.366 \cdot \text{Year}$$

$$2690 = 1.366 \cdot \text{Year}$$

$$2690/1.366 = \text{Year}$$

$$1969.25 = \text{Year}$$

2

Sequences and Difference Equations

The introductory chapter explained the central role that difference equations will play throughout this course. In this chapter you will learn the basic properties of difference equations, and the mathematical terminology and notation that are used in the study of difference equations.

Sequences

We will begin with an example based on a chart printed in *USA Today*[1]. The chart depicted world oil consumption for several years in the past, as well as a predicted level of consumption for the future. The numbers contained in the chart are shown in Table 2.1. These figures are reported in units of millions of barrels of oil per day. So, for 1991, the table entry of 66.6 indicates that on the average the world consumed 66.6 million barrels of oil each day.

This example has an important feature that shows up over and over again in applications: the data are a series of numerical values that go with a series of times, and the times are equally spaced. In this case, we have one value for each year. In other cases we might have data for each day, or for each hour. The common feature is that the data are separated by equal periods of time. We usually refer to data of this type as *discrete*

Year	1991	1992	1993	1994	1995
Oil Used	66.6	66.9	66.9	67.6	68.4

TABLE 2.1
World Oil Consumption in Millions of Barrels per Day

---

[1] From *USA Today*, June 16, 1994

data. The collection of data values is referred to as a *sequence,* and the individual data values are referred to as *terms* of the sequence.

Many many applications are analyzed using discrete data. In its most general meaning, the word *discrete* simply indicates a separation between values. Often, this separation is introduced in the process of recording data: we have a first reading, a definite next reading, and so on. These are discrete data values because they are separated, and can be listed in a definite order. In contrast, it is easy to imagine a situation where data are not separated, at least conceptually. Consider the way air temperature changes over the course of the day. Perhaps at noon the temperature is 70 degrees. You might also have the temperature at 1 minute after noon, or 1 second, or a tenth of a second. There is no one *next* time to consider. Instead, the time can be thought of as a continuous range of possible values, like the points on a line. As we will see later, that is connected with the idea of a *continuous* variable. But for now we will concentrate on sequences of discrete data. Usually these will correspond to measurements made at regularly spaced times, such as daily temperatures or annual rain fall amounts. But discrete data don't have to be associated with regularly spaced times, or with times at all. In one example, we will consider various prices a company can charge for one of its products, and how the price affects sales. In that problem, the data will reflect sales levels at a sequence of prices, not at a sequence of times. The point to keep in mind is this: when you think of discrete data, think of separation. A good image to keep in mind is a list of separate data values.

It is customary in this context to use a shorthand notation consisting of a label (usually a single letter) and a number. For example, we might choose to use the letter C (standing for consumption), and then write C_{1991} for the average daily oil consumption in 1991. Of course for any other year we use a similar notation: C_{1995} stands for the daily consumption in 1995. Reading the data from the table, we see that $C_{1992} = 66.9$ million barrels.

Notice that the number is written in small size type and below the line. It is called a *subscript*. It is important to pay close attention to the position and size of symbols when subscripts are used. Compare C_{1991+4} with $C_{1991} + 4$. For the first one, all of the symbols in $1991 + 4$ are below the line of type, and in the small type size, so these symbols are all part of the subscript attached to the C. Because $1991 + 4 = 1995$, you should recognize C_{1991+4} to mean the same thing as C_{1995}, which is 68.4. In contrast, in $C_{1991} + 4$, only 1991 is part of the subscript attached to C. This time, you are supposed to add 4 to the numerical value of C_{1991}. Since $C_{1991} = 66.6$, that means $C_{1991} + 4 = 66.6 + 4 = 70.6$. Be sure you understand the difference between C_{1991+4} and $C_{1991} + 4$. To practice, write down in the space below the numerical value of C_{1994-2} and $C_{1994} - 2$. The answers are given in a footnote, but don't peek until after you write down your own answers.[2]

You will be using subscripts throughout this course. The examples above illustrate the importance of reading subscript notation carefully. It is also very important to write carefully. It can be very hard to tell in handwritten work what is part of a subscript, and what is not. There can be a big difference in the meaning of two very similar looking expressions. This is illustrated by the example above. For C_{1991+4} refers to the oil

---

[2] Answers: $C_{1994-2} = C_{1992} = 66.9$; and $C_{1994} - 2 = 67.6 - 2 = 65.6$

consumption 4 years after 1991; $C_{1991} + 4$ concerns adding 4 to the oil consumption in 1991. In one case, the 4 is added to the year; in the other case it is added to the amount of oil consumed. These are clearly completely different ideas.

For the examples above, we have used years such as 1991 and 1994 as subscripts, but there are other alternatives. We might have chosen simply to number the data values. In that approach, C_1 is the first data value, or 66.6, C_2 is the second, and so on. We will see in the next section that it is often convenient to number the starting data value with a 0 instead of a 1. In that case, C_0 is the consumption for 1991, C_1 the value for 1992, and so on. This choice permits a simple verbal description of the numbering scheme: *the subscript on each term indicates a number of years after 1991*. That is, C_1 denotes the daily consumption one year after 1991, C_2 stands for the daily consumption two years after 1991, and so on. It even makes some kind of sense to say that C_0 is the consumption zero years after 1991—that is, in 1991.

There is nothing magical about the letter C in the preceding paragraph. Any letter can be used as part of the shorthand for a sequence of numbers. Usually, we try to pick a letter that helps us remember something about the data. In a problem involving population data, we might use the letter p. For example, if the data gives the US population in 1950, 1960, 1970, and so on, we could abbreviate the data as p_0, p_1, etc.

So far, the subscripts have all been given as numerical values. Often it is useful to use a variable as a subscript. This idea can be illustrated using the oil data again. Consider the following three statements:

C_2 is the daily oil consumption 2 years after 1991

C_3 is the daily oil consumption 3 years after 1991

C_4 is the daily oil consumption 4 years after 1991.

These statements form an easily recognized pattern, and communicate quite well what is meant by the C notation. Indeed, after reading the three statements, everyone can agree what C_5 ought to mean. This same pattern can be condensed into a single statement:

C_n is the daily oil consumption n years after 1991.

Here, the point of using n instead of a specific number is to convey the idea that the statement holds for every number. Replace n with 2, or with 3, or with 234. In each case the resulting statement is true. This kind of notation will be used repeatedly. Rather than list several examples and hope that a pattern is apparent, we will make just one statement involving a variable. By replacing that variable with various numbers, you can create your own list of examples.

This idea can be illustrated using an example from Chapter 1. Suppose the following statement is made:

a_n is the amount of carbon dioxide in the atmosphere (in parts per million) n years after 1965.

To help understand what that means, rewrite the statement changing each n to a number, say 7.

a_7 is the amount of carbon dioxide in the atmosphere (in parts per million) 7 years after 1965.

So a_7 gives the parts per million of carbon dioxide in 1972. Repeat this process with several different numbers until you see what the pattern is. With a little practice, you will be able to understand what the pattern is just by looking at the single statement with the variable.

This completes the introduction of the concept of discrete data and sequences, as well as the shorthand notation using subscripts. In the next section these ideas are used to introduce difference equations.

Difference Equations

The oil consumption table was printed in *USA Today* in 1994, before the figures for 1994 or 1995 were available. The oil consumption shown for those years are projections, or estimates. How can estimates like those be made? Generally speaking, the answer involves patterns in the known data. If the data follow some recognizable pattern, we imagine that the pattern will continue into the future and make a projection on that basis. For the *USA Today* article, patterns were probably found in many kinds of data that contribute to oil consumption, such as population growth, industrialization, and economic development patterns. To keep this discussion simple, we will only look at the data displayed in the table, and only consider a very simple pattern. In discussing the simple pattern, we will pretend that the USA Today data are true values, ignoring the fact that the last two values are actually projections.

For a moment, look only at the first two data values. Notice that the consumption rose from 66.6 in 1991 to 66.9 in 1992, an increase of 0.3. One very simple assumption you could make is that the same increase will occur every year. This assumption would be stated in the following form:

> The average daily oil consumption each year is .3 more than in the preceding year.

It is easy to see from the data that this pattern is not correct. At least, it does not exactly fit the data. If it did, the oil consumption for 1993 would be $66.9 + .3 = 67.2$, which is not the figure shown in the table. However, we do not demand that the pattern fit the data perfectly. After all, a pattern is a kind of mathematical model, and mathematical models always involve simplification and approximation. In this course we will see many different kinds of models, some of which fit real data quite well. Our aim now, though, is not to find a highly accurate model for the oil data. Rather, the oil example is intended only to illustrate some notation and terminology that will be used throughout the course. The basic conceptual idea is this: it is often possible to find patterns that approximate a data set. We will study the patterns, eventually comparing them to the original data.

So let us consider the pattern described above. The first data value we have is 66.6. According to the pattern, the next value should be .3 more or 66.9. Then the next value should be 67.2, then 67.5, and so on. We can continue in this way to generate one number after another as long as we choose. At this point, recognizing that these numbers are not exactly the same as the data, we should introduce a new label. We will use the letter c.

Then the previous remarks can be summarized as follows:

$$c_0 = 66.6$$
$$c_1 = 66.6 + .3 = 66.9$$
$$c_2 = 66.9 + .3 = 67.2$$
$$c_3 = 67.2 + .3 = 67.5$$

There is a pattern here. Changing the way the equations are written will make the pattern clearer. The first equation says that $c_0 = 66.6$. But 66.6 appears again in the second equation. Let us write c_0 in that equation, in place of 66.6. This emphasizes that we are adding .3 to the initial data value (c_0), rather than emphasizing exactly what that data value is (66.6). In the same way, the 66.9 that appears in the third equation can be replaced with c_1, and the 67.2 in the last equation can be replaced with c_2. Then the equations become

$$c_0 = 66.6$$
$$c_1 = c_0 + .3$$
$$c_2 = c_1 + .3$$
$$c_3 = c_2 + .3$$

As in previous examples, there is a clear pattern here. It is much more compactly expressed using variable subscripts:

$$c_{n+1} = c_n + .3 \tag{1}$$

This one equation contains the same information as the last three equations above. For example, make $n = 1$, and the equation says $c_2 = c_1 + .3$. But the equation with n says more, because it can be used with any number in place of n. If n is changed to 100, we get $c_{101} = c_{100} + .3$, meaning *the one hundred and first data value equals the one hundredth value plus .3*.

There is one pattern at work here, and we have seen three different ways to communicate what the pattern is. The pattern can be described verbally:

Each data value is .3 greater than the preceding data value.

The pattern can be illustrated with several equations:

$$c_1 = c_0 + .3$$
$$c_2 = c_1 + .3$$
$$c_3 = c_2 + .3.$$

Or the pattern can be expressed in the single equation

$$c_{n+1} = c_n + .3.$$

This final approach is the most compact. It is a kind of shorthand for describing a pattern. Like any shorthand, it is only useful after it becomes so familiar that it is understood

immediately. That requires practice. The shorthand will be used repeatedly throughout
the text. At first, it will be good practice to take the time each time you see such an
equation to substitute several different values for the variable in the subscript and to write
out a verbal description.

For example, if you see this equation:

$$p_{k+1} = 4p_k - 3$$

you should replace the variable in the subscript (in this case, k) with several numbers.
Replacing k with 1 gives $p_{1+1} = 4p_1 - 3$ or $p_2 = 4p_1 - 3$. Replacing k with 2 gives
$p_{2+1} = 4p_2 - 3$ or $p_3 = 4p_2 - 3$. Writing several equations in this way should help you
get a feel for what the pattern is. Then you should write a description of the pattern in
words:

*Each data value is found by multiplying the preceding data value by 4 and
subtracting 3.*

Finally, pick a starting value for the data, and use it to compute several more data
values. If the starting value is 2, the verbal description says to multiply that by 4 and
subtract 3. This gives 5 as the next data value. Repeat the process: multiply the 5 by
4 and subtract 3; 17 is the next data value. Using the equations rather than the verbal
description gives the same results. If the starting value is $p_1 = 2$. Using $k = 1$ in the
equation leads to $p_2 = 4p_1 - 3 = 4 \times 2 - 3 = 5$.

As you can see, there is a lot of information packed into an equation like Eq. (1),
and sometimes it takes a bit of effort to unpack all that information. When equations of
this type are used, it is very important to state verbally what the variables stand for. For
Eq. (1), we should state:

The value c_n is the model value for daily oil consumption n years after 1991.

Please note that there is a difference between c in this statement and the C that was
used at the start of the chapter: C_n represents a true data value, while c_n stands for a
value that comes from the pattern. The same letter is used because both represent oil
consumption, but we use the distinction between upper case (that is, capital) letters and
lower case letters to keep separate the actual data and the model. Of course, the whole
point is to choose a model that gives a good approximation to the real data, so that c_n and
C_n are nearly equal. In fact, comparing c_n and C_n is a good way to see how accurate
the model is, and that is one of the reasons to use the two symbols c and C.

In much of this text, we will be concerned mainly with models, and we won't have to
use one kind of symbol for the actual data and another for the model values. But when it
is necessary to keep these two distinct, you will need to be sensitive to subtle differences
in notation. In other courses you may see other devices used to get at the same idea.
Sometimes, when a variable like x is used to represent actual data, the symbol $\hat{x}$ will
stand for the model value.

The simple pattern expressed in Eq. (1) is characteristic of a very large class of patterns
that will form the core of this course. It is called a *difference equation*. In the case of
Eq. (1), the pattern shows how any term in the sequence can be computed from the
preceding term. This is a general feature of difference equations: a difference equation

always shows how to compute one term of a sequence from one or more preceding terms. The equation itself can be stated in several different forms. For example, one alternative to Eq. (1) is

$$c_n = c_{n-1} + .3 \tag{2}$$

Compare the two equations. Each says that the term on the left side of the equal sign is equal to the preceding term plus .3. But they use variables in a slightly different way. In Eq. (2), n is the subscript of the new term that is being computed. In Eq. (1) n is the subscript of the old term that is used for the calculation. There is no general reason to prefer one form over the other. You should be comfortable with both, and understand how they each describe the same pattern.

Here is another alternative to Eq. (1):

$$c_{n+1} - c_n = .3 \tag{3}$$

In this equation, the left-hand side represents the difference between two successive terms of the sequence. In fact, this version emphasizes the fact that the difference between two successive terms is *constant*. Pick a pair of terms next to each other in the sequence, and no matter which pair you pick, the difference between the terms always has the same numerical value. This idea is important for two reasons. First, it is this idea that gives us the name *difference equation*. Second, the appearance of constants in a model is of fundamental importance in science and mathematics. Mathematicians refer to such constants as *invariants*; physicists talk about *conservation* principles. But the idea is the same: something that can be depended on not to change. It is the foundation of many kinds of analysis. We will see several kinds of application where the models we use express our belief that something remains constant.

There is one more comment that should be made about the various difference equations that have been displayed here. Return for a minute to Eq. (2). Do you see that it doesn't work for $n = 0$? In words, we can't compute c_0 from the preceding term, because there is no preceding term! In fact, Eq. (2) only makes sense for $n = 1, 2, 3$, etc. In the same way, Eq. (1) only makes sense for $n = 0, 1, 2, 3$, etc. When difference equations are used, it is generally necessary to keep in mind which values are permitted for the subscript variable (in this case n). Notice how many things we need to keep straight in this modeling problem: the real data, the model, the notation, the subscript variable, and the difference equation. To emphasize this even more, all of this information is restated in a compact way below:

> C_n represents average daily world oil consumption in millions of barrels n years after 1991. A model for the data is given by the numbers c_n, where $c_0 = C_0$ and for $n > 0$ $c_n = c_{n-1} + .3$.

This kind of information is typical of what is needed to understand the meaning of a difference equation. It provides a context within which the equation is related to a real problem. As you work with difference equation problems, you should try to make a habit of writing out a brief description like this when possible. Usually, you can't write it all down at once, just as we didn't get it all down at once in this chapter. Instead, as you work with a problem, little bits and pieces will show up at different times. When you choose a letter for the original data, you get one piece. When you formulate a pattern as a difference equation, you get another. But at the end, when you are ready to write a description of how you solved the problem, you should begin with a summary statement like that above. This is an organizational tool. Even though it is not the way you figured the problem out, it is a good way to record the problem and its solution for future reference. Then, if you reread the problem solution at a future date, or if someone else reads it, the context for the solution is made clear at the beginning.

Now we have introduced the basic tool for the course: difference equations. To complete this chapter, we will briefly discuss the three types of analysis mentioned in the Introduction: numerical, graphical, and theoretical. To focus the discussion we consider two questions that are typical in using models:

> According to the model, what will the daily consumption be in 1999?

and

> According to the model, when will the daily consumption reach 70 million barrels?

The Numerical Approach

The difference equation Eq. (1) can be used to generate model data for as many years as we like. As shown earlier,

$$c_0 = 66.6,$$

$$c_1 = 66.6 + .3 = 66.9,$$

$$c_2 = 66.9 + .3 = 67.2.$$

We can continue this process to find answers for the two questions. First, to determine the daily consumption in 1999, recall that c_n refers to n years after 1991. Since 1999 is 8 years after 1991, we need the figure for c_8. Just continue the calculations for 6 more steps:

$$c_3 = 67.2 + .3 = 67.5,$$

$$c_4 = 67.5 + .3 = 67.8,$$

$$c_5 = 67.8 + .3 = 68.1,$$

$$c_6 = 68.1 + .3 = 68.4,$$

$$c_7 = 68.4 + .3 = 68.7,$$

$$c_8 = 68.7 + .3 = 69.0.$$

The model predicts a daily consumption of 69.0 million barrels of oil in 1999.

For the second question we can continue the data even further until we see a daily oil consumption of 70 or more:

$$c_9 = 69.0 + .3 = 69.3$$

$$c_{10} = 69.3 + .3 = 69.6$$

$$c_{11} = 69.6 + .3 = 69.9$$

$$c_{12} = 69.9 + .3 = 70.2$$

The model predicts that daily world oil consumption will go above 70 million barrels per year 12 years after 1991, or in 2003.

These calculations are exemplary of the numerical approach to using the model. Notice the pattern of calculations. We keep performing the same operation over and over, each time using the result of the last computation as the starting point for the next. This type of activity is referred to as *recursion*, and sometimes a difference equation is called a recurrence for that reason. The process of computation is sometimes described as computing values of oil consumption recursively. Whatever name you give it, it is a fundamental activity for exploring how a difference equation model behaves. It is also easy to do on a graphing calculator.

The Graphical Approach

In many cases, using diagrams or figures can help us gain additional insights about relationships. In the study of difference equations, one standard kind of diagram is a bar chart. Fig. 2.1 is a bar chart for the real data at the start of the chapter.

This gives some visual impressions of the data, but there are too few numbers to gain much in the way of insight. Next, we will look at a bar chart representation for the model, but this time include many more values. See Fig. 2.2. Here we immediately gain a new insight: in the model, the tops of the bars form a straight line. In later chapters the significance of this observation will be explored further. But one immediate conclusion can be drawn: if we have experience that shows that oil consumption graphs do not usually follow straight lines, the model we are using here will probably not be very accurate. More will be said about this a little later.

A commonly used alternative to the bar chart is called a line graph. Here, we use one point (corresponding to the center of the top of the bar) in place of the entire bar, and

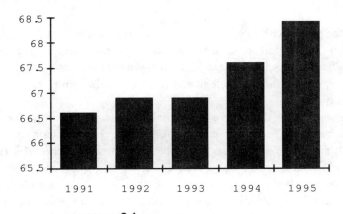

FIGURE 2.1
Bar Graph for Oil Consumption Data

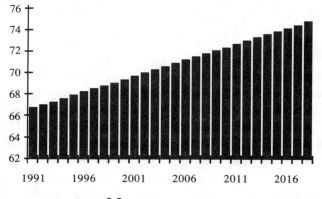

FIGURE 2.2
Bar Graph for Oil Consumption Model

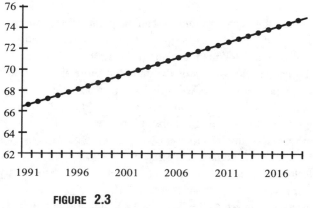

FIGURE 2.3
Line Graph for Oil Consumption Model

then connect the points with lines. A graph of this type is shown in Fig. 2.3. In the line graph it is even more apparent that the model data fall on a straight line. Although the points are emphasized in this figure, with so many points it is more customary to make them smaller, or to leave them out altogether. In that case the figure would simply appear as one straight line.

Using either the bar or line graph, it is possible to answer again the same questions we considered before. That is, we can read off either chart the daily oil consumption level for 1999, and find out the year when the consumption reaches 70. However, the numerical approach is much more accurate for answering questions of this type. The graphical approach is better suited to identifying trends and visual patterns in the data or model. Because the graphical approach conveys impressions about an entire data set, it can also be very useful for getting a first estimate for a problem. Then, a numerical approach can be used to refine the estimate.

The Theoretical Approach

In the previous sections we have used Eq. (1). This is a recursive equation for the model: in order to find c_n for a particular value of n, we first have to find all of the preceding values. Here is a better way to find values of c_n:

$$c_n = n \cdot .3 + 66.6 \qquad (4)$$

For example, to find c_8 we just set $n = 8$

$$c_8 = 8 \cdot .3 + 66.6 = 2.4 + 66.6 = 69.0$$

It is just as easy to compute c_{100}

$$c_{100} = 100 \cdot .3 + 66.6 = 30 + 66.6 = 96.6$$

Eq. (4) is an example of a *functional equation*. It allows us to compute c_n directly from n without first computing all the preceding terms of the sequence. In this situation we say that Eq. (4) gives c_n *as a function of n,* meaning just that c_n can be directly computed as soon as we know n. It is also said in this case that the value of c_n is *determined* by the value of n.

At this point you are probably wondering where Eq. (4) came from. The answer will have to wait until the next chapter, where you will learn how to find a functional equation for difference equations like Eq. (1). In later chapters you will learn how to find functional equations for other kinds of difference equations, as well.

Finding a functional equation is just one aspect of the theoretical method. As another sample, here is how the theoretical approach is used to find when the daily oil consumption reaches a level of 70. In Eq. (4), set c_n to 70 and leave the other side of the equation unchanged:

$$70 = n \cdot 0.3 + 66.6$$

This equation can be solved for n using the methods of algebra:

$$70 - 66.6 = n \cdot 0.3$$

$$3.4 = n \cdot 0.3$$

$$3.4/0.3 = n$$

$$11.333\ldots = n$$

The final value for n can be found by dividing 3.4 by 0.3 on a calculator (the numerical approach) or by multiplying the top and bottom of the fraction by 10 to get 34/3 and converting to a mixed fraction $11\frac{1}{3}$ (the theoretical approach). As is often the case, the numerical approach gives an approximate answer, and the theoretical approach gives an exact answer. Either way, in this problem we see that daily oil consumption reaches 70 more than 11 years and less than 12 years after 1991. So the first year that it exceeds 70 will be 12 years after 1991, or 2003.

These examples of the theoretical approach reveal something of its power, and illustrate how one theoretical result can lead to another. There is another advantage to the theoretical approach—it has greater generality than either the numerical or the graphical approach. As an example, instead of asking when the consumption will reach 70, what if we ask when it will reach 80? Or 90? The theoretical approach allows us to answer all of these questions at once. Using methods of algebra very much like those above, it is possible to find a new equation. Without going into all the steps, the new equation turns out to be

$$n = (c_n - 66.6)/0.3 \tag{5}$$

Now to find when a value of 80 is reached, just substitute 80 for c_n

$$n = (80 - 66.6)/0.3 = 13.4/0.3 = 44.666 \cdots$$

showing that it will be 45 years after 1991. And any other number than 80 could be handled as easily. When we say that the theoretical method is general, it means that it can be adapted to solve many closely related problems at once.

It should also be pointed out that what has been done here is to express n as a function of c_n. Notice that in Eq. (5) the n is all by itself on one side of the equation. That allows the value of n to be directly calculated using the value of c_n. In words, we compute the number of years past 1991, given the level of oil consumption. This reverses the situation of Eq. (4) in which c_n is all by itself on one side of the equation. Using that equation, the oil consumption level is computed directly from the number of years past 1991. This illustrates the idea of *inverting* a function.[3] It is an idea we shall meet often. We will first have an equation with one variable standing alone, and another variable mixed up with numbers and operations. Then we will find a new equation in which it is the second variable that stands alone. In the next chapter you will see how to do this for equations like Eq. (4). That will show the steps that led to Eq. (5).

Back to Reality

The last several sections have focused mainly on the model defined in Eq. (1). But how good a model is it? Does it match the real data we started with? As usual, we can study this question using each of the three methods. Numerically, we can compare the true data C_n with the model values c_n by computing the differences $C_n - c_n$. This leads to the data in Table 2.2. The first two values agree exactly because we designed the model that way. After that, the errors grow steadily. Still, it is not possible to say in the abstract whether this model is good or bad. It depends on the purpose for which it is to be used. This may be a good enough model for some purposes. The point of the table is simply to illustrate a numerical approach to studying how well the model agrees with the data.

A graphical approach to the same question might be to graph both the original data and the model results on one graph. That is shown in Fig. 2.4. Again, it is visually

---

[3] It may be helpful to think of *inverting* as another way to say *reversing*. In Eq. (4) we start with a year and use it to figure out an oil consumption level. Eq. (5) allows us to reverse this process—we can start with a consumption level and figure out a year. It is this reversal that you should think of in connection with inverting a function.

Year	1991	1992	1993	1994	1995
n	0	1	2	3	4
C_n	66.6	66.9	66.9	67.6	68.4
c_n	66.6	66.9	67.2	67.5	67.8
Difference	0.0	0.0	−0.3	0.1	0.6

TABLE 2.2
Data and Model Comparison

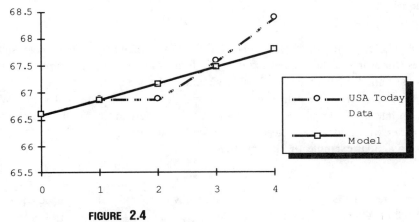

FIGURE 2.4
Line Graphs for Oil Consumption Data and Model

clear that the two graphs seem to be separating as they are followed to the right, but the significance of the differences depends on the problem context.

The theoretical approach might focus on a slightly different question: how could the model be modified to fit the data better? For example, what if we use a different amount in place of 0.3? If the model says that $c_{n+1} = c_n + .25$, is that better? This type of exploration leads naturally to the question of choosing the *best* value to replace 0.3 with. There is a theoretical method for finding the answer to that question, but we won't go into that here. However, most of this course is designed around questions of that type. The models we use are grouped together based on *forms* of difference equations. For example, the following equations all have the same form:

$$c_{n+1} = c_n + .3$$

$$c_{n+1} = c_n + .4$$

$$c_{n+1} = c_n + .5$$

We can condense them into a single line by using a variable for the number at the end of each equation:

$$c_{n+1} = c_n + d$$

This is another kind of shorthand notation. When you read

Let us look at equations that have the form

$$c_{n+1} = c_n + d,$$

you should be thinking:

For example, the d might be 4; that would give the equation $c_{n+1} = c_n + 4$. Or, d might be 1.5, and then the equation would be $c_{n+1} = c_n + 1.5$. What if d is a negative number? Say $d = -2$, then $c_{n+1} = c_n - 2$.

In this way you can get an idea of what is meant by the compact statement $c_{n+1} = c_n + d$. The equation

$$c_{n+1} = c_n + d$$

defines a whole family of closely related difference equations. They have similar properties (for example, they all have graphs that form straight lines), and similar methods of analysis work for all of them. In this situation, the variable d is referred to as a *parameter*. In any actual model, it will have a specific constant numerical value. When we want to talk about properties shared by all of the models in the family, we represent the parameter with a variable instead of any specific number.

Here is a similar example. In the equation

$$c_{n+1} = r \cdot c_n$$

r is a parameter. The equation gives the form of a whole family of difference equations. Some of the equations in this family are

$$c_{n+1} = 2c_n$$

$$c_{n+1} = 1.78c_n$$

$$c_{n+1} = .1c_n$$

We will study several different families of models. Each family is identified by a particular type of difference equation with one or more parameters. When we try to apply a difference equation of this type to a real problem, one common problem is choosing the best possible values for the parameters. Another important question is how to recognize which kind of model will work well for which kind of problem. Over the course you will be introduced to several kinds of models that work well in many problem areas. You will learn what properties of a model type make it especially suitable for problems of a certain type. In the process, you will see how applied mathematicians develop models for real problems.

As a preview of some of these ideas, the next section introduces a few more difference equations.

Sample Difference Equations

The difference equation for the oil consumption model has the form

$$c_{n+1} = c_n + .3$$

It is part of a family of closely related difference equations, all of the form

$$c_{n+1} = c_n + d$$

where d is a parameter. Models featuring this kind of difference equation exhibit what is called arithmetic or linear growth. These equations will be studied in depth in the next chapter.

As a second example, consider a bank account that collects interest. The interest is paid once per month and amounts to one half of a percent of the balance just before the interest payment is made. If the amount after n interest payments is a_n, what will the amount be after one more payment? As an example, suppose the balance is \$27. Then the interest paid is one half percent of \$27. This is computed as $.005 \times 27$. After the interest is paid, the new balance will be $27 + .005 \times 27$. More generally, if the balance is represented by the variable b, the interest payment will be $.005 \times b$ and the new balance will be $b + .005 \times b$. This leads to the difference equation

$$a_{n+1} = a_n + .005 \times a_n$$

As we shall see later, this is an example of geometric growth.

A famous example of a difference equation is

$$f_{n+1} = f_n + f_{n-1}$$

In words, each new f is found by adding the two preceding f's. If the first two numbers in the sequence are 1 and 1, the next will be 2, then 3, then 5, and so on. This pattern was described by the mathematician Leonardo Fibonacci in 1202, and the sequence is named in his honor the Fibonacci numbers. The original context discussed by Fibonacci was a puzzle involving rabbits that reproduce according to the following pattern. Each month a pair of rabbits produces one pair of children, a male and a female. Each new pair produces children of its own, one pair per month, starting 2 months after being born. The puzzle is this: if you start with one pair of new children, how many rabbits will there be after 10 months? After 15 months? After any number of months? Fibonacci's solution was to show that, if f_n is the number of rabbits after n months, then $f_0 = f_1 = 1$ and for $n > 0$

$$f_{n+1} = f_n + f_{n-1}$$

Although Fibonacci became interested in the pattern in connection with a puzzle, mathematicians have found that the Fibonacci numbers reveal a large number of interesting patterns. What is more, these numbers arise naturally in biology. As one example, we consider the problem of counting ancestors for a common bee[4]. Now, male bees are produced asexually by the queen, so each male has a mother but no father. Each queen bee, by contrast, has both a father and a mother. Starting with a queen, we can chart the ancestors in a diagram as in Fig. 2.5. In the diagram, each black circle is a male, and each white circle is a female. The circle at the top of the diagram is the queen bee

---

[4] This example is taken from *Concrete Mathematics,* by Ronald L. Graham, Donald E. Knuth, and Oren Patashnik. Addison Wesley, Reading, Mass., 1989. See page 277.

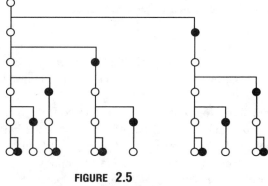

FIGURE 2.5
Ancestry for a Queen Bee

we are interested in. The white circle directly below is the mother of the queen, while the black circle off to the right is the father of the queen. In fact, for any white circle, there is a white circle directly below, indicating the mother, and a black off to the side for the father. In contrast, for black circles, there is only one parent, shown by a white circle directly below. We can continue this diagram for as many generations of bees as we wish. Observe that the queen has 2 parents, 3 grandparents, 5 great-grandparents, 8 great-great-grandparents, and so on. It is this pattern of numbers, 2, 3, 5, 8, etc., that we are interested in. To be specific, let a_n be the number of ancestors that the queen has in the nth preceding generation. Then a_1, the number of ancestors in the first preceding generation (that is the parents), is 2, $a_2 = 3$, $a_3 = 5$, and so on. The question is, how can we compute a_n for any n? The answer is that the a's are just the Fibonacci numbers.

An explanation for this result can be found in the diagram. In Fig. 2.6 parts of the ancestry chart are highlighted to show the ancestors of the queen's mother and paternal grandmother. There are two points to consider. First, since the mother and grandmother are also queen bees, their ancestry charts look just like the one we started with. Second,

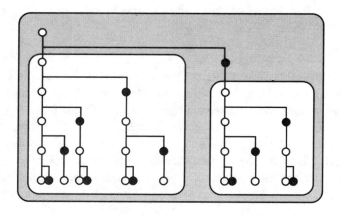

FIGURE 2.6
Ancestry of Mother and Grandmother Bees

we can find all of the ancestors for our queen by looking at the ancestors of the mother and grandmother. For example, the number of ancestors for our queen 4 generations back can be found by adding the mother's ancestors 3 generations back to the grandmother's ancestors 2 generations back. How many ancestors does the mother have 3 generations back? The answer is a_3, because all queens have the same number of ancestors at each generation. How many ancestors does the grandmother have 2 generations back? a_2. So, we can see that $a_4 = a_3 + a_2$. The same argument works for any number of generations. This shows that $a_{n+1} = a_n + a_{n-1}$, which is the same difference equation as the one found by Fibonacci. This example illustrates how difference equations can be found by examining patterns.

Sometimes difference equations are just made up without any connection to an application. Here is one with a funny pattern:

$$a_{n+1} = a_n/a_{n-1}$$

Pick a couple of numbers to start the pattern, say $a_1 = 4$ and $a_2 = 7$, and work out the next 8 or 9 numbers in the pattern, following the difference equation, and you will see what the pattern is.

As a final example, we will develop a difference equation that can be used to model how drugs are metabolized from the body, or how impurities are washed out of a lake. To simplify the explanation, we will look here at a water tank that holds 10 gallons. Suppose someone accidentally spills a pollutant into the tank. For the sake of discussion, we will imagine that 4 pounds of salt fall into the tank, and completely dissolve. In an attempt to purify the water, we drain the tank and refill it with pure water. Unfortunately, not all of the water drains out. The drain is located in such a way that one gallon is left at the bottom and doesn't go out of the drain. So when we put in fresh water, we only put in 9 gallons, and that mixes with the one salty gallon that was left in the tank. If we repeat the process several times, the salt is reduced further and further. The difference equation will describe how much salt is left in the tank after each draining and refilling.

Initially, there are 4 pounds of salt in the water. When we remove 9 gallons from the tank, that is 9/10 of the total. The remaining one gallon is one-tenth of the total, so it contains one-tenth of the salt. That is .4 pounds. Now when we refill the tank and the water mixes up, there are .4 pounds of salt dissolved in the 10 gallons. We remove 9 gallons, leaving behind one-tenth of the water, and with it, one-tenth of the .4 pounds of salt. That is, we leave .04 pounds of salt. The same pattern always applies. If there are s pounds of salt dissolved in the ten gallons in the tank, when we drain the tank, we leave behind one-tenth of the water, and so one-tenth of the salt. That means that there will be $.1s$ pounds of salt. After refilling the tank with pure water, the same $.1s$ pounds of salt remain. Starting with s pounds of salt, if we drain and refill the tank once, there will then be $.1s$ pounds of salt. This gives the difference equation

$$s_{n+1} = .1s_n$$

where s_n is the amount of salt after draining and refilling the tank n times.

This type of difference equation arises in many many applications. If there is a lake with some pollution in it, we can think of the amount of water flowing into the lake and out of the lake each day as being like draining part of the tank and refilling it. The difference equation for the pollution left in the lake will be of the same form as the one for the tank. For medical applications, the water tank concept provides a good first approximation to the way the body removes a drug from the blood stream. In each hour, a certain fraction of the blood is purified, leaving some of the drug behind. If the body purifies about nine-tenths of the blood, then about one-tenth of the drug remains in the blood at the end of the hour. The difference equation for this situation is

$$d_{n+1} = .1d_n$$

where d_n is the amount of drug in the body after n hours.

A modification of this equation applies when more medicine is added on a regular basis. If the patient takes a pill every hour, and adds .5 milligrams of the medicine, then the difference equation becomes

$$d_{n+1} = .1d_n + .5$$

This shows that each hour all but $1/10$ of the drug already in the body is removed, and that .5 new milligrams of the drug are added. This model can be used to understand how medicines can build up in the body. It also has uses in setting drug doses.

Summary

The main point of this chapter has been to introduce the idea of difference equations, and the terminology and notation that are used with them. In addition, the basic ideas of developing and applying mathematical models involving difference equations were discussed. A particular example involving oil consumption data was referred to throughout the chapter to illustrate the ideas that were presented. Numerical, graphical, and theoretical techniques were illustrated as approaches to two typical questions that arise in modeling. One of the fundamental questions in this regard concerns how well a model fits real data. Often a model can be improved by making a slight modification to one or more parameters. This leads to the idea of studying families of models with very similar difference equations, an approach that will be followed throughout the course. Several different kinds of difference equations were given as examples at the end of the Chapter. These show some of the kinds of patterns that difference equations can lead to.

Exercises

Reading Comprehension

1. Explain what is meant by each of the following terms, introduced in the reading: sequence of numbers, discrete data, subscript, terms, difference equation, invariant, recursion.

2. What is the advantage of using a variable for a subscript? Why not just use numerical subscripts?

3. Explain why the following difference equations all express the same pattern: $a_n = a_{n-1}/3$, $a_{n+1} = a_n/3$, $a_n - a_{n-1}/3 = 0$.

4. For each part of this problem, write a mathematical equation that expresses the written statement:

   a. In a set of data, the 4th number is equal to the 3rd number plus .65.
   b. In a set of data, the 5th number is equal to .85 times the 4th number.
   c. In a set of data, each number is 2.38 more than the preceding number.
   d. In a set of data, each number is found by multiplying the two preceding numbers.
   e. In a set of data, each number is 20% less than the preceding number.

5. The following is given: $a_{n+1} = (a_n - 2)^2$ and $a_1 = 5$. A student is asked to figure out a_4, and makes the following computations:

$$
\begin{aligned}
a_2 &= (a_1 - 2)^2 &=& (5 - 2)^2 &=& 9 \\
a_3 &= (a_2 - 2)^2 &=& (9 - 2)^2 &=& 49 \\
a_4 &= (a_3 - 2)^2 &=& (49 - 2)^2 &=& 2,209
\end{aligned}
$$

Is this student using a difference equation or a functional equation? Is the method recursive? Explain.

6. The following is given: $a_n = 3n^2 - 2n + 5$. A student is asked to figure out a_4 and a_7, and makes the following computations:

$$
\begin{aligned}
a_4 &= 3 \cdot 4^2 - 2 \cdot 4 + 5 &=& 48 - 8 + 5 &=& 45 \\
a_7 &= 3 \cdot 7^2 - 2 \cdot 7 + 5 &=& 147 - 14 + 5 &=& 138
\end{aligned}
$$

Is this student using a difference equation or a functional equation? Is the method recursive? Explain.

7. Explain what a parameter is, and give an example showing how a parameter is used.

Mathematical Skills

1. For each part of this problem, a difference equation and one or more initial values are given. Use the difference equation to compute additional terms as specified.

   a. If $a_0 = 1,000$ and $a_{n+1} = a_n + 0.006a_n$, compute a_1, a_2, and a_3.
   b. If $f_1 = 1$, $f_2 = 1$, and $f_{n+2} = f_n + f_{n+1}$, compute f_3, f_4, and f_5.
   c. If $a_1 = 1$ and $a_{n+1} = a_n + 2$, compute a_1, through a_{10}.

2. For each part of this problem a sequence of numbers is shown. See whether you can find a pattern for each sequence. Describe the pattern in words, and, if possible, with either a difference equation or a functional equation.

   a. $2, 4, 6, 8, 10, \ldots$
   b. $1, 3, 5, 7, 9, \ldots$
   c. $1, 4, 9, 16, 25, \ldots$
   d. $1, 3, 6, 10, 15, \ldots$
   e. $1, 2, 4, 8, 16, 32, \ldots$
   f. $1, 4, 5, 9, 14, 23, \ldots$
   g. $1, 2, 6, 24, 120, 720, \ldots$

3. For the difference equation $b_{n+1} = b_n/b_{n-1}$ there is a simple pattern starting with any two initial values. Make up your own values of b_1 and b_2 and use the difference equation to figure out the next several b's until you can see the pattern. It might take 6 or more steps for you to recognize the pattern. Then pick different starting values of b_1 and b_2 and repeat the process. Do this for several more starting choices of b_1 and b_2. Finally, write a verbal description that fits all of these patterns.

4. Several familiar patterns of numbers are described by difference equations. For each equation below, work out enough terms of the sequence to see what pattern is generated.

   a. $a_{n+1} = a_n + 2, a_1 = 2$
   b. $a_{n+1} = a_n + 5, a_1 = 5$
   c. $a_{n+1} = a_n + 10, a_1 = 10$
   d. $a_{n+1} = -a_n, a_0 = 1$
   e. $a_{n+1} = 2a_n, a_1 = 2$

5. Functional equations are given below for each of the patterns in the preceding problem. Match each functional equation with the correct difference equation.

   a. $a_n = (-1)^n$
   b. $a_n = 5n$
   c. $a_n = (2)^n$
   d. $a_n = 2n$
   e. $a_n = 10n$

6. For each of the difference equations in the two preceding problems, figure out a_{10} using (a) the difference equation and (b) the functional equation. Which is easier to use? Which would you want to use to find a_{1000}?

7. In the discussion of the world oil consumption example, the notation c_n was used to represent the oil consumption n years after 1991. This problem is concerned with finding the correct n for a given year. For example, given the year 1993, which c_n gives the oil consumption? The answer is c_2 because with $n = 2$, this gives the oil consumption 2 years after 1991. Which c_n gives the oil consumption for each of the following years?

   a. 1992   b. 1995   c. 2000
   d. 2050   e. 1991   f. 1984

8. Referring to the preceding problem, this problem is concerned with finding the year that corresponds to a given n. For example, c_8 represents oil consumption for 1999, because 1999 is $n = 8$ years after 1991. Tell which year each of the following refers to:

   a. c_5 b. c_{25} c. c_0 d. c_{-2} e. c_{-10}

Problems in Context.   This set of problems refers to the carbon dioxide data discussed in Chapter 1. For that discussion, an equation was expressed in the form

$$\text{Carbon Dioxide Level} = 1.366 \cdot \text{Year} - 2365 \tag{6}$$

Using this equation, you will generate data and then follow many of the steps of analysis used in this chapter's discussion of oil consumption.

1. The first data point in the carbon dioxide example was for the year 1965. Using c_n to stand for the carbon dioxide level n years after 1965, find c_n for $n = 1$ through 10. [For each n, figure out the correct year, and use Eq. (6).]

2. Find the change from each year to the next. That is, calculate $c_2 - c_1$, $c_3 - c_2$, $c_4 - c_3$, etc. What pattern do you observe?

3. Express the pattern from the preceding problem as a difference equation. That is, write it in the form $c_{n+1} = ?$, but in place of the question mark write down something that involves c_n.

4. Write out a paragraph like the one in the box on page 17 describing c_n and C_n, but this time you should be writing about the carbon dioxide data. For C_0 use 319.19.

5. Make a bar graph and a line graph for the data you worked out in question 1.

6. Use a graphical and numerical approach to estimate the year when the carbon dioxide level will reach 350.

7. Compare the data in Table 1.1 with the values you have been working with in this problem. Make a table similar to Table 2.2.
    For the next three problems some algebra is needed. These problems are recommended for students who have studied algebra previously and feel comfortable with the subject. The correct methods will be explained in the next chapter.

8. Can you determine the functional equation for c_n? It should be of the form $c_n = ?$, but in place of the question mark there should be something that involves n.

9. Use the equation from the preceding problem to express n as a function of c_n. This time the form should be $n = ?$, and the question mark should be something involving c_n.

10. Use the equation in the preceding problem to find the year when the carbon dioxide level will reach 350. Compare this to your answer from problem 6.

Solutions to Selected Exercises

Reading Comprehension

4. a. $a_4 = a_3 + .65$
    b. $a_5 = .85 a_4$
    c. $a_{n+1} = 2.38 + a_n$
    d. $a_{n+1} = a_n \cdot a_{n-1}$
    e. $a_{n+1} = a_n - .20 \cdot a_n$

5. This is a difference equation. The student uses a_1 to get a_2, then uses a_2 to get a_3, then a_3 to get a_4. That is a recursive method.

6. This is a functional equation. To find a_4, the student changes the n to a 4 and computes a_4 directly, without using any of the other a_n values. Similarly with a_7. This is not a recursive process because the result of the first step is not used in the second step.

Mathematical Skills

1. a. $a_1 = a_0 + .006a_0 = 1,000 + 6 = 1,006.$ $a_2 = a_1 + .006a_1 = 1,006 + 6.036 = 1,012.036$

   b.   Use $n = 1$ so that $f_{n+2} = f_{1+2} = f_3$. This gives $f_3 = f_{1+2} = f_1 + f_{1+1} = f_1 + f_2 = 1 + 1 = 2$. Similarly, $f_4 = f_3 + f_2 = 2 + 1 = 3$.

2. a.   Each number is 2 more than the preceding number. A difference equation is $a_{n+1} = a_n + 2$. A functional equation is $a_n = 2n$. How do you find that? You have to look for a pattern of the right type. For a difference equation, the pattern tells how to go from a_1 to a_2, then from a_2 to a_3, and so on. In this problem, the question is how to get from 2 to 4, then from 4 to 6, then from 6 to 8 and so on. The answer is *add 2 each time*, and that gives the difference equation. For the functional equation the idea is different. This time we have to get from 1 to a_1, from 2 to a_2, from 3 to a_3, and so on. So for this particular problem I need to explain how to get from 1 to 2, from 2 to 4, from 3 to 6, and so on. Once you understand the question, you just have to try to think of the right pattern.

   b.   Again, the pattern is that each number is 2 more than the one before it. The difference equation is the same as before: $a_{n+1} = a_n + 2$. Of course, there are different starting values in this problem and the preceding one, so, although the difference equation is the same for both, the number patterns are different. That means the functional equations will be different. Remember that we need to go from 1 to a_1 from 2 to a_2, and so on. The pattern we are after might be symbolized this way: $1 \rightarrow 1,\ 2 \rightarrow 3,\ 3 \rightarrow 5,\ 4 \rightarrow 7$. The pattern is: double the first number and subtract 1. The equation is $a_n = 2n - 1$.

   c.   This time the functional equation might be more obvious than the difference equation. Did you notice that $1 = 1 \times 1,\ 4 = 2 \times 2,\ 9 = 3 \times 3$, and so on? The functional equation is $a_n = n \times n$. The difference equation is more obscure. You might have noticed how the odd numbers show up in the pattern. Start at 1, add **3** to get 4, add **5** to get 9, add **7** to get 16. Each time you go to the next number, you add the next odd number. The difference equation would be something like this: $a_{n+1} = a_n +$ the next odd number. But what is the next odd number? The answer is closely related to the functional equation from the preceding problem. The difference equation can be written $a_{n+1} = a_n + (2n + 1)$ This is admittedly pretty obscure. It shows that some simple patterns are not easy to express using difference equations.

   d.   These numbers follow a different kind of pattern: $1,\ 1 + 2,\ 1 + 2 + 3,$ $1 + 2 + 3 + 4$, and so on. A difference equation for this pattern can be written in the form $a_n = a_{n-1} + n$ with $a_1 = 1$. There is a functional equation but it is not obvious. In Chapter 5 you will learn how to find functional equations for this type of pattern.

   e.   Here, each number is twice the preceding number. The difference equation can be written either as $a_{n+1} = 2a_n$ or $a_{n+1} = a_n + a_n$. The functional equation can be written $a_n = 2^n$ with $a_0 = 1$. We will study this kind of pattern in Chapter 9.

f. This is a pattern like the Fibonacci numbers—each number is the sum of the two preceding numbers. The difference equation can be written $a_{n+1} = a_n + a_{n-1}$. The first two numbers, 1 and 4, were not picked according to any pattern. But all the numbers after the 4 follow the pattern of the difference equation. There is a functional equation for this pattern but it is too complicated to present here.

g. The pattern here is $1, \ 1 \cdot 2, \ 1 \cdot 2 \cdot 3, \ 1 \cdot 2 \cdot 3 \cdot 4, \ 1 \cdot 2 \cdot 3 \cdot 4 \cdot 5, \ 1 \cdot 2 \cdot 3 \cdot 4 \cdot 5 \cdot 6, \ldots$ This is most easily represented by the difference equation $a_n = a_{n-1} \cdot n$ with $a_1 = 1$.

3. If you start with $b_1 = 3$ and $b_2 = 5$, the next number will be $b_3 = b_2/b_1 = 5/3$; then $b_4 = b_3/b_2 = (5/3)/5 = 1/3$; then $b_5 = (1/3)/(5/3) = 1/5$; $b_6 = (1/5)/(1/3) = 3/5$; $b_7 = (3/5)/(1/5) = 3$; and $b_8 = 3/(3/5) = 5$. From that point on, the pattern repeats: $3, 5, 5/3, 1/3, 1/5, 3/5$. If the first two numbers are a and b, then the pattern that will occur is $a, b, b/a, 1/a, 1/b, a/b$ repeated over and over.

4. a. The even numbers
   b. Counting by 5s
   c. Counting by 10s
   d. Switching back and forth between 1 and -1
   e. Powers of 2: $2^1, 2^2, 2^3, \ldots$

5. a. matches d
   b. matches b
   c. matches e
   d. matches a
   e. matches c

7. $1992 \rightarrow c_1$, $1995 \rightarrow c_4$, $2000 \rightarrow c_9$, $1991 \rightarrow c_0$, and $1984 \rightarrow c_{-7}$.

8. $c_5 \rightarrow 1996$, $c_{25} \rightarrow 2016$, $c_0 \rightarrow 1991$, $c_{-2} \rightarrow 1989$

Problems in Context

1. For c_1 the year is 1966. The carbon dioxide level from the equation is $1.366 \cdot 1966 - 2365 = 320.556$. The values for the first 10 c's are given below:

n	1	2	3	4	5
c_n	320.556	321.922	323.288	324.654	326.020

n	6	7	8	9	10
c_n	327.386	328.752	330.118	331.484	332.850

2. Each year the difference is 1.366.

3. $c_{n+1} = c_n + 1.366$.

4. C_n represents the amount of carbon dioxide in the atmosphere, in parts per million, n years after 1965. A model for the data is given by the numbers c_n, where $c_0 = 319.19$ and for $n > 0$ $c_n = c_{n-1} + .366$.

5. The graphs are shown below:

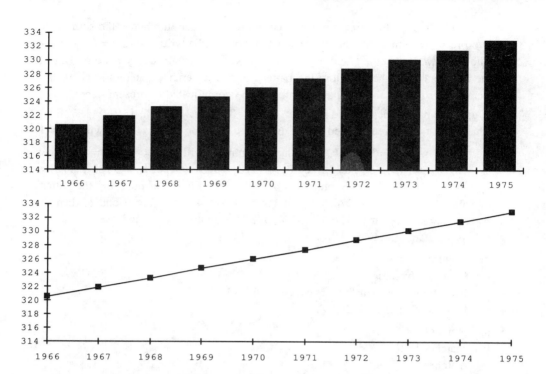

6. In the graph below, the line graph has been extended to show where a carbon dioxide level of 350 will be. It appears that level will be reached in about 1988. Using the equation and the year 1988 the carbon dioxide level is found to be $1988 \cdot 1.366 - 2365 = 350.6$. That is pretty close, but a little too high. Putting in 1987.5 gives $1987.5 \cdot 1.366 - 2365 = 349.9$, which is closer to the exact figure.

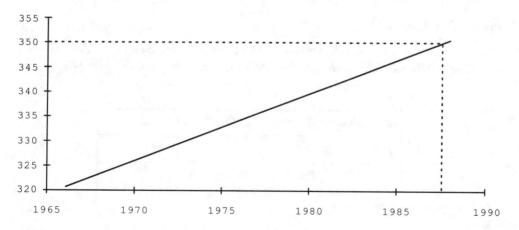

7. The data from Table 1.1 only include four years. For each of those years the model value from the equation and the original data value are included in the table below. The errors are all less than 1.2, which is not very significant for data values on the order of 330.

Year	1965	1970	1980	1988
n	0	5	15	23
C_n	319.9	325.3	338.5	351.3
c_n	319.19	326.02	339.68	350.61
Difference	0.71	-0.72	-1.18	0.69

8. $c_n = 1.366n + 319.19$

9. $n = (c_n - 319.19)/1.366$

10. $n = (350 - 319.19)/1.366 = 22.555$ is the number of years after 1965. This gives the year as 1987.555, which is pretty close to the approximate answer 1987.5 found earlier.

3

Arithmetic Growth

In the last chapter the following difference equation was considered as a model for daily world oil consumption year by year:

$$c_{n+1} = c_n + .3 \tag{1}$$

In this chapter you will see that this is just one example of *arithmetic growth*. Additional examples of this kind of model will be presented, and a general procedure for finding a functional equation will be described. The functional equation is more convenient than the difference equation for answering questions about the behavior of the model.

The Arithmetic Growth Difference Equation

The general idea of arithmetic growth can be stated verbally as follows:

> *Growth of a variable concerns the way the variable changes over time. Under the assumption of Arithmetic Growth, equal periods of time result in equal increases of the variable.*

For example, let the variable represent daily world oil consumption for each year. An arithmetic growth assumption means that in any year the daily oil consumption is expected to increase by the same amount as any other year. If daily oil consumption increases by .3 million barrels from 1991 to 1992, then it must increase by an equal amount in any other year. For the oil consumption model, the assumed increase we first considered was .3. But in general in an arithmetic growth model the increase can be any amount. For this reason, we represent the increase by a parameter, d. This leads to a general difference equation

$$a_{n+1} = a_n + d \tag{2}$$

Here, the letter a is used to emphasize that this equation applies in a wider array of problems than just the oil consumption case where we used the variable c.

Here are some examples. In a study of a flu epidemic, the equation

$$p_{n+1} = p_n + 500$$

might be used to indicate that the number of people who have been infected is going up by 500 per month; the parameter d is 500. The equation

$$f_{n+1} = f_n + 10$$

could be used to represent the fine for an overdue book at the library. The parameter d is 10 and indicates that the amount of the fine increases by 10 cents a day. And we have already seen the equation

$$c_{n+1} = c_n + .3$$

for the oil consumption model. In that case, the parameter d is .3. In all of these examples, the equation has the same form as Eq. (2); however, a particular numerical value appears in place of d, and each equation uses a different variable, p, a, and c.

The assumption of arithmetic growth, as described verbally in the box, leads to a difference equation of the form of Eq. (2). For the equation to make sense, the meaning of a_n should be made clear. For example, in the oil consumption problem, we defined c_n to be the average daily oil consumption n years after 1991. Then, assuming arithmetic growth, c_n must increase by the same amount each year. The amount of that increase is d. Actually, the arithmetic growth assumption implies more than the difference equation. It implies that the oil consumption should increase by the same amount each half year, or each month. This broader idea of arithmetic growth will be taken up in greater detail later. The point of emphasis here is that, if you agree to use an arithmetic growth model, as described verbally, then you are led to the difference equation Eq. (2).

This illustrates an important idea—using a common sense verbal description to devise a difference equation. Here the verbal idea of equal growth in equal periods of time leads to a difference equation of the form shown in Eq. (2). In other chapters we will consider different kinds of verbal descriptions, and the difference equations they inspire.

Although the expression *arithmetic growth* suggests that something is getting bigger, it is sometimes used to describe a variable that is getting smaller over time. In this case, equal periods of time would result in equal decreases in the variable, and the parameter d in Eq. (2) would be a negative number. For example, suppose you are modeling the reserves of a nonrenewable resource, such as the fuel on board a satellite. If the satellite uses up the same amount of fuel each month, the supply of fuel will decrease. This would be a good application for an arithmetic growth model, even though the variable is shrinking rather than growing. Perhaps it would be reasonable to consider this to be an example of negative growth. In any case, we will include arithmetic decay (meaning shrinkage) among our examples of arithmetic growth.

The arithmetic growth model is one of the simplest that can be used. For this reason, it is a good starting point for many kinds of problems. In the next section we will consider three different examples of modeling problems that can be started in this way. Later, we will return to the models with mathematical tools that are more advanced, and which provide better accuracy. But for now, we will be content to study arithmetic growth.

Examples

The first example is population growth. An instance of this is the tracking of school-age populations by school boards. If the population is increasing, additional schools may have to be built. If the population decreases, it may become too expensive to keep all existing schools open, particularly since state and federal funding is often based on the number of students attending school. Over the short term, say for a five-year period, an arithmetic growth model may be accurate enough to make some reasonable predictions about future population figures. It is also a fairly natural model to consider. If you find that the number of elementary school students has increased by about 5,000 for each of the last two or three years, it is easy to imagine that the population might grow in the same way in future years. This would be expressed in a difference equation as follows. First, we need to define a variable for the population size. Let p_n be the number of school-aged children in the district in n years. The difference equation is

$$p_{n+1} = p_n + 5,000$$

By itself, this equation does not give us predicted values for future years. We also have to know the starting value, p_0. If the current population size is $p_0 = 100,000$ then after one year the model predicts a population of size $p_1 = 105,000$.

The second example involves *interest*, as paid on loans and savings accounts. In most real situations, some sort of compound interest is used, and the result is *not* arithmetic growth. However, in a very special simplified kind of interest calculation, arithmetic growth does occur. This is called *simple interest*.

Here is an example of simple interest. Suppose that you borrow $1,000 from a relative to help pay for your education. You agree to pay 5% interest per year. The interest for one year is 5% of $1,000, or $50, so that at the end of a year you owe $1,050. Unfortunately, you are unable to pay off the loan that year, and another year goes by. So you add another $50 of interest, bringing the total owed to $1,100. Each year that goes by increases the amount owed by another $50. If a_n is the amount owed at the end of n years, then $a_0 = 1,000$ and the difference equation is $a_n = a_{n-1} + 50$. This is of the same form as Eq. (2), and shows that simple interest corresponds to arithmetic growth of the amount owed.

The final example concerns a model that is used in business applications. In many situations, the price that is charged for an item has an effect on how many items will be sold, particularly for items that are not necessities. A student may choose to purchase a soda between classes, but she can certainly get along without the soda. If the price is very low, say 10 cents, it is likely that many students will choose to buy a soda. If the price is very high, say five dollars, it is unlikely that many sodas will be sold. Between these extremes, it is a reasonable expectation that the number of sodas that can be sold will decrease as the price increases. In this kind of situation, the number of items that can be sold is called the *demand* for the item. Using this terminology, the preceding remarks can be restated as follows: the demand goes down as the price goes up.

To investigate the relationship between price and demand, sales surveys are often performed. The basic idea of a sales survey is to determine how many items can be sold at various prices. Students in a finite math class conducted a survey of this type for canned soda one semester. They found they could sell 141 cans at a price of 40 cents per can, and for each nickel increase in the price of the soda, the number they could sell was reduced by about 12. Of course, the survey data did not fit this pattern exactly. The pattern is a simplified description of the actual price-and-demand data; it gives a *model* for the effect of price on demand. This model fit the data fairly well for prices from 40 cents up to about a dollar.

We can formulate this pattern as an arithmetic growth model. It is different from previous examples in one important respect. Here, we are not interested in how the demand changes over time. Rather, we consider the way demand changes in response to hypothetical changes in price. In this context, arithmetic growth means that equal changes in the price of a soda result in equal changes in the number that can be sold. If raising the price from 40 to 45 cents results in a loss of 12 sales, then raising the price from 60 to 65 cents will also result in a loss of 12 sales. That is what arithmetic growth means. To put this into the framework of difference equations, we can begin with a base price of 40 cents per can and refer to the expected number of sales at the base price as $s_0 = 141$. Now imagine raising the price a nickel at a time, and estimating from the survey the expected number of sales at each price. Let s_1 be the sales at 45 cents a can, s_2 the number of sales at 50 cents a can, and so on. Then the pattern observed in the data says that each s is 12 lower than the preceding s. In other words, s_1 is 12 less than s_0; s_2 is 12 less than s_1; and so on. Using the kind of careful description used earlier, we can describe the model as follows.

> At a base price of 40 cents, $s_0 = 141$ cans of soda can be sold; s_n is the number of cans of soda that can be sold after raising the price by a nickel n times. A model for this situation is given by the difference equation $s_{n+1} = s_n - 12$.

This is an example of arithmetic growth where the parameter d is negative. The difference equation has the same form as before, and can be analyzed using the same methods that apply to the other examples. This illustrates that arithmetic growth can be used in models that predict how changes in one variable (price) can affect another variable (number of sales), as well as in models where we wish to predict how changes occur over time.

The examples in this section all illustrate the idea of arithmetic growth. Each one is described by a difference equation that is of the form shown in Eq. (2). The next section will discuss numerical, graphical, and theoretical properties of arithmetic growth models.

Numerical, Graphical, and Theoretical Properties

Numerics. Numerically, an arithmetic growth model can be explored easily, once the starting value is specified. In the school population example, the difference equation is $p_n = p_{n-1} + 5,000$. If we start with $p_0 = 100,000$, then we can compute $p_1 = 100,000 + 5,000 = 105,000$, $p_2 = 105,000 + 5,000 = 110,000$, and so on. However, if the population starts at 65,000, then the successive values will be 70,000, 75,000, etc. Look back at the examples of the preceding section. Can you see the importance of the starting values in each of those examples? It should be clear that a starting value is essential for the use of any difference equation. Once a starting value is given, the rest of the values are easily computed as far ahead as we wish. Without the starting value, the difference equation can't be used to tell us what any of the following values will be. For this reason, choosing the starting value is an important part of developing a model which involves a difference equation. The starting value is also given a special name—it is called the *initial value*.

Graphics.   The graphical representation of an arithmetic growth model as either a bar graph or a line graph produces a straight line. For the bar graph, it is the tops of the bars that line up, while for a line graph the individual points for the sequence values all fall on a line. The value of the parameter d is reflected in the steepness of the line. If d is positive, the line slopes up to the right, whereas for negative d the line slopes down to the right. The bigger d is, the more steeply the line slopes. In the next chapter you will learn more specifics about the sloping of the lines. The fact that the graphs always involve straight lines explains why arithmetic growth models are sometimes referred to as *linear* models.

Theory.   Do you recall the distinction between functional equations and difference equations discussed in Chapter 2? The difference equation is used *recursively*; we have to use the initial value to find the next value, then use that to find the next, and so on. To find the 100th value in the sequence, we have to compute all of the 99 values that come before it. A functional equation is direct. We can use it to find the 100th data value in a single computation. That is what makes functional equations so useful.

A theoretical analysis allows us to find a functional equation that can be applied in any case of arithmetic growth. To illustrate with an example, the following table shows the first several values of the school population, as described in the example on p. 39. We begin with $p_0 = 50{,}000$.

$$p_0 = 50{,}000$$

$$p_1 = p_0 + 5{,}000$$

$$p_2 = p_1 + 5{,}000 = p_0 + 2 \times 5{,}000$$

$$p_3 = p_2 + 5{,}000 = p_0 + 3 \times 5{,}000$$

$$p_4 = p_3 + 5{,}000 = p_0 + 4 \times 5{,}000$$

There is a clear pattern here, which can be condensed into the following equation:

$$p_n = p_0 + 5{,}000n \tag{3}$$

Although the equation can be understood simply as a pattern of data, it also follows in a logical way from an understanding of the difference equation. Remember that p_n is the population after n years. Each year the population goes up by 5,000. Clearly, in n years it will go up by $5{,}000 \times n$. If the starting value is p_0, after n years it will be at $p_0 + 5{,}000 \times n$.

The p_0 that shows up in these equations would actually be a number in any specific case. To apply the equation in the population example, we need to know that $p_0 = 50,000$. Then Eq. (3) becomes $p_n = 50,000 + 5,000n$. But we leave p_0 in the general form of the equation, as a reminder of the role played by the initial value of the sequence. Notice that the p_0 is being treated here in much the same way that d was treated earlier. It is a parameter. It is thought of as being a specific number in any particular application of the equation, but we realize that it will be different numbers in different applications.

To return to the main point of this discussion, observe that Eq. (3) is a functional equation for p_n. It gives p_n as a function of n, because, as soon as you replace n with a particular number, the equation tells exactly how to compute p_n. You can now compute p_{100} directly using $n = 100$ without first computing the preceding 99 values.

This whole line of reasoning applies for any arithmetic growth model. If the difference equation is $a_{n+1} = a_n + d$, then term number n will be reached after adding d repeatedly n times. Starting from a_0 this results in a value of $a_n = a_0 + dn$. This is the general form for functional equations for all arithmetic growth models. We formulate it as a general principle for future reference.

> If an arithmetic growth model has an initial value of a_0 and obeys the difference equation $a_{n+1} = a_n + d$, then for any $n \geq 0$, $a_n = a_0 + dn$.

This can be applied immediately to get a functional equation for any arithmetic growth model. Just fill in the values of the parameters a_0 (the starting value) and d (the constant amount added at each step of the difference equation). For the simple interest problem, the initial value was \$1,000 and the amount added each year was \$50. The functional equation is then $a_n = 1,000 + 50n$. In a similar way, in the demand model the number of sodas that can be sold at an initial price of 40 cents is 141. That number decreases by 12 as you go to each higher price. Therefore, $s_n = 141 - 12n$.

This last example is a little confusing because n is such a strange variable—the number of five cent increases. A much more natural variable to use would be the price itself. In fact, this is a good place to introduce an idea that will be important later: in many applications it is useful to consider difference equations for and relationships between several different variables at the same time. Here, we will illustrate this idea by looking at two variables, the price per can of soda and the number of cans that can be sold. We have already worked out the functional equation for s_n, the number of cans sold after n increases in price. Now let us look at the price itself. Let p_n be the price after n nickel increases. Since the starting price is 40 cents, we have $p_0 = 40$. Then, with each nickel increase, the price goes up by 5. So, the prices follow an arithmetic growth model, and we can conclude that $p_n = 40 + 5n$. Now the model makes a little more sense. For any n, we can compute both the price p_n and the demand s_n. As an example, with $n = 5$ we have a price of $p_5 = 65$, and we can sell $s_5 = 81$ sodas at that price.

The functional equation that goes with an arithmetic growth model is a powerful tool for studying the model. With it, you can easily determine what will happen in the future, or determine when in the future something of interest will happen. For this reason, it is a good idea to figure out the functional equation whenever an arithmetic growth model is used. In fact, once you have determined that an arithmetic growth model is appropriate, there is a routine outline that you should follow. This outline is illustrated in the following example. The statements in **boldface** are the steps of the outline.

Example Problem. A cellular phone company has equipment that can service 100 thousand customers. In 1990 they had 70 thousand customers. Over the last few years the number of customers has been increasing by about 4,500 per year. Assuming that the growth continues at the same pace, what will the situation be in the year 2000? 2010? How long will it be before additional equipment will be needed?

Step 1: Formulate an arithmetic growth model. In this case, the idea that the number of customers goes up each year by the same amount, 4,500, identifies this as an arithmetic growth model. What is it we wish to study? The number of customers. So make up a name for the variable: *Let c_n be the number of customers, in thousands, n years after 1990.* Notice that this statement does three things: it gives a letter for the data values, it indicates that the data will be in units of thousands of customers, and it establishes the initial year or starting year for the model. You should reread the problem and clarify in your own mind why these choices were made. Why is the starting year 1990? Why are units of thousands of customers used?

Step 2: Formulate the arithmetic growth difference equation. Since the number of customers goes up by 4,500 per year, the difference equation should say that each c is 4,500 more than the preceding c. Be careful. Note that 4,500 customers is four and one-half *thousands* or 4.5 thousands of customers. The difference equation is

$$c_{n+1} = c_n + 4.5$$

Step 3: Formulate the initial value and any restrictions on n. The problem says there were 70 thousand customers in 1990. That gives $c_0 = 70$ (Why?) The difference equation makes sense for any $n \geq 0$. (Why?)

Step 4: Draw a graph. The visual portrayal of the model gives you another way to look at things. Sometimes you can recognize mistakes or errors because something about the difference equation or numerical results does not agree with the graph.

Step 5: Formulate the functional equation. The functional equation tells us directly what c_n will be. If the number of customers goes up by 4.5 thousand each year, then after n years it will have increased by $4.5n$. Since it started at 70, that gives $c_n = 70 + 4.5n$. If you prefer working with formulas, use $c_n = c_0 + nd$ where $c_0 = 70$ is the initial number of customers and $d = 4.5$ is the amount of increase each year. The end result is the same: $c_n = 70 + 4.5n$.

Step 6: Answer any specific questions. Restate the questions in terms of the model. What will the situation be in the year 2000? That will be 10 years after 1990, so $n = 10$. Then, $c_{10} = 70 + 4.5 \cdot 10 = 115$ means there will be 115 thousand customers in the year 2000. The year 2010 corresponds to $n = 20$. In 2010 there will be $c_{20} = 70 + 4.5 \cdot 20 = 160$ thousand customers, according to the model. When will there be 100 thousand customers? First we ask, for what n does $c_n = 100$? This problem can be approached graphically and numerically, or we can use algebra. For the latter, use the functional equation for c_n and solve for n:

$$c_n = 70 + 4.5n$$

$$100 = 70 + 4.5n$$

$$100 - 70 = 4.5n$$

$$30 = 4.5n$$

$$\frac{30}{4.5} = n$$

$$6.6666 \cdots = n$$

That means that after 6 and 2/3 years we expect there to be 100 thousand customers. That will occur during the year 1996.

The exact order of these steps is not critical. What is important is to understand the overall approach of formulating the model in regular English, then as a difference equation, and then using a functional equation. In the exercises, you will again have a chance to work with this kind of problem.

Proportional Reasoning and Continuous Models

The idea of arithmetic growth leads to two additional ideas that are important in modeling. The first is the idea of proportional reasoning. Here is an example:

> At a price of 40 cents a can, 120 cans of soda can be sold. If the price is raised to 55 cents, then only 90 cans will be sold. If the price is set at 60 cents per can, how many do you think can be sold?

Take a minute now to answer this question just using common sense.

Did you predict 80 cans would be sold at 60 cents per can? Most people do, arguing as follows:

> Raising the price by 15 cents caused a reduction of 30 in sales. That means that each time you raise the price a nickel you lose 10 sales. So, by raising the price from 55 to 60 we expect to lose another 10 sales, giving a result of 80 cans. That is, if the price is set at 60 cents per can, we expect to sell 80 cans.

This kind of reasoning is very common. It is based on the assumption that two quantities change in such a way that their ratio always remains constant or invariant. In this example, we are assuming that the ratio *(price increase)/(sales loss)* gives the same value when the price increase is 20 cents as it does when the price increase is 15 cents. Proportional reasoning can always be expressed in terms of ratios. The connection of this kind of proportional reasoning to arithmetic growth models is quite direct: an arithmetic growth assumption will always lead to the same results as using proportional reasoning.

So far, we have seen that assuming arithmetic growth leads to a particular kind of difference equation. As mentioned before, the original arithmetic growth assumption says a little bit more than the the difference equation that it leads to. Whereas the difference equation might say that the same growth occurs every year (for example), the principle of arithmetic growth indicates that the same growth must also occur every half-year, every month, every week, and so on. In fact, if we know how much growth occurs in a year, we can actually figure out how much growth occurs in any other period of time.

To get an idea of why this is so, consider again the simple interest model. As developed on p. 39, the amount a student owes for a loan increases by $50 every year. That is what gave us the difference equation. But the arithmetic growth assumption also tells us that there must be a $25 increase for each *half* of a year. After all, the increase in the first half of the year must be the same as in the second half of the year, and the total increase over the year is $50. Using similar arguments, we can figure out the increase for any fraction of a year. This is another instance of proportional reasoning. The end result is this: the functional equation that we derived for arithmetic growth makes sense even when the values of n are not whole numbers. In the interest example, the amount owed after n years was found to be $1,000 + 50n$. This result was obtained based on the assumption that n is a whole number. But even for fractional values of n this makes sense. For example, if $n = 3.5$, we find the amount owed is $1,000 + 50 \times 3.5 = $1,175$, and this is the amount that would be owed after three and one-half years. As so often before, we again encounter an important general principle in an example. We will see frequently the idea that a functional equation derived thinking of n as a whole number still makes sense when n is replaced with fractional values. When a variable (such as n) is thought of as being restricted to whole number values, it is described as a discrete variable. Even if the values are not whole numbers, the variable can be discrete if it is restricted to a particular set of separate values. For example, if we make measurements every tenth of a second, the time values are .1, .2, .3, .4, etc. These are elements of a discrete set. In contrast, when a variable is allowed to take on all possible fractional values (possibly restricted to fall between some given maximum and minimum values), the variable is said to be continuous. One good way to test whether a variable is discrete or continuous is to ask whether each value is followed by a specific next value. If n stands for a whole number, then $n = 8$ is followed by a definite next value $n = 9$. However, if n stands for any fraction between 0 and 1, the value $n = .5$ is not followed by a definite next value. It could be .6, but .51 is also a possibility, as are .501, .5001, .50001, and so on. There is no closest next value. That is the idea of a continuous variable.

This issue came up from considering the functional equation $a_n = 1,000 + 50n$, which came out of an analysis in which n was supposed to be a whole number. The functional equation still makes sense if we make n a continuous variable, allowing n to be any fraction. Then, both the amount of time (n) and the amount of money a are continuous variables, linked by the functional equation. In the next several chapters you will study many properties of continuous variables related by simple functional equations.

Beyond Arithmetic Growth: Oil Reserves

The development of a mathematical model is a cumulative process. A model for one phenomenon is used as a basis for modeling a related phenomenon. This process continues, leading to ever more involved models. To illustrate this process, we will refer once again to the oil consumption model.

Given a model for average daily oil consumption, it is a simple matter to proceed to a model for oil reserves. Using 1991 as the baseline year, as before, let us consider the total oil reserves that exist in each succeeding year. According to the International Petroleum Encyclopedia for 1994, the world oil reserves in 1991 amounted to 999.1 billion barrels = 999,100 million barrels. How much oil was left in 1992? In 1993? If you subtract what was consumed in a year from the oil reserves at the start of the year, the result gives the oil reserves at the start of the next year. This leads directly to a new difference equation.

Let us agree to use r_n to represent the oil reserves n years after the start of 1991. This gives r_0 as the oil reserves for 1991, so $r_0 = 999,100$. What is r_1, the oil reserves for 1992? Recalling that c_0 is the daily consumption during 1991, observe that over an entire year the world will consume 365 days worth of oil, or $365c_0$. Accordingly, $r_0 - 365c_0$ gives the reserves for the next year, that is, r_1. In equation form, $r_1 = r_0 - 365c_0$ Reasoning similarly, $r_2 = r_1 - 365c_1$, $r_3 = r_2 - 365c_2$, and so on. This pattern gives a difference equation for the reserves:

$$r_{n+1} = r_n - 365c_n$$

Here we see again a situation where it is useful to have two different variables in the model. Combining the reserves model with the consumption model gives us the following, richer, description.

> c_n represents average daily world oil consumption in millions of barrels n years after 1991. r_n represents the total supply of oil available at the start of the year, n years after 1991. Our model includes the following initial values and difference equations:
>
> $$c_0 = 66.6$$
>
> $$r_0 = 999,100$$
>
> $$c_{n+1} = c_n + .3$$
>
> $$r_{n+1} = r_n - 365c_n$$

We can use this model to compute both the annual consumption and the reserves for any year after 1991. If we are mainly interested in the reserves, we can streamline the process a little by using the functional equation $c_n = 66.6 + .3n$. Just replace c_n with $66.6 + .3n$ in the difference equation $r_{n+1} = r_n - 365c_n$, and we obtain a difference equation for r alone

$$r_{n+1} = r_n - 365(66.6 + .3n)$$

So, building on the model for oil *consumption*, we have formulated a difference equation for oil *reserves*.

An obvious question to ask, now, is: *When will the oil run out?* That is, in how many years will the reserves reach 0? In terms of the model, the problem is to find the n for which $r_n = 0$. This is very similar to questions we have considered before. Numerical and graphical methods can be applied, and you will do that in the exercises. A preferred approach would be to come up with a functional equation for the reserves. The methods for deriving this equation are not yet available to you, but the final result turns out to be

$$r_n = 999,100 - 24,254.25n - 54.75n^2$$

You will be asked to verify this for a few small values of n in the exercises, too. In future chapters you will learn how to obtain the functional equation for r_n and how to use it to figure out when the reserves will reach 0.

The point of this example is to illustrate how models can be built on top of other models. From a simplistic model for oil consumption we are led to a more complicated model for oil reserves. A completely reasonable idea for modeling the reserves using the consumption model leads quickly to much more complicated mathematical questions. In a real application, any conclusion about when the oil will run out would carry the following qualification: *This conclusion assumes a linear model for consumption.* By using alternate consumption models we can obtain quite different conclusions. We will explore this example again in future chapters.

Summary

In this chapter we have discussed arithmetic growth. Arithmetic growth models all share many common features, including the form of their difference and functional equations. They all have straight-line graphs, and they all embody proportional thinking. Arithmetic growth led us into the idea of continuous variables. We also saw how one model can be built on another in the example of oil reserves.

Exercises

Reading Comprehension

1. Write a short essay, about one page in length, on the topic of arithmetic growth. Tell what features are shared by all arithmetic growth models, and give details about graphical properties, difference equations, and functional equations of these models. What are the limitations of using arithmetic growth models to approximate real phenomena? How can you tell whether a particular application might be an appropriate place to use an arithmetic growth model?

2. Write brief paragraphs to explain each of the following concepts:*simple interest, initial value, demand, proportional reasoning, continuous variable.* Include examples in your paragraphs as appropriate.

3. Why are arithmetic growth models often referred to as linear models?

4. In the reading the development of mathematical models is described as being cumulative in nature. What is meant by that? Give an example.

Mathematical Skills

1. For each difference equation find the corresponding functional equation:
   a. $a_{n+1} = a_n + 2;\ a_0 = 1$
   b. $a_{n+1} = a_n + 2;\ a_0 = 5$
   c. $a_{n+1} = a_n + 2;\ a_0 = -312$
   d. $b_{n+1} = b_n - 1.3;\ b_0 = 100$
   e. $p_n = p_{n-1} + .8;\ p_0 = 11.3$
   f. $a_{n+1} = a_n + 2;\ a_3 = 12$

2. Find the graph for each part of the preceding problem. It is recommended that you use a graphing calculator or a computer graphing tool for this problem.

3. For each part, a functional equation is given. Find the corresponding difference equation, and the initial value a_0.
   a. $a_n = 15 - 3n$ b. $a_n = 15 + 3n$
   c. $a_n = 20 - 3n$ d. $a_n = 20 + 3n$
   e. $a_n = 20 - 5n$ f. $a_n = 20 + 5n$
   g. $a_n = 20n - 3$ h. $a_n = 20n + 3$

4. For each part of problem 3 find a_8.

5. In part (a) of problem 3 find the n for which $a_n = 3$.

6. In part (b) of problem 3 find the n for which $a_n = 3$.

7. For each part of problem 3 find an equation for n as a function of a_n. [For example, for part a, since $a_n = 15 - 3n$, we have $3n = 15 - a_n$ so $n = 5 - a_n/3$. This equation expresses n as a function of a_n.]

Problems in Context

1. A weather balloon carries a battery-powered radio transmitter which sends weather data back to the ground. When the balloon is sent up, the battery carries a charge of 30 units. It uses up 2.4 units of charge per hour. Let q_n represent the charge on the battery n hours after the balloon is sent up.
   a. Using a numerical method, find q_1, q_2, and q_3.
   b. What is the difference equation for q_n?
   c. What is the functional equation for q_n?
   d. What will the charge be 4 hours after launch?
   e. Find an equation expressing n as a function of q_n.
   f. The radio transmitter cannot continue to work once the charge on the battery falls below 4 units. How many hours will that take?

2. A student borrows $5,000 from his aunt. He promises to pay the money back as soon as possible, with simple interest. The interest will be calculated at one half of a

percent per month. Let p_n be the amount of money the student will have to pay the aunt if he makes the payment after n months. For example, if he makes the payment after a year, that is 12 months. At half a percent per month, the interest will be 6 percent in 12 months. Now 6 percent of $5,000 is $0.06 \times 5,000 = 300$. So if the loan is paid back after 12 months, the amount that has to be paid is $5,300, the original loan of $5,000 plus the $300 interest. That is p_{12}. In contrast, p_6 is what the student has to pay if he makes the payment after 6 months.

    a.   What is p_0? Does that make sense?

    b.   What is the difference equation for p_n?

    c.   What is the functional equation for p_n?

    d.   Use the functional equation to figure out how much must be paid if the payment is made after 18 months.

    e.   Use the functional equation to find an equation for n as a function of p_n. This is the inverse equation.

    f.   Use the inverse equation to find how long it will be before the student owes twice the amount she originally borrowed. That is, find n so that $p_n = \$10,000$.

3. A scientist studying the spread of a new disease in a small town decides to use an arithmetic growth model. She estimates that 3,700 people have the disease at the start of her study, and that there are 45 new cases each day. Follow steps similar to those in the last two problems to analyze the arithmetic growth model for this epidemic.

4. This is a continuation of the preceding problem. The scientist has also found that about 3 percent of the people who get the disease require treatment with a special medicine. The local hospital had 500 doses of the medicine on hand at the start of the study. According to the model, how long will it be before this medicine is used up?

5. (Continuation of the preceding problems.) Suppose that the small town in the study is isolated—very few people arrive or leave. Given what you know about the way diseases spread, do you think an arithmetic growth model is reasonable? Consider both predictions made over a short period of time, and those over much longer periods of time.

6. Oil Reserves. Review the discussion of the oil reserves model on p. 46. There, the following functional equation was given:

$$r_n = 999{,}100 - 24{,}254.25n - 54.75n^2$$

Use this equation to compute r_n for several different values of n. To check that the answers are correct, also compute r_n using the difference equation $r_{n+1} = r_n - 365(66.6 + .3n)$ that is based on the oil consumption model. The starting value is $r_0 = 999{,}100$. The point of this exercise is to provide you with some evidence that the functional equation above is correct. Later you will see how the equation was obtained.

7. Using the functional equation for r_n in the previous problem, use numerical techniques to study two questions: What will the oil reserves be in the year 2000? And

when will half the oil that was available in 1991 be used up?

8. Consider again the exercises about the epidemic of a disease in a small town. Modify the previous discussion as follows. Suppose that the spread of the disease is increasing, and that the researcher observes 45 new cases on the first day of the study, 50 new cases on the next day of the study, and 55 new cases the day after that. She models the number of new cases per day using an arithmetic growth model. Here, let c_n be the number of new cases of the disease in day n of the study. So $c_1 = 45$, $c_2 = 50$, and so on. Study the way the number of new cases each day increases. Then use numerical methods to figure out how many days the hospital's medicine supply will last.

Group Activity

1. Look through newspapers or magazines for a graph that appears to be nearly a straight line for all or part of the data. Create an arithmetic growth model for this graph, and write a short report on the model. Your report should define what is being modeled, what assumptions are being made, and over what range of data the model is expected to be valid. Also discuss the kinds of predictions that can be made using your model, and why these predictions might be useful.

2. Using the data from your news article, or from the carbon dioxide example (see Fig. 1.1 on page 2), find the best line you can for the data. Do this by eye. Simply put a ruler on the graph and draw a line as close as possible to the data points. What is the difference equation for the points on this line? What is the functional equation?

3. Again using the data from the preceding problem, compute the difference between each data value and the preceding data value. Are these differences fairly constant? If so, compute the average difference and use it to define an arithmetic growth difference equation. Compare the model based on this difference equation with what you found in the preceding questions.

Solutions to Selected Exercises

Mathematical Skills

1. a. $a_n = 1 + 2n$

   c. $a_n = -312 + 2n$

   e. $p_n = 11.3 + .8n$

   f. Starting with $a_3 = 12$, can you find a_2? a_1? a_0? The difference equation tells us that the sequence of a's goes *up* by 2 each time. So if $a_3 = 12$, a_2 must have been 10, $a_1 = 8$, and $a_0 = 6$. Therefore, $a_n = 6 + 2n$. As a check, this gives $a_3 = 6 + 2 \times 3 = 12$, which is what we expected.

3. There are three different ways to approach these problems: (1) use the general form of the functional equation to identify the parameters a_0 and d; (2) use the functional equation to work out a few terms of the sequence so that you can see the pattern of

the difference equation; and (3) use algebra to derive the difference equation. Each method will be illustrated below.

a.   Using method 1: compare these equations:

$$a_n = a_0 + nd$$

$$a_n = 15 - 3n$$

The two equations will be the same if $a_0 = 15$ and $d = -3$. So the initial value for this problem is $a_0 = 15$, and the difference equation is $a_{n+1} = a_n - 3$.

c.   By method 2: Using the equation $a_n = 20 - 3n$ we can compute several numbers in the sequence. Starting with $n = 0$ the equation gives $a_0 = 20 - 3 \cdot 0 = 20$. Similarly, $a_1 = 20 - 3 \cdot 1 = 17$; $a_2 = 20 - 3 \cdot 2 = 14$; and $a_3 = 20 - 3 \cdot 3 = 11$. So, the pattern is 20, 17, 14, 11, ... It should be clear that the difference equation for this pattern is $a_{n+1} = a_n - 3$, and the value of $a_0 = 20$ was already found.

e.   Using method 3: The difference equation begins with a_{n+1} so that is what we use in the functional equation. The result is $a_{n+1} = 20 - 5(n + 1) = 20 - 5n - 5$. But $20 - 5n = a_n$, so $a_{n+1} = a_n - 5$. That gives the difference equation. The initial value is given by $a_0 = 20 - 5 \cdot 0 = 20$.

g.   Answer: $a_{n+1} = a_n + 20$; $a_0 = -3$.

5. The functional equation is $a_n = 15 - 3n$, and we want $a_n = 3$. So set the left side of the functional equation equal to 3: $3 = 15 - 3n$. You can solve this by trial and error, just guessing different values of n and seeing if they work. Or, use algebra:

$$3 = 15 - 3n$$

$$1 = 5 - n$$

$$n + 1 = 5$$

$$n = 4$$

Check your answers, if possible. Here, we can go back to the functional equation and check for $n = 4$: $a_4 = 15 - 3 \cdot 4 = 3$. That is what the problem required.

7. For part c:

$$a_n = 20 - 3n$$

$$a_n + 3n = 20$$

$$3n = 20 - a_n$$

$$n = \frac{20 - a_n}{3}$$

The answer for part g is $n = (a_n + 3)/20$

Problems in Context

1. a.   The starting charge is 30. Each hour the charge goes down by 2.4. So after one hour the charge is 27.6, this is q_1; after another hour the charge is 25.2, which is q_2; and after another hour the charge is 22.8, which is q_3.

   b. $q_{n+1} = q_n - 2.4$.

c. We start at 30. q_n is found by reducing the original amount by 2.4 per hour for n hours. The total reduction is $2.4n$, so $q_n = 30 - 2.4n$.

d. The answer is $q_4 = 30 - 2.4 \cdot 4 = 20.4$.

e. Using algebra on the functional equation

$$q_n = 30 - 2.4n$$

$$q_n + 2.4n = 30$$

$$2.4n = 30 - q_n$$

$$n = \frac{30 - q_n}{2.4}.$$

f. We want to know when $q_n = 4$ or less. Set $q_n = 4$, and use the equation from the previous problem: $n = (30 - 4)/2.4 = 10.833333$. So after 10.8333 hours (or 10 hours and 50 minutes), the charge will be less than 4, and the transmitter will stop working.

3. Following the outline starting on page 43: First, define the variables. Let s_n be the number of sick people n days after the start of the study. Next, assuming 45 new cases per day, the difference equation is $s_{n+1} = s_n + 45$. Next, observe that the starting value is $a_0 = 3,700$. Now we can make a graph. It is recommended that you use a graphing calculator or a computer, but it is possible to do it by hand. Observe that the first several data values will be 3,700, 3,745, 3,790, 3,835, 3,880, etc., always increasing by 45 cases per day. The functional equation for this problem is $s_n = 3,700 + 45n$.

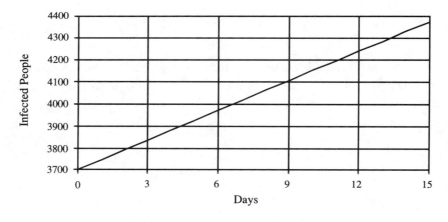

FIGURE 3.1
Graph for problem 3.

4. In this problem, there are a lot of different parts to keep straight. At first, it is normal to feel a little lost or overwhelmed. When that happens, a good strategy is to look at specific cases and get a feel for the numbers. In this particular problem,

you can do this by systematically working out how much medicine is used up after 1 day, then after 2 days, and so on. At the very start, with 3,700 people infected, how much medicine is required? 3,700 doses? No, only 3 percent of the people need the medicine. What is 3 percent of 3,700? It is computed as $.03 \cdot 3,700 = 111$ doses. After one day, another 45 cases come in. How many of them will need the medicine? Right, 3 percent. Now 3 percent of 45 is $.03 \cdot 45 = 1.35$. Of course, you will not use up 1.35 doses of medicine that day. But the idea of 3 percent of patients needing the medicine is an average. So some days you might give one dose, others you will give two doses, but in the long run, it will average out to 3 percent. For this reason,'we will act as if 1.35 doses were given the first day. Then the total medicine used up is $111 + 1.35 = 112.35$. Do you see that each day another 1.35 doses will be used? And that therefore there is an arithmetic growth model for the amount of medicine used up? Let d_n be the number of doses used up by end of day n. We know that $d_0 = 111$. Then $d_1 = 111 + 1.35$, and d_2 will be the d_1 amount plus 1.35, and so on. The difference equation is $d_{n+1} = d_n + 1.35$. That means the functional equation is $d_n = 111 + 1.35n$. We want to know when this will reach the total 500 doses available. That is, we want to solve $111 + 1.35n = 500$. As usual, you can use graphical and numerical methods to get an idea of what the answer will be. For an exact answer, we use algebra: $1.35n = 389$; $n = 389/1.35 = 288.15$. This shows that the medicine will last 288 days, and will be used up by the next day, assuming the epidemic continues to grow according to the arithmetic growth model.

6. Here are several sample calculations

$$r_0 = 999,100 - 24,254.25 \cdot 0 - 54.75 \cdot 0^2 = 999,100$$

$$r_1 = 999,100 - 24,254.25 \cdot 1 - 54.75 \cdot 1^2 = 974,791$$

$$r_2 = 999,100 - 24,254.25 \cdot 2 - 54.75 \cdot 2^2 = 950,373$$

$$r_3 = 999,100 - 24,254.25 \cdot 3 - 54.75 \cdot 3^2 = 925,845$$

Using the difference equation, we start with $r_0 = 999,100$, $r_1 = 999,100 - 365(66.6 + .3 \cdot 1)$. This gives the same answer as the functional equation. In fact, all of the answers check out.

7. The first part is a matter of direct use of the functional equation. The year 2000 is 9 years after 1991, so the oil reserves at that time will be

$$r_9 = 999,100 - 24,254.25 \cdot 9 - 54.75 \cdot 9^2 = 776,377$$

For the other part of the question, use trial and error guessing different n's. We are looking for an oil reserve that is half the original amount of 999,100, that is, for a reserve of 499,550. We already found that $r_9 = 776,377$ which is larger than 499,550. So we need to try an n that is larger than 9. How about $n = 15$? That gives

$$r_{15} = 999,100 - 24,254.25 \cdot 15 - 54.75 \cdot 15^2 = 622,968$$

That is still too big. Try $n = 20$.

$$r_{20} = 999{,}100 - 24{,}254.25 \cdot 20 - 54.75 \cdot 20^2 = 492{,}115$$

This is a little less than the answer we are looking for. Continuing in this way, you can test different values of n. This trial-and-error approach is often a feature of numerical methods. In this problem, we find $r_{19} = 518{,}505$, which is more than half of the 1991 oil supply. So it will be between 19 and 20 years, according to this model, before the oil is half gone.

4

Linear Graphs, Functions, and Equations

In the last chapter, we found that the graphs of arithmetic growth models always appear as straight lines. We also found that the functional equations for arithmetic growth models always appear in the same form. This form defines what is called a *linear equation*, so called because the graph is a straight line. In this chapter we will study linear equations a little more carefully, covering ideas connected with functions, graphs, and solving equations. Additional important topics are *continuous* models and *proportional reasoning*.

Linear Functions and Equations

The general form for the functional equation of an arithmetic growth model is

$$a_n = a_0 + d \cdot n$$

In this equation, a_0 and d are parameters. That is, they stand for numerical values that can change from one problem to another, but are always thought of as remaining fixed or constant. For one of the examples we used, $a_0 = 66.6$ and $d = .3$ giving us

$$a_n = 66.6 + .3n \tag{1}$$

Recall that this equation expresses a_n as a function of n because it permits us to compute a value of a_n as soon as we know n. To be more specific, we refer to this as a *linear* function because the functional equation is a linear equation.

A word is in order here about terminology, and in particular about the terms *equation*, *expression*, and *function*. It is a common error to use the word *equation* for just about anything that can be written down with algebraic symbols. Many students refer to all of the following as equations:

$$y = 3x + 4 \qquad 3x + 2 \leq 5y \qquad 7x^3 + 5xy - 2$$

To be technically correct, only the first of these is an *equation*, because only the first has an equal sign. The middle example should properly be called an inequality, while the third is simply an expression. An expression has no equal sign or inequality. It can be thought of as a recipe for computing something. Indeed, if you replace every variable in an expression with a number, you end up with something that can be computed. In this chapter, the focus is on examples that are *linear*. To use the terms properly, $y = 3x + 4$ should be called a linear *equation,* while $3x + 4$ should be called a linear *expression.*

Both equations and expressions are closely connected to the concept of *function*. The accepted mathematical definition of a function is closest conceptually to a set of (x, y) data points. So when we say that a_n is a function of n, it is most correct to think of the graph, where each point is of the form (n, a_n). Normally, we use an equation to find these data points. For the equation $a_n = 66.6 + 3n$, we can start with any n we like, say $n = 3$, and compute $a_3 = 66.6 + 9 = 75.9$. That gives $(3, 75.9)$ as one of the points that make up the function. In cases like this, it is often convenient to refer to either the equation itself, or to the expression that appears on the right side of the equation, as the function. For example, you will sometimes see statements like

For the function $y = 3x + 4$ ...

or

Note that $3x + 4$ is a function of x ...

In each case, it is understood from the context that the function being considered is made up of points (x, y) and that for each point, if you are given the x, you can compute the y as $3x+4$. Although technically it is not completely correct to refer to expressions or equations as functions in this way, it usually does not lead to any confusion. Throughout this book, the convention will be to use *expression* and *function* more or less interchangeably, and to use *equation* to emphasize the idea that there is an equal sign present. To illustrate this distinction using the earlier examples, $y = 3x + 4$ will be called a linear *equation,* while $3x + 4$ will be called a linear *function.*

So far, linear equations and functions have been discussed only in connection with specific examples. Let us proceed to a definition of the concept of linearity. It is based on the mathematical operations used to perform a calculation. When we use Eq. (1) to compute a_n, what operations are performed? The variable n is multiplied by one constant, and added to another. These simple operations define the concepts of linear function and equation:

> Linear functions and linear equations involve just two types of operations: (1) addition/subtraction, and (2) multiplication/division by constants.

For example, consider the operations that have to be performed in

$$z = 3(x - 7y) + w/5$$

In each multiplication at least one of the two quantities multiplied is a number, and the only division that appears is division by 5, a number. There are also several additions and subtractions. This is a linear equation. It defines z as a linear function of x, y, and w. In contrast, the following examples are not linear equations

$$xy + 2 = z$$
$$x^2 + y^2 = z^2$$
$$y = -2 + \sqrt{4 - 3x}$$
$$y = \frac{3x + 4}{4x - 3}$$

In each of these there are operations other than addition, subtraction, and multiplication or division by constants. For example, in the second equation there appears an x^2. This is an abbreviation for $x \cdot x$ and so requires multiplication of two non-constant quantities. You should try to decide why each of the other equations is not a linear equation.

The definitions of linear equation and linear function above are formulated in terms of numerical operations. There is an alternative description that emphasizes visual appearance:

> A linear function is a combination of variables and constants in which (1) variables have no exponents, (2) variables are not multiplied by other variables, (3) variables do not appear in denominators, (4) variables are not acted on by other operations, such as square roots and logarithms.

This is not really a complete definition since it doesn't specify exactly what the other operations in item (4) are, but it provides a useful way to think about linear functions.

A linear equation may involve as many variables as we choose, but for most of the course we will be especially interested in linear equations in two variables, like Eq. (1). There, one of the variables is n and the other is a_n. As you are aware, it is possible to rewrite one equation in many different forms by applying the rules of algebra. The equation $3x + 4y = 2$, for example, can also be written as $4x + 3y = 2$, $3x + 4y - 2 = 0$, $3x = 2 - 4y$, and in many other ways. Among all the possible ways to express a linear equation in two variables, there are a few standard formats that are usually used. For example, in Eq. (1) the constant 66.6 would normally be moved to the end of the equation, giving the form

$$a_n = .3n + 66.6 \tag{2}$$

The significance of this and of other standard forms will be discussed later in this chapter.

Continuous Models

An important aspect of Eq. (2) is the fact that n is thought of as a whole number. Do you remember the discussion in Chapter 2 of discrete data? There we emphasized the idea of data sequences with a first value, a second value, and so on. The variable n was introduced as a label or counter for the sequence values, and that is why n made sense for whole number values: $n = 1$ for the first data value, $n = 2$ for the second, and so on. But in many applications it is reasonable to think about fractional values for n, especially when the data values are for a series of times. Suppose, for example, a model for the AIDS epidemic uses a_n as the number of infected people after n months. Then $a_{3.5}$ would mean the number of infected people after three and one-half months. In fact, we can reasonably interpret an a with any decimal value for the subscript: $a_{4.687}$ would mean the number of infected people after 4.687 months. However, this notation is rarely used. Instead of expressing the decimal as a subscript, it is written in parentheses. That is, $a(4.687)$ would be used instead of $a_{4.687}$.

The preceding example illustrates the idea of a continuous variable. If we only allow n to be a whole number, the model that is based on n is a discrete model. If we allow n to be any real number (meaning any decimal, or any point on a number line), then n is a continuous variable, and may be part of a continuous model. Continuous and discrete models can also be distinguished by the kinds of graphs that are used. In the discrete case, the graphs are defined by a series of individual separate points, as in Fig. 2.1 and Fig. 2.3. In the continuous case, we imagine that there are actually data points for each n on the horizontal axis. If we could individually graph all of these points, they would crowd together completely covering a line without gaps or separations. This is the image that inspires the term *continuous*.

Which is better, discrete or continuous? For our purposes, the discrete approach will be easier to understand when it comes to describing how changes occur. It is the context in which we formulate difference equations. However, when we combine the theoretical approach with a difference equation, the end result is frequently a functional equation which makes sense with continuous variables. This is what happened in the case of the oil consumption model. The difference equation was formulated by thinking about how the consumption next year might depend on the consumption this year. In that setting, and in particular in the equation

$$c_{n+1} = c_n + .3$$

it was natural to think of n as a whole number. That is, we think of n as a discrete variable. Eventually we obtained the functional equation

$$c_n = .3n + 66.6$$

Now it is natural to allow n to be any decimal, so that we can talk about the oil consumption after 3.4 years, for example. When we use the functional equation, it is reasonable to think of n as a continuous variable. As we proceed, we will follow this same progression many times: starting with a discrete model and a difference equation; using it to find a functional equation; and then using the functional equation as part of a continuous model.

This discussion of continuous variables appears here to prepare for the ensuing consideration of linear equations. In the remaining sections of the chapter you will learn about algebraic and graphical aspects of linear equations, as well as applications and the use of proportional reasoning. In all of these topics, continuous (rather than discrete) variables will be used.

Algebra and Solving Linear Equations

Consider again an arithmetic growth model for oil consumption:

$$c_n = .3n + 66.6$$

What is the daily oil consumption when $n = 10$? The answer to this question is simply a matter of computation: Replace n by 10 and carry out the operations on the right side of the equation. In contrast, the following question is not so easily answered: When will the daily oil consumption equal 75? In this case, we are specifying c_n and asking for the value of n. Replacing c_n with 75 leads to

$$75 = .3n + 66.6$$

The task is to find the n for which this becomes a true statement. Finding that n is referred to as solving the equation.

There is a general method for solving linear equations, and usually, the result is a single answer. We will see later that other kinds of equations can be more difficult to solve, and may produce multiple answers.

The Importance of Solving Equations. The general idea of solving an equation is one of the most fundamental concepts in algebra. Why is it so important? The answer brings up the topic of functions again. Notice that in the first question, the one that is easy to answer, we are given n and seeking c_n. Since the equation gives c_n as a function of n, that is an easy task. In the second question the roles are reversed. We are asked to find the n that produces a certain value of c_n. This is sometimes referred to as *inverting* the function, since we want to begin with the result of the function computation and figure out what the starting point was. It occurs repeatedly with functions. For example, suppose we have a model that predicts the ozone depletion as a function of time. The model will predict for any time what the level of ozone depletion will be. Then someone will ask when a particular critical ozone level will be reached. That question calls for inverting the function in the model: it specifies the ozone depletion and asks for the time. For another example, we may have a model that predicts how many voters will choose our candidate as a function of the amount of money that will be spent on advertising. If you spend 2 million dollars, the model predicts how many votes you will get. But what the candidate wants to know is, how much must be spent to win? You know how many votes are needed, and want to determine how much money it will take. That calls for inverting the function in the model. In all of these examples, if the function in the model is given by an equation, then the process of inverting the function amounts to solving an equation. That is why solving equations is such an important part of algebra.

The Process of Solving Equations. When algebra is used to solve an equation, there is one basic process that is used over and over again: replace the equation you have with a simpler equation without changing the solution. Here is an example. If the original equation is $3x + 4 = 10$, then we replace it with the simpler equation $3x = 6$. We can argue that the new equation has the same solution as the original as follows. If $3x + 4$ and 10 are equal, then subtracting 4 from them both will produce equal results. That is $3x$

and 6 must be equal. That is another way of saying that the same value of x solves both equations. For linear equations, there are only a few necessary operations that simplify equations. You can

- algebraically rearrange either side
- add the same amount to both sides
- subtract the same amount from both sides
- multiply both sides by the same amount
- divide both sides by the same amount (but not by zero)

Using these operations, any linear equation can be reduced to the simple form *variable* = *number*. Here is a somewhat involved example:

$$2(3x - 5) + 6 = 2 - 4x + 2(x - 4)$$
$$6x - 10 + 6 = 2 - 4x + 2x - 8$$
$$6x - 4 = -2x - 6$$
$$6x - 4 + 2x = 2x - 2x - 6$$
$$8x - 4 = -6$$
$$8x - 4 + 4 = -6 + 4$$
$$8x = -2$$
$$8x/8 = -2/8$$
$$x = -1/4$$

In this example, each step has been done separately to emphasize what operations are being performed. Normally, when you solve an equation, you will be able to do several steps at once and shorten the process significantly.

No Solution; Every Number a Solution.   It is possible to have linear equations without any solution, or for which every number is a solution. Although these equations rarely occur in practice, it is not always obvious when they do occur. For example, the following equation looks like a typical linear equation.

$$3x + 2 = 5 + 3(x - 4)$$

A few steps of algebra lead to the equation

$$2 = -7$$

This is an impossible situation, and is never true no matter what x is. So the original equation has no solution. On the other hand, this equation

$$2 - 4x = 4(3 - x) - 10$$

is true for every value of x. Try a few. If you make $x = 0$ the equation becomes $2 = 12 - 10$, which is certainly true. What do you get if $x = 1$? $x = -1$? In all these cases the resulting equation is true. Now no number of examples can *prove* that this equation is true for every x. However, this conclusion *can* be reached by using algebra. Applying the same kind of steps as in the preceding examples, we reach the equation $2 = 2$. Since this is true for all values of x, so is the original equation. As already stated, equations of this type rarely actually occur in practice, although they do appear once in a while, especially as a result of an error. It is important to understand what they mean when they do occur. Then, if an error has been made, it will be easier to spot.

More than One Variable. So far the examples have all involved just one variable. The same kinds of operations can be applied in equations with more than one variable. In that case, it is often useful to simplify the equation to a form with one of the variables isolated on one side of the equation, and all other terms on the other side of the equation. As an example, we will again use the oil consumption model

$$c_n = .3n + 66.6$$

This equation gives c_n as a function of n. Using algebra, we can replace it with one that has n isolated on one side of the equation:

$$c_n = .3n + 66.6$$
$$c_n - 66.6 = .3n$$
$$(c_n - 66.6)/.3 = n$$
$$n = (c_n - 66.6)/.3$$

This process is described as solving for n in terms of c_n. Notice that the final equation gives n as a function of c_n, because as soon as the value of c_n is substituted, the value of n can be immediately computed. As described before, solving the equation for n in this way amounts to *inverting* the original equation, the one giving c_n as a function of n. The new equation for n as a function of c_n is very useful. Using it, we can easily determine for any level of oil consumption, when that level will be reached. Following the convention mentioned earlier for replacing subscripts with quantities in parentheses, we can express the original equation in the form

$$c(n) = .3n + 66$$

where we think of n as any whole number, fraction, or decimal. In the same way, the inverse equation can be written

$$n(c) = (c - 66.6)/.3$$

This emphasizes that n is expressed as a function of the consumption c.

Graphs of Linear Equations

We have seen in earlier work how to create a line graph for the discrete model

$$c_n = .3n + 66.6$$

In that discussion, we used whole number values for n and plotted a point for each n. If we now think of n as a continuous variable, it would be impossible to plot a point for every possible value of n. However, methods of higher mathematics can be used to show that the result of plotting all possible points would be a smooth continuous curve, and in the case of a linear equation, a straight line. Graphing on a computer is performed by plotting many points so closely spaced that the appearance of a smooth curve results. The case of a straight line is special. We need only plot two points, and then draw a line through them using a straight edge.

What is more interesting is to develop a qualitative feel for the connections between parts of the equation and aspects of the graph. This insight can be useful in two ways. First, given a graph showing a line, we might want to determine the equation of the line. Second, if we have an equation, we might want to form an idea of the appearance of the line. Examples of each case will be given in more detail. Before moving on to those examples, we introduce a slight change in notation. To emphasize the idea that the variables are continuous, we will represent them as single letters, without using a subscript. In the context of a specific problem, the letters will be chosen to help remind us what the variables stand for. For the case of oil consumption, we will use c for consumption and n for number of years. For general discussions that are not connected with a specific problem, x will be used for the variable on the horizontal axis and y for the variable on the vertical axis.

Special Forms for Linear Equations.   There are three different forms that are often used for linear equations.   The first is the one we mentioned earlier.   For the oil consumption example, it was

$$c = .3n + 66.6$$

The more general form is

$$y = mx + b \tag{3}$$

In this form the constants m and b have special significance. The number b indicates the point at which the line crosses the y axis. It is called the *y-intercept*. For example, the graph of the equation $y = 3x + 5$ crosses the y axis at 5 (Fig. 4.1). Similarly, the graph of $y = 3x - 5$ crosses the y axis at -5 (Fig. 4.2). Notice that in the equation $y = mx + b$, b is the value of y that results when x is set to 0. An x value of 0 always gives a point on the y axis.

Slope.   The number m also has a special meaning. It is called the slope and tells how steeply the line slopes up or down as you trace it from left to right. The numerical value of the slope tells how much the line moves up for every unit you move to the right. For the equation $y = 3x + 5$, the slope is 3. This indicates that the line moves up 3 units for every unit you move to the right. This is illustrated in Fig. 4.3. For the line $y = -3.2x + 5$, the slope of -3.2 indicates that the line moves *down* 3.2 units for each unit to the right (Fig. 4.4). When the slope is a fraction, it has an alternate interpretation. For $y = \frac{3}{4}x + 2$, the line moves up 3 units for every 4 units to the right (Fig. 4.5). In this context, think of $3/4$ as 3 *over* 4, and think of the slope as *up* 3 *over* 4.

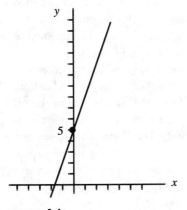

FIGURE 4.1
Positive *y* intercept

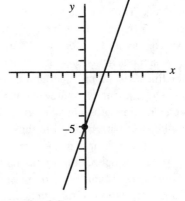

FIGURE 4.2
Negative *y* intercept

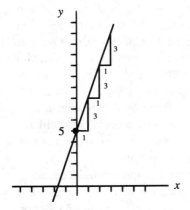

FIGURE 4.3
Positive slope

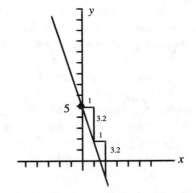

FIGURE 4.4
Negative slope

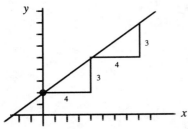

FIGURE 4.5
Slope is a fraction

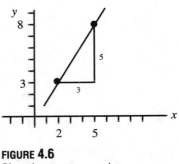

FIGURE 4.6
Slope between two points

   The slope of the line can be computed using any two points of the line. Draw a triangle as in Fig. 4.6. Then moving from left to right along the line, the distance you go up is called the rise (it is negative if you go down). The distance you go to the right is called the run. The slope is the rise over the run. From the point $(2,3)$ to the point $(5,8)$ the rise is 5 (from 3 to 8) and the run is 3 (from 2 to 5), so the slope is $5/3$. This is another way of saying that for every 3 units you go to the right, you also go 5 units up.

A slope is actually a rate. In a purely geometric discussion, a slope of 3.2 indicates that the line rises 3.2 units per unit traveled to the right. In the context of a specific application, the slope acquires whatever units of measurement are used on the two axes. For the oil consumption problem, the equation $c = .3n + 66.6$ relates the daily consumption c in millions of barrels of oil to the number n of years after 1991. The slope .3 indicates that daily oil consumption goes up by .3 (million barrels of oil) for every 1 (year) increase in n. The slope therefore has units of millions of barrels of oil per year. The statement *daily oil consumption is increasing at the rate of .3 million barrels of oil per year* is a direct translation of the mathematical statement that the slope is .3.

The Slope–Intercept Form. As explained in the preceding discussion, when an equation is expressed in the form $y = mx + b$, we can immediately see the slope and the y–intercept of the line. For example, the graph of $y = 2.4x - 7$ has a slope of 2.4 and a y–intercept of -7. With this in mind, $y = mx + b$ is called the *slope-intercept* form of a linear equation. It gives an immediate visual understanding of the graph. Start at the point b on the y axis, and proceed on a diagonal going up m units for every unit you go right. You can actually generate a sequence of points following the recipe *up m over 1* over and over. For a numerical example, if the equation is $y = \frac{1}{2}x + 3$, start out at 3 on the y axis, go up 1 and over 2 to a new point, then up 1 and over 2 to another new point, etc. This is illustrated in Fig. 4.7. If you think about this process, you will see that it is intimately connected to the idea of arithmetic growth. The vertical positions of the points grow arithmetically as you generate them from left to right.

The slope-intercept form of a line allows you to visualize the graph given the equation. It is also useful for the reverse problem: determine the equation given the graph. For example, in Fig. 4.8 a line has been drawn. What is the equation? A visual inspection

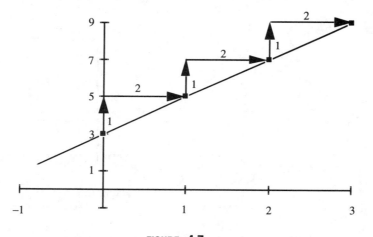

FIGURE 4.7

Up 1 and Over 2

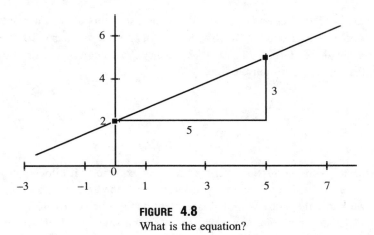

FIGURE 4.8
What is the equation?

of the graph reveals that the y intercept is 2. A little careful measurement of the triangle that is shown in the figure will convince you that the slope is $3/5$. So the equation can be immediately written down as $y = \frac{3}{5}x + 2$.

The Point–Slope Form. Sometimes you have a graph in which it is inconvenient to find the y intercept. In Fig. 4.9 the axes have not been drawn in the usual locations. That is, they do not meet at the 0 point of each axis. Because our notion of y intercept assumes that x is 0 on the vertical axis, it is an error to think of the y intercept as the point where the line crosses the vertical axis for this figure. Instead, the true y intercept, and the y axis itself, would actually be far to the left of the figure. It is still possible to determine the equation in a simple way. First, using the two points shown, $(73, 45)$ and $(83, 51)$, the slope is easily computed. The graph goes over 10 (from 73 to 83) and up 6 (from 45 to 51) so the slope is $6/10$ or $.6$. Now imagine another point on the line at an

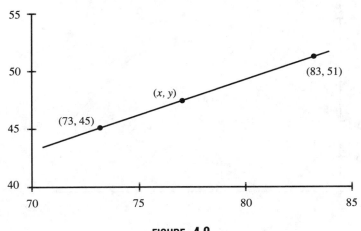

FIGURE 4.9
No y intercept

unknown location (x, y). Let us recompute the slope using this unknown point and the first point, $(73, 45)$. The rise is $y - 45$ (from 45 to y), and the run is $x - 73$ (from 73 to x). This gives the slope as $(y - 45)/(x - 73)$. Since we already know the slope is .6, we conclude that for any point on the line,

$$\frac{y - 45}{x - 73} = .6$$

or

$$y - 45 = .6(x - 73).$$

This is referred to as the *point-slope* form of a linear equation. It allows you to write down the equation for a line as soon as you know the slope and one point of the line. In this case, we know the slope is .6 and one point of the line is $(73, 45)$. For another example, Fig. 4.10 shows a graph for soda prices (p) and sales (s). If the price is set at 80 cents, 200 sodas can be sold. If the price is raised to 90 cents, only 160 sodas will be sold. The slope between these two points is $-40/10 = -4$ in units of sodas per penny. That is, the sales drop by 4 sodas per penny increase in price. Knowing the slope (-4) and one point $(80, 200)$, we can immediately write the equation: $s - 200 = -4(p - 80)$.

The general form of the point-slope equation is usually written

$$y - y_0 = m(x - x_0) \tag{4}$$

In this equation, x_0, y_0, and m are parameters. In any actual application there will be numerical values for these parameters. Just as in the example $x_0 = 73$, $y_0 = 45$, and $m = 0.6$. In using the general form Eq. (4), think of the parameters as fixed numbers and of the variables x and y as representing many different points all on the line.

Like the slope–intercept form, the point–slope form can be used to visualize the graph given the equation. For example, in $y - 5 = .12(x - 4)$ we can recognize immediately that the point $(4, 5)$ is on the line, and the slope is .12. As before, we can generate a series of points by starting at $(4, 5)$ and repeatedly going up .12 and over 1.

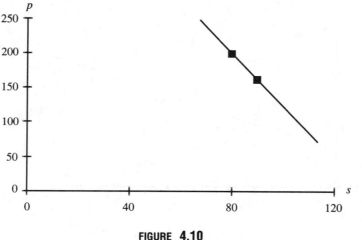

FIGURE 4.10
Sales vs. Price

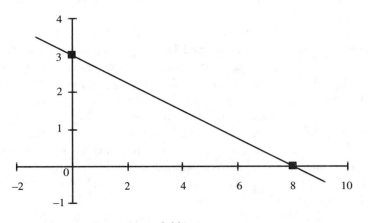

FIGURE **4.11**
Using Both x and y Intercepts

The Two-Intercept Form. There is a third form of a linear equation that is especially useful if we know where the line crosses both axes. In Fig. 4.11 the line shown crosses the x axis at 8 and the y axis at 3. From this information the following equation can be written down immediately:

$$\frac{x}{8} + \frac{y}{3} = 1$$

The reasoning for this equation is a little different from the previous cases. Observe that this *is* a linear equation, so the graph must be a straight line. It is clear that the illustrated x intercept satisfies the equation, for at the x intercept, $x = 8$ and $y = 0$. By similar reasoning the illustrated y intercept also satisfies the equation. This shows that the graph of the line must go through the two intercepts shown, and so must be exactly the line that is illustrated. This example uses what is called the 2–intercept form of a linear equation. The general expression of this form is

$$\frac{x}{a} + \frac{y}{b} = 1 \tag{5}$$

Here, the letters a and b are parameters. In using this form, a should be replaced by the x intercept of the line and b should be replaced by the y intercept of the line. Note: there is no 2–intercept form for a line that goes through the point $(0,0)$. For such a line, the x and y intercepts are both 0. With $a = 0$ and $b = 0$, Eq. (5) would not make sense.

Activity. You have now read about three different forms for linear equations. Before proceeding with the reading, it would be a good idea to go back and review those three forms. In your notes for this section, write out the three forms using parameters (like m and b, or m, x_0, and y_0). For each equation, write out in words what each parameter tells you about the graph. Give a specific example of each equation, using numbers in place of the parameters, and show the graph of each example.

Fitting a Line to Data

It is almost always the case that a linear model only approximates the data that appear in a problem. In that case, the linear model is chosen to come as close as possible to the data points. How is this done? One approach is to draw the line that appears best to the eye, and then determine the equation using the methods of this chapter. The equation derived in this way then becomes part of the linear model.

As an example of this approach, consider again the oil consumption data used before, and repeated below in Table 4.1. The graph for the individual data points is shown in Fig. 4.12. The line shown in the graph was simply drawn by eye to fit the data. Likewise, the two points shown on the line were estimated by eye using the labels on the axes. Using these points, the slope of the line is $(67.78 - 66.6)/(3 - .5) = 1.18/2.5 = .472$. Then the equation of the line is $c - 66.6 = .472(n - .5)$ where c stands for consumption and n stands for number of years after 1991. The linear equation can be rearranged to $c = .472(n - .5) + 66.6 = .472n + 66.364$. Note that this is very similar in form to the first equation we used for consumption: $c = .3n + 66.6$. The slope and intercept are slightly different, and reflect our attempt to get the line close to all the data. The point

Year	1991	1992	1993	1994	1995
Oil Used	66.6	66.9	66.9	67.6	68.4

TABLE 4.1
World Oil Consumption in Millions of Barrels per Day

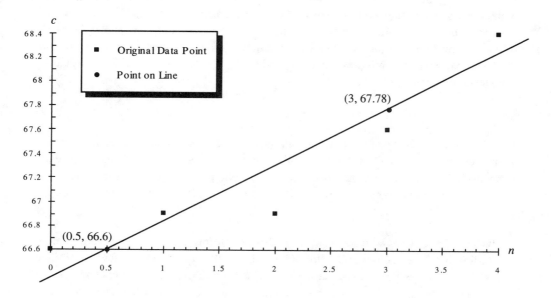

FIGURE 4.12
Best line by eye

of this example is to illustrate the process of starting with a graph and determining the corresponding equation. The new equation once again expresses c as a function of n, and can be used in the same way that the earlier equation was used. It can be thought of as the best linear model for the data.

In this discussion the line was picked by visual appearance. In a later chapter the idea of choosing the best line to fit a set of data will be reconsidered in a more analytical way.

Formulating Linear Models

The properties of linear equations and their graphs can be used to formulate linear models. Because there are several ways to write a linear equation, there is no one way to formulate a linear model. The key point is to recognize that a linear model is called for, and then to use the given information in an appropriate way. To illustrate this process, the example problem from Chapter 3 will be redone below.

Example Problem. A cellular phone company has equipment that can service 100 thousand customers. In 1990 they had 70 thousand customers. Over the last few years the number of customers has been increasing by about 4,500 per year. Assuming that the growth continues at the same pace, what will the situation be in the year 2000? 2010? How long will it be before additional equipment will be needed?

Step 1: Recognize this as a linear model. In this case, the idea that the number of customers goes up each year by the same amount, 4,500, identifies this as an arithmetic growth model, and we know that those always lead to linear models. That means we could formulate a difference equation and eventually find a linear equation, just as in the previous chapter. That is one effective approach. Here, we will use an alternative approach, one that does not involve difference equations.

What is it we wish to study? The number of customers for each year. So make up a name for the variables: *Let c be the number of customers, in thousands, and let y be the year.* Notice that this statement gives units for the variables, as well as defining their meanings.

Now suppose each year the number of customers increases by the same amount. That is an indication that this is a linear model. You can see this visually by graphing several data points, or can reason it out using the idea of slope. The number of customers increases at the rate of 4,500 customers per year. A rate is a slope. Here the rate is the same each year, so on the graph the slope will be same between any two points. That means the graph is a straight line and indicates a linear model.

In some problems, the linear model will only approximately fit the given data. In that case, you might decide a linear model is appropriate by graphing the data and observing that the points are nearly arranged in a straight line.

Step 2: Identify the given information in terms of points, slopes, and intercepts. In 1990 there were 70 thousand customers. This gives a data point: $(y, c) = (1990, 70)$. Note that I have made y the first variable because I am envisioning a graph with years along the horizontal axis. The increase in customers is given as 4,500 per year, or in the units we are using for c, as 4.5 thousands of customers per year. That is a slope. You can also find the slope by finding another data point. If the increase is 4.5 per year, then in 1991 there must be 74.5 thousand customers. That gives the point $(1991, 74.5)$. Now that two points have been found, the slope can be calculated as the rise over the run. By either method, the slope is 4.5.

Step 3: Draw a graph. The graph shows points and intercepts pictorially. It also gives you a visual portrayal of the way the variables are related. If you are a visual thinker, the graph makes using the slope–intercept or point–slope forms clearer.

Step 4: Find an equation for the variables. Now that we know one point $(1990, 70)$ and the slope (4.5), we can use the point–slope form of linear equation. Remember that we are using y and c as the variables instead of x and y. The equation is

$$c - 70 = 4.5(y - 1990)$$

In other problems, the given information might be in a different form. If two points are given, you can compute a slope. It may be that the y intercept is given. Then you can use the slope–intercept form. In some cases you will have several data points given, and will want to choose the best fitting straight line, as described earlier.

Step 5: Answer any specific questions. Restate the questions in terms of the model. What will the situation be in the year 2000? That is, how many customers will there be. In terms of the model, the question is this: when $y = 2000$, what is c? To find the answer, substitute 2000 for y in the linear equation, and solve for c. The equation you must solve is $c - 70 = 4.5(2000 - 1990)$, and the answer is $c = 115$. That is, in 2000 the model predicts 115 thousand customers. If we are going to do this for several different years, it is convenient to rewrite the equation so that c is expressed as a function of y. The new equation is $c = 4.5(y - 1990) + 70$ or, more simply, $c = 4.5y - 8885$. In 2010, we compute $c = 4.5 \cdot 2010 - 8885 = 160$. When will there be 100 thousand customers? That is, if $c = 100$ what is y? As before, we can substitute the given variable value, in this case c into the linear equation and solve for the unknown y. Which equation should you use? Either one is correct. But if there will be many questions of this type, it is best to get a new equation that expresses y as a function of c. That gives $y = (c + 8885)/4.5$. Then, with $c = 100$ we find immediately $y = 8985/4.5 = 1996.66667 = 1996$ and 2/3. This should be interpreted to mean that the number of customers will reach 100 thousand two-thirds of the way through 1996.

Proportional Reasoning

A linear model embodies a concept referred to as *proportional reasoning*. This idea was discussed in the previous chapter in an example involving soda prices and sales. In that context, the statement is made that for every increase of 5 cents in the price there will be a decrease of 10 in soda sales. With what you have learned in this chapter, you should now recognize this as a description of a slope: on a graph of soda sales versus price, between any two points the slope is $-10/5 = -2$. Proportional reasoning always involves straight lines and slopes.

Proportional reasoning is a way of relating changes of two quantities. For the soda example, it says that the changes of price and sales are always in the proportion $-10/5$. Notice that this is another example of an invariant. Under proportional reasoning, it is assumed that dividing a change in one variable by the corresponding change in another variable always results in the same number. In a graph showing the first variable on the vertical axis and the second variable on the horizontal axis, dividing the change in the first variable by the corresponding change of the second variable is simply computing the rise over the run, that is, the slope. Now we can see that the underlying assumption in proportional reasoning is that there is a linear relation between the two variables. Put another way, using proportional reasoning in a model connecting two variables is precisely the same as adopting a linear model.

Remember the discussion on page 44 about how soda prices affect the number that can be sold? You were told that 120 cans of soda would sell at 40 cents per can, and that 90 cans would be sold at 55 cents per can. Then you were supposed to use common-sense to figure out how many cans would sell at 60 cents per can. The reasoning that was then presented (and which was supposed to be your common sense approach) is really just proportional reasoning. From 40 to 55 cents a can is a change of 15 cents in price. From 120 to 90 cans of soda is a change of -30 cans that will be sold. These changes form a ratio (or proportion) of $15/-30 = -.5$, and we assume that this is an invariant. That is, we assume that making a change in the price per can always leads to a change in the number that will be sold in the same proportion, a decrease of 2 cans for each one cent increase in price. So, what was called common sense earlier, is just another instance of proportional reasoning.

For most people, proportional reasoning is easy to use when the numbers work out nicely, but becomes more difficult with fractions or decimal figures. As you now know, a proportional reasoning problem can always be expressed in the form of a linear model. This chapter has presented a large amount of information about linear equations and models. When it has been mastered, these are powerful tools for formulating models. Unfortunately, just using common sense to formulate models can lead to dramatic errors. Here is an illustration.

A Common Error.   Returning to an earlier example, let us again consider the connection between price and number of sales for sodas. The information we used previously was that 200 sodas could be sold at a price of 80 cents and 160 could be sold at a price of 90 cents. To predict how many could be sold at other prices, say 87 cents, it would be very handy to have an equation giving the number of sales s as a function of the price p. Using proportional reasoning, we would say that a 10 cent increase in price produced a 40 soda decrease in sales, so that for every penny the price is raised, the sales go down by 4. How can that be expressed as an equation?

Many students will write as an answer

$$p + 1 = s - 4$$

and will interpret it loosely to mean *an addition of* 1 *to price goes with a subtraction of* 4 *from sodas.* This is wrong. The problem is that $=$ is being misinterpreted. It does not mean *goes with*. It means *is the same number as*. Reread the equation in that light. It says that the price, plus one more, is the same number as the number of sodas minus 4. This just isn't true. In the first place, the two sides of the equations are in different units, cents on the left and sodas on the right, so saying they are equal is like comparing apples and oranges. Second, we know that at a price of 80 cents we can sell 200 sodas. That is when $p = 80$ we know $s = 200$. Using proportional reasoning again, at a price of 81 the number of sales would be 196. Here the $p + 1$ is 81, and the $s - 4$ is 196. These numbers go together, but they are certainly not equal. To repeat: $p + 1 \neq s - 4$.

It is clearly not right to translate the idea *adding 1 to the price subtracts 4 from sales* into the equation $p + 1 = s - 4$. This is an example of incorrectly formulating a linear model. At first it seems plausible, but closer inspection shows that it really doesn't make sense. So what is the right approach? The answer is to use all your tools about linear equations. First, recognize that by using proportional reasoning we are really adopting a linear model, meaning a linear equation for p and s. We know two data points, because $s = 200$ when $p = 80$ and $s = 160$ when $p = 90$. To find the equation for the straight line joining these data points, we can calculate a slope and use the point-slope form for a line. That is just what was done earlier, leading up to the equation $s - 200 = -4(p - 80)$. This is the right equation. It is not easy to find in this equation the numbers mentioned in the statement:

> For each 10 cent increase in price there is a 40 cent decrease in sales.

Making direct use of the proportional reasoning idea to derive the right equation is a little bit complicated and difficult to understand. It is much easier to apply a knowledge of algebra and linear equations.

Summary

In this chapter we have discussed linear equations and the use of continuous variables was introduced. The algebraic process of solving linear equations was presented. Connections between three different standard forms of a linear equation and the graphs of the equations were described. The concepts of slope and intercepts were also presented. These ideas were used to determine the equation for a line fit by eye to a set of data. Finally, linear equations were related to proportional reasoning.

Exercises

Reading Comprehension.   Write short essay answers to each question. After you write an answer, compare it to the explanation that is given in the reading.

1. Explain the difference between a linear function and a linear equation.

2. What is a discrete variable? What is a continuous variable? How do we use each type of variable in this course?

3. The four equations shown on page 57 are examples of nonlinear equations. For each equation, give a specific reason that shows it is nonlinear. Example: in the

second equation there appears an x^2, which means $x \cdot x$. Linear equations never have variables with exponents.

4. In the reading there are two different definitions of linear function, one that emphasizes the way the function appears, and one that concerns the operations that are used in computation. Write versions of these definitions in your own words.

5. Describe the process of solving an equation. Include in your answer a list of the operations that are permitted for simplifying an equation.

6. On page 60 there is a very detailed example of solving a linear equation. Explain how each equation after the first was obtained from the preceding equation.

7. Explain how a linear equation with one variable can have no solution. Explain how a linear equation can have an infinite number of solutions.

8. Explain what it means to invert a function. Give an example.

9. Suppose in the equation $T = -0.01h + 72$, the variable h stands for a height off the ground, and the variable T stands for the temperature at that height. Explain what it means to solve the equation for h in terms of T. Why might it be useful to do this?

10. What is slope? What is an intercept? How are these related to linear equations and their graphs?

11. Explain why a slope is a kind of rate. Give an example.

12. Describe three different forms for a linear equation. For each type, give an example and explain how the graph is related to the equation.

Mathematical Skills

1. Solve the following equations for x.
   a. $3x - 4 = 5 + 2x$
   b. $3(x - 5) + 3 = 9x + 6(2 - x)$
   c. $2(x - 3) = 3(x - 2)$
   d. $4(4 - x) + 5(x - 3) = x + 1$

2. For each equation, tell what form the equation is in (point–slope, slope–intercept, or 2–intercept form), and describe how to create a quick graph.
   a. $y = 3x - 4$ b. $y - 2 = .5(x - 3)$
   c. $y = x + 2$ d. $y - 5 = 2x$
   e. $x/2 + y/4 = 1$ f. $y = 2.3x$

3. Change each equation to slope–intercept form.
   a. $3x - 4 = 5 + 2y$
   b. $3(y - 5) + 3 = 9x + 6(2 - x)$
   c. $2(x - 3) = 3(y - 2)$
   d. $4(2 - x) + 5(y - 3) = x + 1$

4. For each part of the preceding problem, use algebra to express x as a function of y.

5. For each part of problem 3, express the line in the 2–intercept form, if possible.

Problems in Context

1. In the Fahrenheit temperature scale, water freezes at 32 degrees and boils at 212 degrees. The centigrade scale is defined so that water freezes at 0 and boils at 100 degrees. There is a linear equation that can be used to convert a centrigrade temperature to Fahrenheit. Find this equation. (Hint: Make a graph showing data points of the form (C, F), where C is the centrigrade temperature corresponding to a Fahrenheit temperature of F. This graph will be a straight line. You are looking for the equation of this line.) Is it possible for a temperature to be the same number using both Fahrenheit and centigrade?

2. Scuba divers are subject to the effects of increasing pressure as they go deeper and deeper into the ocean. As a rule of thumb, divers use the following linear model. At the surface, the water exerts a pressure of 15 pounds per square inch. For every 33 feet of depth, the pressure increases by 15 pounds per square inch, so that at a depth of 33 feet, say, the pressure is 30 pounds per square inch. Write an equation for pressure as a function of depth. What is the pressure at a depth of 100 feet? Suppose an underwater camera can withstand pressures of up to 1,000 pounds per square inch. How deep can the camera go into the ocean safely?

3. A company manufactures backpacks. The total cost to make each backpack, including materials and labor, is $23. In addition, the company has expenses of $12,000 per month for items such as rent, insurance, and power. These expenses do not depend on the number of backpacks made. Using this information, develop a linear model for monthly total costs as a function of the number of backpacks made each month. Write a short report about your model, defining the variables you use, showing the equation for costs, and explaining your reasoning.

4. In a large city, air pollution increases during the day, as auto emissions and other types of pollution enter the atmosphere. One day, the pollution level was 20 parts per million at 8 in the morning, and had increased to 80 parts per million by noon. Develop a linear model for the pollution level as a function of time. The Air Quality Management District is required to publish an unhealthy air alert on any day when the pollution level reaches 150 parts per million. If the linear model is valid from 8 in the morning to 6 at night, will it be necessary to publish an alert?

5. In Chapter 3 starting on page 42 two different variables are used in a model involving soda prices and demand. Starting from a base price of 40 cents per soda, the price is raised by a nickel several times. This gives a sequence of prices, 40, 45, 50, 55, and so on. These prices are represented by the variable p, with p_n the price after n increases. For each n there is also a demand, s_n. For example, s_1 is the number of sodas that can be sold when the price is p_1. As presented in the earlier discussion,

$$s_n = 141 - 12n$$

$$p_n = 40 + 5n$$

In this exercise, you will combine the equations to relate s_n and p_n directly. To simplify the algebra, just write s instead of s_n and p instead of p_n.

a. First, observe that the two equations can be used to find pairs of prices and demands that go together. For example, for $n = 1$ we can see that the price will be 45 and that there will be a demand for 129 sodas. This gives a data point, $(p_1, s_1) = (45, 129)$. Make a graph showing p on one axis and s on the other, and plot (p_1, s_1) as well as several other points. What does this graph suggest about the relationship between p and s?

b. Solve the second equation for n in terms of p. That is, rearrange the equation so that starting with a given price p, you can compute the corresponding value of n.

c. Use the equation for n in terms of p to replace the n in $s = 141 - 12n$. That is, you have a formula for n as a function of p: replace n in the s equation using that formula. [Note that this gives s as a function of p. With this equation, you can easily compute how many sodas can be sold at a given price.]

d. Use the preceding equation to find out how many sodas will be sold if the price is 75 cents.

e. Use your equation to solve for p in terms of s.

f. Use the equation from the preceding question to figure out what price should be charged in order to sell 200 sodas.

g. Simplify the equation for s in terms of p into the slope-intercept form, so that you can find the slope and intercept. Use this information to sketch the graph of p and s. Compare the result with the graph you made in part a.

Group Activities. Many scientists are interested in the subject of global warming. The basic issue concerns whether human activities are influencing the climate of the entire world. One widely held theory says that the amount of carbon dioxide in the atmosphere is increasing as a result of burning various kinds of fuel. According to this theory, the atmosphere will heat up as a result of the increased carbon dioxide levels. How much? And how soon? These questions are studied by developing models and making predictions. The models are very involved, and consider many variables. In this problem, we will look at just one of the variables: the number of automobiles in the US. This variable is used to predict how much gasoline is burned in the US, and that leads to predictions about the amount of carbon dioxide added to the atmosphere. In Table 4.2 are

Year	Automobiles
1940	27.5
1950	40.3
1960	61.7
1970	89.3
1980	121.6
1986	135.4

TABLE 4.2
Millions of Automobiles in the US

data on the number of automobiles in the US[1]. The figures are in units of one million, so that in 1940 there were 27.5 million automobiles. In this activity you will develop and use a linear model for the number of automobiles in the US.

1. Make a graph of the data. This can be done by hand or with a computer or calculator graphing tool. If you do it by hand, use graph paper and be careful to plot the points accurately. Do the points appear to be approximately in a straight line? A visual inspection of this type is an important first step in modeling. It would not be a good idea to use a linear model if the data do not have some appearance of lining up.

2. A second way to check if the data are approximately lined up is to compute the slope from each point to the next point. If these slopes are all identical, then the points must line up exactly. If the slopes are approximately equal, a linear model might be appropriate. Compute these slopes and see how close they all are.

3. Using a ruler, draw a line that comes as close as possible to the data points in your graph. One way to do this is to use a piece of string or thread. Pull the string tight and move it around until it seems to fit the data very well. Then have someone else use a pen to mark one point on the graph at each end of the string. Finally, use the ruler to draw the line on the graph.

4. Find the equation of your line.

5. Use the equation of the line to estimate how many autombiles were in the US in 1990. Also predict how many automobiles will be in the US in 2000.

6. How good was your estimate? Go to the reference desk in the library and ask how to find out the number of automobiles in the US in 1990. Compare the actual figure with the estimate from your model. Does the result give you any more or less confidence in your projection for the year 2000?

7. There is a theoretical way to choose the best line to fit a set of data. It is based on a carefully spelled out explanation of what is meant by *best* in this context. Use a calculator or computer to obtain the equation for the theoretically best fitting line for the automobile data in this problem. Your instructor can explain how to do this. On a graph, compare the calculator or computer's line with the one you came up with earlier. Does one seem better than the other?

Solutions to Selected Exercises

Mathematical Skills

1. a. $x = 9$ b. No solutions
   c. $x = 0$ d. Every number

---

[1] The data in this table were taken from *Earth Algebra*, preliminary edition, by Christopher Schaufele and Nancy Zumhoff. Harper Collins, 1993, page 89.

2. a. Slope–intercept. Start at -4 on the y axis, and plot several points going up 3 and over 1 each time.

   b. Point–Slope. Start at the point $(3, 2)$ and plot several more points going up 1 and over 2 each time. The slope here is .5 or $1/2$. A slope of 1 *over* 2 means you can find points by going up 1 and over 2.

   d. Point–Slope with point $(0, 5)$ and slope 2.

   e. Two–Intercept form. Mark 2 on the x axis and 4 on the y axis, then draw the straight line connecting these points.

   f. Slope–Intercept with slope 2.3 and y intercept 0. Also, Point–Slope with point $(0, 0)$ and slope 2.3.

3. a. $y = (3/2)x - 9/2$ or $y = 1.5x - 4.5$.

   c. $y = (2/3)x$.

4. a. $x = (2/3)(y + 9/2)$ or $x = (2/3)y + 3$.

   c. $x = (3/2)y$.

5. a. $\dfrac{x}{3} + \dfrac{y}{-4.5} = 1$.

   c. There is no Two–Intercept form for this line.

Problems in Context

1. Make a graph with the horizontal axis labeled C and the vertical axis labeled F. When it is 0 centigrade, it is 32 degrees Fahrenheit. This gives one data point $(C, F) = (0, 32)$. In fact, this is the F axis intercept for the line. Another data point is $(C, F) = (100, 212)$. Using the two data points, we can compute a slope. From $(0, 32)$ to $(100, 212)$, there is a rise of $212 - 32 = 180$ and a run of $100 - 0 = 100$. The slope is therefore $180/100 = 1.8$. This gives the equation $F = 1.8C + 32$. Suppose that there is a temperature that is the same in both temperature units. Call that temperature T. That means that (T, T) is one of the points on the graph of centigrade–Fahrenheit data points. Using the equation, that gives us $T = 1.8T + 32$. Solving that equation for T produces the answer: -40. That is, if the temperature is 40 degrees below 0 centigrade, it is also 40 below zero Fahrenheit.

2. The first thing to do is recognize this as a linear model. The pressure increases 15 pounds per square inch (psi) for each 33 feet of depth. If the depth is increased by 66 feet, there will be a 30-psi increase in pressure; if the depth is increased by 11 feet, there will be a 5-psi increase in pressure. This is proportional reasoning, and indicates that a linear model is called for. What variables should be used? Let p stand for pressure and d for depth below the surface, with p in psi and d in feet. Now express the given information in terms of points, slopes, and so on. At the surface, $d = 0$ and $p = 15$. That gives one point: $(d, p) = (0, 15)$. If you graph this with the d along the x axis and the p on the y axis, you will see that you now have a y intercept of 15. The pressure increases by 15 when the depth increases by 33. That is a slope of $15/33$. Or, if you prefer, get another data point by observing that at a depth of 33 feet the pressure must be 30. Then use the points $(0, 15)$ and $(33, 30)$

to compute the slope. With either of these two points and the slope, or using the slope and the intercept, an equation can be found: $p = (15/33)d + 15$. Notice that p is all by itself on one side, so the equation gives p as a function of d, as required. What is the pressure at a depth of 100 feet? That is, when $d = 100$, what is p? The answer is $p = (15/33)100 + 15 = 60.4545$. It is a good idea to leave the units in the calculation. Remember that the slope is in units of psi per foot. Then we have $p = (\frac{15}{33}$ psi per foot$) \cdot 100$ feet $+ 15$ psi. The answer comes out in units of psi, as it should for a pressure. Sometimes, when an error is made, it will be revealed by the fact that units come out wrong. We also want to know what the depth is if the pressure is 1,000 psi. That is, if $p = 1,000$ what is d? Using the equation for p and d and using 1,000 for p we obtain $1,000 = (15/33)d + 15$. Solve for d: $d = (1,000 - 15)(33/15) = 2,167$. This shows that the camera will be safe up to a depth of 2,167 feet.

3. In this problem you are told to use a linear model. The variables are c the total monthly costs and b the number of backpacks made each month. Suppose that $b = 10$. What is the total monthly cost? The backpacks cost $23 each, for a total of $230. Together with the other expenses of $12,000 per month the total is $12,230. This gives one data point. Using similar reasoning you can find another data point, and then the slope, and then the equation of the line for c and b. The final result will be $c = 23b + 12,000$.

4. Let p be the pollution level, in parts per million, t hours after 8 A.M. We have two data points $(0, 20)$ and $(4, 80)$. These can be used to find the equation $p = 15t + 20$. What is the highest the pollution gets? At 6 P.M. $t = 10$ so $p = 170$. Since that is above the 150 level, it will be necessary to publish an alert.

5. a. Plot points like these: $(45, 129)$, $(50, 117)$, $(55, 105)$, $(60, 93)$, etc.

   b. $n = (p - 40)/5$

   c. $s = 141 - 12(p - 40)/5$.

   d. $s = 141 - 12(75 - 40)/5 = 57$.

   e. $p = -5(s - 141)/12 + 40$

   f. $p = -5(200 - 141)/12 + 40 = 15.416666$. Of course, we can only set the price to a whole number of cents. Make the price 15 cents and we will actually sell 201 sodas.

   g. $s = (-12/5)p + 237$ The slope is $-12/5$ or -2.4, and the s intercept is 237. If this graph is drawn on the same axes as in part a, the line should go through all the points that were originally plotted.

5

Quadratic Growth Models

Chapter 3 showed that all arithmetic growth models have several features in common, including the forms of the difference and functional equations, the shapes of the graphs, and methods of application. Indeed, it is in recognition of these commonalities that all arithmetic growth models are considered to be part of the same *family* of models. In this chapter we will extend this idea by studying a second family of models, namely, the quadratic growth models.[1] The subject is introduced by the following example.

A New School Population Example

A school board is studying the way enrollments have been increasing over the past several years. They compile the following data:

Year	1991	1992	1993	1994	1995
Enrollment	20,000	24,000	29,000	35,000	42,000

TABLE 5.1
Enrollments in one School District

Would this situation be a good place to apply an arithmetic growth model? To find out, we look at how much the enrollment increases each year: From 1991 to 1992, the increase is 4,000 students; the next year, from 1992 to 1993, the increase is 5,000, the next year it is 6,000, then 7,000. These are the differences in the enrollment data. They

---

[1] The term *quadratic* is connected with the algebraic methods that apply to these models. Historically, these methods were first described systematically by mathematicians in the Islamic Empire in connection with problems relating the areas and sides of squares and rectangles, also called quadrilaterals. As used today, the algebraic methods include quadratic equations and the quadratic formula, topics you may recall studying in an earlier course.

are not constant, so an arithmetic growth model would not be appropriate here. Instead, the differences themselves are increasing in a regular way. In fact, if you look at the differences, 4,000, 5,000, 6,000, 7,000, you will see that they are growing according to an arithmetic model. Each number in the pattern is 1,000 more than the preceding number. Put another way, the *differences of the differences* are all equal to 1,000. This is an example of a new kind of growth model, called *quadratic growth*.

Quadratic Growth Models

Of course, the numbers in the preceding example were made up. In real applications, it is very rare to see data follow such an exact pattern. But the example was intended to illustrate an idea. We already studied arithmetic growth, in which each number in a pattern increases (or decreases) by a fixed amount from the preceding number. But in many situations, this kind of model does not really apply. The increases or decreases observed in the data are not all about the same. Instead, they are either growing larger or smaller in some regular way. One very simple way for this to happen is what appeared in the example above: the increases themselves increase by a fixed amount. This idea defines what is called a quadratic growth model.

In the example above, the original set of data do not follow an arithmetic growth pattern. But the differences between successive data values *do* follow an arithmetic pattern. This idea is so important that it is worth repeating. But it is easiest to state using the terminology of *differences*, so let us review that terminology. When a set of data are given in a sequence, such as 4, 9, 16, 25, 36, 49, subtracting each number from the one that comes after generates a new set of values, in this case, 5, 7, 9, 11, 13. These new values are called the *differences* for the original data set. The differences are the increases or decreases in the original data set. For the example, from 4 to 9 there is an increase of **5**; from 9 to 16 the increase is **7**; from 16 to 25, **9**; 25 to 36, **11**; from 36 to 49, **13**. These are the same numbers we listed before as the differences for the original data set. When a data set involves decreasing numbers, the differences are negative: for the data 25, 13, 7, 4, 2, 1, the differences are $-12, -6, -3, -2, -1$.

Using the idea of differences, it is now possible to give a concise statement of the concept of quadratic growth:

> In a quadratic growth model, the differences follow a pattern of arithmetic growth.

Read that statement carefully! The data values themselves don't exhibit arithmetic growth — the *differences* do.

There is another way to describe quadratic growth models, using a concept called *second differences*. Second differences are the differences of the differences. We have already seen how you can start with a data sequence and compute the differences. Take the differences of these differences in the same way, and that gives what are called the second differences of the original data values. For example, suppose the original data values are 6, 12, 20, 30, 42, 56. Then the differences are computed to be 6, 8, 10, 12,

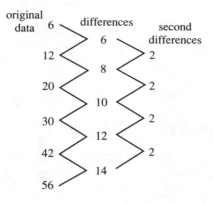

FIGURE 5.1

Constant Second Differences

14. Now look at the differences for these numbers — from 6 to 8, then from 8 to 10, 10 to 12, and so on. Those are the second differences of the original data. And in this example, the second differences are all the same. They all equal 2. This is illustrated schematically in Fig. 5.1.

For any quadratic growth model, the differences follow a pattern of arithmetic growth. That means that the differences of these differences must be a fixed constant. That is the second way to think about quadratic growth.

> In a quadratic growth model, the second differences are constant.

This idea applies to the example above. Because all the second differences are equal to 2 in 6, 12, 20, 30, 42, 56, this sequence is as an example of quadratic growth.

In the foregoing material, the emphasis has been on explaining *what* a quadratic growth model is. To review, quadratic growth is one kind of model that can be considered when an arithmetic growth model is inappropriate. Arithmetic growth can only be used when successive terms in a data sequence increase by a fixed (or nearly fixed) amount. Quadratic growth applies when the successive terms increase not by a constant amount, but by an amount that is itself changing over time, although in a very simple way. Specifically, in quadratic growth, the differences follow a pattern of arithmetic growth. Or, put another way, in quadratic growth, the second differences are constant.

Quadratic growth models are used to make predictions about real phenomena. This involves the same basic methods of analysis that we saw for arithmetic growth. First, the model is usually formulated using a difference equation. All quadratic growth models have similar difference equations. These will be discussed in the next section. Second, based on the difference equation, a functional equation is developed. Finally, the functional equation can be used to answer specific questions about the future behavior of the model. Understanding the graphical and numerical features of quadratic models contributes valuable insight in this task. Functional equations and numerical and graphical aspects will be covered after the discussion of difference equations.

a_0	a_1	a_2	a_3	a_4
20,000	24,000	29,000	35,000	42,000

TABLE 5.2
Enrollment Data with Subscript Notation

Quadratic Growth Difference Equations

As a first step, let us look again at the school enrollment example. We will use subscript notation, with a_0 standing for the enrollment in 1991, a_1 for the enrollment in 1992, and so on. The first few a's are shown in Table 5.2.

Using subscript notation, the pattern of the differences can be shown as follows:

$$a_1 = a_0 + 4,000$$
$$a_2 = a_1 + 5,000$$
$$a_3 = a_2 + 6,000$$
$$a_4 = a_3 + 7,000$$

To emphasize a particular pattern in this example, relate each of the differences to 4,000, the starting difference. That is, express 5,000 as $4,000 + 1,000$, 6,000 as $4,000 + 2,000$, and so on. That produces the following pattern.

$$a_1 = a_0 + 4,000$$
$$a_2 = a_1 + 4,000 + 1,000$$
$$a_3 = a_2 + 4,000 + 2,000$$
$$a_4 = a_3 + 4,000 + 3,000$$

Based on this pattern, you should have no difficulty predicting the appropriate equation for any a_n. For example, a_7 should be given by

$$a_7 = a_6 + 4,000 + 6,000$$

As in previous explorations of this kind, we can express the pattern that has been found using a single equation, by using a variable for one of the subscripts. In

$$a_7 = a_6 + 4,000 + 6,000$$

use n in place of the subscript 6. Then the 6,000 becomes $n \cdot 1,000$ or $1,000n$ and the subscript of 7 becomes $n + 1$. The new equation is

$$a_{n+1} = a_n + 4,000 + 1,000n$$

That is the difference equation for the school enrollment example.

The constants that appear in this difference equation are related to the differences of the original data: 4,000, 5,000, 6,000, 7,000. This is an arithmetic growth pattern that starts at **4,000** and goes up by **1,000** with each successive value. The constants 4,000 and 1,000 are the same ones that appear in the difference equation

$$a_{n+1} = a_n + 4,000 + 1,000n$$

Next look at the second differences. The original data are 20,000, 24,000, 29,000, 35,000, 42,000; the differences are 4,000, 5,000, 6,000, 7,000; and the second differences are 1,000, 1,000, 1,000. These second differences are all the same, as they must be for a quadratic growth model. But more significantly, the constant second difference of **1,000** is again one of the constants from the difference equation. These same relationships can be observed in any quadratic growth model.

This can be illustrated by considering some variations on the first example. In the original version, the differences started at 4,000 and went up by 1,000 each year. Here is a new data set: 20,000, 26,000, 33,000, 41,000, 50,000. This time the differences are 6,000, 7,000, 8,000, 9,000; they start at 6,000 and go up by 1,000 each time. Working through the same steps we used earlier, we can obtain the difference equation for the new data set. The pattern would look like this:

$$a_1 = a_0 + 6,000$$

$$a_2 = a_1 + 6,000 + 1,000$$

$$a_3 = a_2 + 6,000 + 2,000$$

$$a_4 = a_3 + 6,000 + 3,000$$

and the difference equation would be

$$a_{n+1} = a_n + 6,000 + 1,000n$$

As before, the two numbers that appear on the right side of the difference equation are **6,000**, the starting point for the differences, and **1,000**, the amount the differences increase by at each step, which is also constant second difference for the original data.

For another variation, suppose the original data were 20,000, 26,000, 33,500, 44,000, 59,000. Then the differences would be 6,000, 7,500, 10,500, 15,000. This time, the differences start at **6,000** and go up by **1,500** each year. In this case the pattern would be

$$a_1 = a_0 + 6,000$$

$$a_2 = a_1 + 6,000 + 1,500$$

$$a_3 = a_2 + 6,000 + 3,000$$

$$a_4 = a_3 + 6,000 + 4,500$$

To make this a little clearer, rewrite it in this form:

$$a_1 = a_0 + 6,000$$

$$a_2 = a_1 + 6,000 + 1,500$$

$$a_3 = a_2 + 6,000 + 1,500 \cdot 2$$

$$a_4 = a_3 + 6,000 + 1,500 \cdot 3$$

The difference equation for this variation is

$$a_{n+1} = a_n + 6{,}000 + 1{,}500n$$

The constants in this equation, **6,000** and **1,500**, have the same significance that the constants had in the earlier example: the differences of the data can be described as starting at **6,000** and increasing at each step by **1,500**, which is the constant second difference for the original data.

As all of these variations suggest, the general form for a quadratic growth difference equation is

$$a_{n+1} = a_n + \text{a fixed number} + \text{another fixed number} \cdot n$$

In this form, the fixed numbers are *parameters*. In different instances of quadratic growth they will be different numbers, but they remain constant as long as a single model is being discussed. As for arithmetic growth, we will represent these parameters with variables, d for the first one and e for the other. The general form of the quadratic growth difference equation is then

$$a_{n+1} = a_n + d + en \qquad (1)$$

This is worth highlighting for future reference.

In any quadratic growth model, there is a difference equation of the form

$$a_{n+1} = a_n + d + en$$

where d and e are parameters that represent constant numerical values. The differences for the model begin with the value d and grow by e with each step. The parameter d is equal to $a_1 - a_0$. The parameter e is the constant second difference for the model.

Notice how closely this difference equation resembles the arithmetic growth difference equation. There is just one additional term, en. This also gives a different way of defining what is meant by quadratic growth. Remember that for arithmetic growth, the difference between two successive terms is a fixed number. For quadratic growth, the difference between two successive terms has the form $d + en$. In other words, it is a linear function of n. So,

---

A quadratic growth model is one in which the difference between a_{n+1} and a_n is a linear function of n.

The importance of the difference equation in quadratic growth is mainly involved with formulating the model. Expressing the difference equation in the standard form helps us understand what the parameters of the model are. These parameters reappear in the functional equation for quadratic growth, and that is the equation that really is useful in making predictions.

Functional Equations.   In the preceding section difference equations were considered for quadratic growth models. Remember that difference equations are *recursive*, they let you compute the next data value once you know the preceding value. The only way to compute a particular data value using the difference equation is step by step: start out with a_0, then compute a_1, then a_2, then a_3, and so on, until you reach the one you want. In contrast, functional equations are direct. If you want to compute the 25th data value, you simply substitute 25 for n and immediately compute a_{25}.

The functional equations for quadratic growth models all share a common form. What that looks like can be worked out using patterns. To illustrate this idea we will work out a functional equation for the school enrollment example considered earlier. Recall that the enrollments start at $a_0 = 20{,}000$, and that the difference equation is

$$a_{n+1} = a_n + 4{,}000 + 1{,}000n$$

Starting with $a_0 = 20{,}000$, we can systematically work out a_1, then a_2, and so on. For example, $a_1 = 20{,}000 + 4{,}000 + 1{,}000 \cdot 0$, or more simply, $20{,}000 + 4{,}000$. We won't work this out as a final answer of 24,000, because that will hide the pattern that we need to find. Proceeding,

to get a_2, take the value we just found for a_1 and add $4{,}000 + 1{,}000 \cdot 1$. That gives the equation

$$a_2 = 20{,}000 + 4{,}000 + 4{,}000 + 1{,}000 \cdot 1$$

Similarly, $a_3 = a_2 + 4{,}000 + 1{,}000 \cdot 2$, so take what we already have for a_2 and add $4{,}000 + 1{,}000 \cdot 2$. To make it clearer what the pattern is, when we write this next result, we will keep the 4,000's together, and keep the parts that have 1,000 together. This is what it looks like:

$$a_3 = 20{,}000 + 4{,}000 + 4{,}000 + 4{,}000 + 1{,}000 \cdot 1 + 1{,}000 \cdot 2$$

This process can be continued as long as we wish. The following pattern results.

$$
\begin{aligned}
a_0 &= 20{,}000 \\
a_1 &= 20{,}000 + 4{,}000 \\
a_2 &= 20{,}000 + 4{,}000 + 4{,}000 && + 1{,}000 \cdot 1 \\
a_3 &= 20{,}000 + 4{,}000 + 4{,}000 + 4{,}000 && + 1{,}000 \cdot 1 + 1{,}000 \cdot 2 \\
a_4 &= 20{,}000 + 4{,}000 + 4{,}000 + 4{,}000 + 4{,}000 && + 1{,}000 \cdot 1 + 1{,}000 \cdot 2 + 1{,}000 \cdot 3
\end{aligned}
$$

This pattern can be expressed a little more compactly by combining all the 4,000's together and all the terms with 1,000 together. That is, write $4{,}000 \cdot 3$ instead of $4{,}000 + 4{,}000 + 4{,}000$, and write $1{,}000(1 + 2 + 3)$ instead of $1{,}000 \cdot 1 + 1{,}000 \cdot 2 + 1{,}000 \cdot 3$. Making that change, the equations above become

$$
\begin{aligned}
a_0 &= 20{,}000 \\
a_1 &= 20{,}000 + 4{,}000(1) \\
a_2 &= 20{,}000 + 4{,}000(2) && + 1{,}000(1) \\
a_3 &= 20{,}000 + 4{,}000(3) && + 1{,}000(1 + 2) \\
a_4 &= 20{,}000 + 4{,}000(4) && + 1{,}000(1 + 2 + 3)
\end{aligned}
$$

Now it should be clear that there is a pattern that can be used to write an equation for any a_n. For example, the predicted equation for a_7 would be

$$a_7 = 20{,}000 + 4{,}000(7) + 1{,}000(1 + 2 + 3 + 4 + 5 + 6)$$

That prediction proves to be correct, and, in fact, the pattern is generally valid. Using the pattern, we can extend the set of equations above as far as we like. Below we have extended them to a_7.

$$
\begin{aligned}
a_0 &= 20{,}000 \\
a_1 &= 20{,}000 + 4{,}000(1) \\
a_2 &= 20{,}000 + 4{,}000(2) \;+\; 1{,}000(1) \\
a_3 &= 20{,}000 + 4{,}000(3) \;+\; 1{,}000(1 + 2) \\
a_4 &= 20{,}000 + 4{,}000(4) \;+\; 1{,}000(1 + 2 + 3) \\
a_5 &= 20{,}000 + 4{,}000(5) \;+\; 1{,}000(1 + 2 + 3 + 4) \\
a_6 &= 20{,}000 + 4{,}000(6) \;+\; 1{,}000(1 + 2 + 3 + 4 + 5) \\
a_7 &= 20{,}000 + 4{,}000(7) \;+\; 1{,}000(1 + 2 + 3 + 4 + 5 + 6)
\end{aligned}
$$

An Amazing Shortcut.   Although this gives us a direct method for calculating a_n, it is not very convenient to use. Suppose you wanted to compute a_{50}. The pattern says to start with 20,000 plus $4{,}000 \cdot 50$, and that is easy enough. But next there is a 1,000 and all the whole numbers from 1 through 49, in parentheses.[2] Just writing this down is awkward, and nobody would want to actually do all that addition. Fortunately, there is a shortcut that can be applied to problems of this type.[3]

Here is how the shortcut works. To add up all the numbers from 1 to 49, multiply 49 by the next whole number, 50, and divide by 2. The answer is $49 \cdot 50/2 = 1{,}225$. Similarly, to add up all the numbers from 1 to 100, just multiply 100 by 101, and divide by 2: $100 \cdot 101/2 = 5{,}050$. You can see that with a calculator this shortcut is easy to use even with very large numbers. The total from adding up all the numbers from 1 to 1,000 is $1{,}000 \cdot 1{,}001/2 = 500{,}500$.

This shortcut is an amazing simplification. It almost seems too good to be true. In fact, how do we know it *is* true? Of course, it is easy to check whether it gives the right answer for small numbers. We can add up $1 + 2 + 3 + 4$ directly to find the answer 10, and see that the shortcut answer $4 \cdot 5/2$ is correct. But how do we know that it works for gigantic numbers like 1,000 or 10,000 or a million? One answer is that a deductive proof (similar to what many students study in high school geometry) is known. The main concept involved in this proof will be presented at the end of the chapter. For the present, you are asked to accept that the shortcut works as described so that it can be used in the continued study of quadratic growth functional equations.

Before returning to that subject, one additional digression is justified. The idea of adding up many many numbers is sometimes awkward to describe. We have referred before to adding up all the whole numbers from 1 to 1,000, for example. But it is simply not practical to write out all 1,000 numbers with plus signs. For this reason, there are two shorthand notations that are commonly used for sums.

---

[2] Check the pattern carefully: the last number added in the parenthesis is one less than the subscript on the left side of the equation, so for a_{50}, the last number in the parentheses is 49.

[3] The shortcut has been known for centuries, and has been discovered independently by many mathematicians from different cultures. The prominent 19th century German mathematician Karl F. Gauss is supposed to have figured it out at the age of 11 when his mathematics class was assigned to add all of the whole numbers from 1 to 100.

The first notation uses three dots to indicate continuation of a pattern. In this approach, rather than writing $1 + 2 + 3 + 4 + 5 + 6 + 7$ out in full, the shorter $1 + 2 + 3 + \cdots + 7$ is used. Similarly, $1 + 2 + 3 + \cdots + 100$ is understood to mean adding up all the whole numbers from 1 to 100. The first few terms are included to show what the pattern is, and the last number tells where to stop adding. So $2 + 4 + 6 + \cdots + 100$ means to add all the *even* whole numbers from 2 to 100. Using this shorthand notation, the shortcut for adding whole numbers can be written

$$1 + 2 + 3 + \cdots + n = n(n + 1)/2$$

A second very common shorthand uses the Greek letter Σ (sigma) to stand for a sum. It will be too great a distraction to discuss this notation here. So we will defer that topic to the end of the chapter. Now we can turn again to the discussion of functional equations in quadratic growth.

A Functional Equation for a_n. Remember that we were studying the pattern of values for a_n in the school enrollment model. We had worked out the following equations:

$$
\begin{aligned}
a_0 &= 20{,}000 \\
a_1 &= 20{,}000 + 4{,}000(1) \\
a_2 &= 20{,}000 + 4{,}000(2) \quad + \quad 1{,}000(1) \\
a_3 &= 20{,}000 + 4{,}000(3) \quad + \quad 1{,}000(1 + 2) \\
a_4 &= 20{,}000 + 4{,}000(4) \quad + \quad 1{,}000(1 + 2 + 3) \\
a_5 &= 20{,}000 + 4{,}000(5) \quad + \quad 1{,}000(1 + 2 + 3 + 4) \\
a_6 &= 20{,}000 + 4{,}000(6) \quad + \quad 1{,}000(1 + 2 + 3 + 4 + 5) \\
a_7 &= 20{,}000 + 4{,}000(7) \quad + \quad 1{,}000(1 + 2 + 3 + 4 + 5 + 6)
\end{aligned}
$$

Now we can apply the shortcut for adding whole numbers. For example, in the last equation, instead of $1 + 2 + 3 + 4 + 5 + 6$ we can write $6 \cdot 7/2$. Applying the shortcut in a similar fashion to each of the last few equations in the set above, we get

$$a_4 = 20{,}000 + 4{,}000(4) + 1{,}000(3)(4)/2$$

$$a_5 = 20{,}000 + 4{,}000(5) + 1{,}000(4)(5)/2$$

$$a_6 = 20{,}000 + 4{,}000(6) + 1{,}000(5)(6)/2$$

$$a_7 = 20{,}000 + 4{,}000(7) + 1{,}000(6)(7)/2$$

There is again an easily recognized pattern here. You will have no trouble guessing that $a_{73} = 20{,}000 + 4{,}000(73) + 1{,}000(72)(73)/2$. Using the variable n, the pattern can be expressed in the general form

$$a_n = 20{,}000 + 4{,}000(n) + 1{,}000(n - 1)(n)/2$$

That is the functional equation for a_n in the school enrollment model.

With this functional equation we can predict the enrollment for future years. Remember that a_0 is the enrollment for 1991. For 1999, which is 8 years later, we have $a_8 = 20{,}000 + 4{,}000(8) + 1{,}000(7)(8)/2 = 80{,}000$.

The functional equation for this example reveals the form that always appears for a quadratic growth model. Look again at the equation:

$$a_n = 20{,}000 + 4{,}000(n) + 1{,}000(n-1)(n)/2$$

Do you see that the **4,000** and **1,000** are the same constants that appeared in the difference equation? The remaining constant, 20,000, is the initial value for the model, because the enrollment started at $a_0 = 20{,}000$. In a similar way, the initial value of the model and the two parameters for the difference equation always appear in the functional equation. To illustrate this, here are the variations we considered earlier. First, if the enrollment starts out at $a_0 = 20{,}000$ and the difference equation is

$$a_{n+1} = a_n + \mathbf{6{,}000} + \mathbf{1{,}000}n$$

then the functional equation would be

$$a_n = 20{,}000 + \mathbf{6{,}000}(n) + \mathbf{1{,}000}(n-1)(n)/2 \tag{2}$$

In the next variation, let $a_0 = 20{,}000$ again, but this time use the difference equation

$$a_{n+1} = a_n + \mathbf{6{,}000} + \mathbf{1{,}500}n$$

The functional equation is then

$$a_n = 20{,}000 + \mathbf{6{,}000}(n) + \mathbf{1{,}500}(n-1)(n)/2$$

Both of these functional equations can be developed using patterns for the first few values of a_n, just as we did for the original example. In fact, it would be a good idea to work through the pattern yourself, on a piece of scratch paper, at least once. Just follow the same steps that were presented above starting on page 85, but using different parameters, say $a_0 = 20{,}000$, and with the difference equation $a_{n+1} = a_n + \mathbf{6{,}000} + \mathbf{1{,}000}n$. As you work through this process, keep asking yourself how the pattern changes when you use the new parameters instead of the original ones. This will show you that Eq. (2) is correct, and help you understand how the functional equation arises.

It will also illustrate how to jump directly from the difference equation to the correct functional equation. This procedure is succinctly described in terms of the parameters d and e introduced earlier. When the difference equation is expressed in the form

$$a_{n+1} = a_n + d + en$$

the functional equation will be

$$a_n = a_0 + dn + e(n-1)(n)/2$$

In fact, the entire development of a functional equation can be performed using d and e for the parameters in the difference equation, rather than specific numbers. Starting with the difference equation

$$a_{n+1} = a_n + d + en$$

and expressing the starting value for the progression as a_0, we can set $n = 0$ to get $a_1 = a_0 + d$. Similarly, with $n = 1$ in the difference equation, we obtain $a_2 = a_1 + d + e \cdot 1 = a_0 + d + d + e \cdot 1$. Retracing the steps we used before (starting on page 85), but with a_0, d and e in place of numerical values, we obtain the following pattern:

$$
\begin{aligned}
a_1 &= a_0 + d \\
a_2 &= a_0 + d + d & & + e \cdot 1 \\
a_3 &= a_0 + d + d + d & & + e \cdot 1 + e \cdot 2 \\
a_4 &= a_0 + d + d + d + d & & + e \cdot 1 + e \cdot 2 + e \cdot 3
\end{aligned}
$$

This format makes it easy to see what was added to each equation to get the next one, so you can verify that the difference equation is being properly followed. However, it is more convenient to use a compressed format. On each line, combine all of the d's into one term, and combine all of the multiples of e into another term. That produces

$$a_1 = a_0 + d$$

$$a_2 = a_0 + d \cdot 2 + e$$

$$a_3 = a_0 + d \cdot 3 + e(1 + 2)$$

$$a_4 = a_0 + d \cdot 4 + e(1 + 2 + 3)$$

$$a_5 = a_0 + d \cdot 5 + e(1 + 2 + 3 + 4)$$

$$a_6 = a_0 + d \cdot 6 + e(1 + 2 + 3 + 4 + 5)$$

See how similar this is to what we had earlier? By now you can guess what will happen next. Wherever we see something like $1 + 2 + 3 + 4$, the shortcut is used. Then the pattern becomes

$$a_1 = a_0 + d$$

$$a_2 = a_0 + d \cdot 2 + e$$

$$a_3 = a_0 + d \cdot 3 + e(2)(3)/2$$

$$a_4 = a_0 + d \cdot 4 + e(3)(4)/2$$

$$a_5 = a_0 + d \cdot 5 + e(4)(5)/2$$

$$a_6 = a_0 + d \cdot 6 + e(5)(6)/2$$

The pattern that is revealed in these equations can be expressed as a functional equation for a_n:

$$a_n = a_0 + dn + e(n-1)n/2 \tag{3}$$

This gives the functional equation in a parametric form that can be used with any quadratic growth model. For future reference we summarize the results below.

> If a quadratic growth model has an initial value of a_0 and obeys the difference equation $a_{n+1} = a_n + d + en$, then the functional equation is
>
> $$a_n = a_0 + dn + e(n-1)n/2$$

An Example. Once you know the difference equation for a quadratic growth model, the functional equation can be found by using the formulation in the box. To illustrate, here is an example.

Problem: Find the functional equation for the pattern

$$5, 27, 59, 101, 153, 215, \ldots$$

The first step is to find the difference equation. Computing the differences

$$22, 32, 42, 52, 62$$

you can see that the second differences are all equal to 10. This indicates that the numbers follow a quadratic growth model.[4] What is more, since the differences start at 22 and increase by 10 for each step, the parameters for this problem are $d = 22$ and $e = 10$. That tells us the difference equation should be

$$a_{n+1} = a_n + 22 + 10n$$

We also observe that $a_0 = 5$. Now the problem is to find a functional equation for a_n. This is the point at which to use the formulation in the box. We know the parameters are $a_0 = 5$, $d = 22$, and $e = 10$. That means we can immediately write down

$$a_n = 5 + 22n + 10(n - 1)n/2$$

With this equation, we can find any one of the a's. For example, $a_3 = 5 + 22 \cdot 3 + 10 \cdot 2 \cdot 3/2 = 101$, which agrees with the data originally given for the problem. It is just as easy to compute $a_{20} = 5 + 22 \cdot 20 + 10 \cdot 19 \cdot 20/2 = 2{,}345$.

This sample problem illustrates how you can proceed from a pattern of numbers to a functional equation, if the pattern exhibits quadratic growth. As it is presented, the problem is completely divorced from any real world interpretation. The starting point is a meaningless set of numbers, and the ending point is the functional equation. Of course in real life problems, the numbers are data values, and the functional equation is not an end in itself, but is used to make predictions and answer questions. What will the enrollments be in five years? When will the enrollment reach 50,000? So the true application of the material just presented occurs in the middle of a larger problem. After the data are collected, after you decide that a quadratic model applies, then you can find the functional equation and use it to answer questions about the model. The fact is that quadratic growth models do come up in real problems. In the next section we will take a look at some situations where quadratic growth models are useful.

---

[4] That is hardly surprising in an example at this point in the chapter, but in some other context, if you are trying to decide whether a set of data might follow a simple model, looking at differences in this way is a good start. If the differences themselves are constant, you will know that an arithmetic growth model is correct. If the differences are not constant but the second differences are constant, that indicates a quadratic growth model. If no differences are constant, some other type of model may be required.

Applications of Quadratic Growth

In this section we will look at three situations where quadratic growth models are used. The first, like the opening example of the chapter, concerns data patterns with equal or nearly equal second differences. The second situation involves problems that can be analyzed using networks. The third arises when it is of interest to add up successive data values in an arithmetic growth model.

Constant Second Differences. The opening example of this chapter is not a genuine application. The numbers were invented to illustrate the concept of quadratic growth. But the example does provide a context for understanding one way that quadratic growth models can be used. In real data, it is very rare to see second differences that are exactly constant. But if the second differences appear nearly constant, that is a clue that a quadratic growth model might provide a good approximation. An example of this is given in the Group Activities at the end of the chapter.

A famous historical example of this application of quadratic growth involves Galileo's study of motion at the start of the 17th century. Galileo studied falling objects and balls rolling down ramps.[5] In modern terms we would describe his data as follows. Lay out a measuring tape along the path of motion, and every second record the moving body's position along the tape. So, for a rolling ball, the starting point on the tape would be at 0. After one second it might be at 10 on the tape, after another second it would be at 40, and so on. These successive positions on the tape, 0, 10, 40, and so forth, form a data sequence of the kind we have studied in this course.

In Galileo's studies he found a common pattern. The differences in the data grew in proportion to the odd numbers. If the first difference was one foot, then the succeeding differences would be 3 feet, 5 feet, 7 feet, etc. Or, if the first difference was one yard, then the succeeding differences would be 3 yards, 5 yards, 7 yards, and so on. Whatever he found for the difference of the first two data values, the difference of the next two would be three times as great, then 5, then 7, and would continue in the pattern of odd numbers. Of course, none of the measurements were exact. The differences did not come out exactly in the pattern of odd numbers. But they were always close enough to that pattern to allow Galileo to recognize it.

Using the methods discussed earlier in this chapter, we would describe Galileo's discovery this way. In his motion data, the differences of the values are approximately proportional to the arithmetic growth pattern 1, 3, 5, 7, 9, etc. The second differences in the data will then be approximately constant. For example, if the differences are approximately 10, 30, 50, 70, and so on, then the second differences would all be about 20. Similarly, if the ball travels 2 feet in the first second, the differences will be 2, 6, 10, 14 (the doubles of the odd numbers) and the second differences will all be 4. With second differences that are approximately constant, we recognize that quadratic growth

---

[5] This description of Galileo's work is based on the account in *Physics*, by Robert Resnick and David Halliday, Wiley, New York, 1955, part 1, pages 48–49.

will be a good choice for a model. In fact, quadratic models are used in physics today for the motion of falling bodies.

As this example illustrates, in some situations the use of a quadratic model is a matter of choice. If the data have second differences that are approximately constant, the analyst may choose to adopt a model in which the second differences are truly constant. That means choosing a quadratic growth model. The model will be an approximation to the real data, and if the approximation is accurate enough for the intended purposes, then the assumption of constant second differences is justified. So, for these examples, we might say that quadratic growth is just one possible approximation to a real problem.

Network Problems.   Quadratic growth can also be dictated by the nature of a problem. This is observed in a variety of problems featuring networks. A network can be thought of as a collection of objects connected together in some way. They may be cities connected by roads or airline routes, telephones connected by phone lines, computer processors connected by data pathways, or satellites connected by radio links. As an illustration, we will consider here a network of computers connected together to exchange email. A very naive approach would be to connect each pair of computers directly with a separate wire, say a telephone line.[6] That might require a lot of phone lines for large numbers of computers. But how many? Suppose there are 100 computers, how many phone lines are needed? What if there are 1,000 computers? This problem can be analyzed using difference equations, and we will see that the type of equation that is developed is a quadratic growth difference equation.

It is easy enough to count the number of connections needed when the number of computers is small. For two computers, only one telephone line is needed. For three computers, three lines are needed. How many are needed for four computers? A good way to find the answer is by drawing a diagram. Put down one dot for each of the four computers, draw a line from each dot to every other dot, and count the lines. Do that now. The answer is on p. 94, but don't look at it until after you have worked it out for yourself.

For larger numbers of computers, this process of counting soon becomes impractical. Difference equations offer an alternative. The main idea is this: it is much easier to count the number of *new* telephone lines required to add one new computer to the network, than it is to count the total number of phone lines. If there are 10 computers already on the network and one new computer is to be connected, how many new lines are needed? The new computer just needs to be connected to each of the existing ones, 10 in all. Then every old computer is connected to the new computer, and of course all the old computers

---

[6] This is not a realistic way to connect large networks of computers, but it is an easily visualized example that illustrates the idea of quadratic growth in network problems. A more realistic problem is simply to enumerate the number of different pairs of computers. This might be of interest in analyzing the email traffic within the network. As a first step in monitoring the flow of data it would be natural to keep track of how much mail passed from any one computer to each of the others. This can be recorded for each pair of computers. How many pairs are there? Analyzing that question is mathematically identical to counting how many wires are needed to connect each pair of computers.

were already connected to each other. To repeat the conclusion: with 10 computers on the network, ten new lines are needed to add one computer.

This reasoning can be applied to any size network. If there are 50 computers on a network, and one computer is added, how many new telephone lines are needed? _____ How many if there are 100 computers on the network? _____ How many if there are n computers on the network? _____

By focusing on how the number of phone lines increases as one new computer is added, we are finding just the kind of relationship needed to formulate a difference equation. Of course, we need to have some variables first, so let us say that t_n is the number of telephone lines needed for n computers. To be more complete,

> If n computers are connected in a network, t_n is the number of telephone lines required to provide a direct connection from each computer to every other.

Our direct counting above showed that $t_2 = 1$, $t_3 = 3$, and $t_4 = 6$.

What is the difference equation for t_n? Let us look at some examples. Suppose there are 100 computers on a network. We don't know yet what t_{100} is, but we do know what happens if we add one new computer: another 100 phone lines will be needed. That means the number of phone lines for 101 computers (t_{101}) is 100 more than the number of phone lines needed for 100 computers (t_{100}). This is expressed in equation form as

$$t_{101} = t_{100} + 100$$

Similarly, without knowing how many phone lines are required for 50 computers on a network, we can state that 50 additional lines would be needed for a 51 computer network. So

$$t_{51} = t_{50} + 50$$

The general statement of this pattern is

$$t_{n+1} = t_n + n \tag{4}$$

This is an example of a quadratic growth model. For this example, the parameter $d = 0$ and $e = 1$. Then the general equation

$$t_{n+1} = t_n + d + en$$

becomes

$$t_{n+1} = t_n + 0 + 1 \cdot n = t_n + n$$

and that is the same form as Eq. (4). Now we can immediately formulate the functional equation. Using Eq. (3) with $d = 0$ and $e = 1$, we find

$$t_n = t_0 + n(n-1)/2$$

The only problem is, what is t_0? With 0 computers, it is clear that we don't need any phone lines, so it would be reasonable to say that $t_0 = 0$. That would give the equation

$$t_n = n(n-1)/2 \tag{5}$$

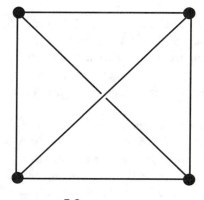

FIGURE 5.2
6 Lines Needed for 4 Computers

However, we need to be a little careful, because the difference equation really only makes sense for n greater than or equal to 1. Let's just check whether the equation works for small values of n.

According to the Eq. (5), for $n = 2$ we should have $t_2 = 2 \cdot 1/2 = 1$. That is the right answer. For $n = 4$, the equation says $t_4 = 4 \cdot 3/2 = 6$, and that is also the correct answer. This shows that the choice of $t_0 = 0$ really is correct for our difference equation, and that Eq. (5) correctly gives the functional equation for this problem.

Now we can answer some of the questions we raised before. If there are 100 computers in the network, how many telephone lines would be needed to connect each pair directly? The answer is $t_{100} = 99 \cdot 100/2 = 4,950$. What if there are 1,000 computers? You might guess that with 10 times as many computers you would need 10 times as many lines. But the functional equation gives $t_{1,000} = 999 \cdot 1,000/2 = 499,500$. That is more than 100 times more telephone lines! This shows one of the reasons that mathematical models are important. In many situations, using *common sense* just doesn't lead you to reliable conclusions. In this example, we have seen how a difference equation approach leads indisputably to a quadratic growth model. Our knowledge of these models gives us a functional equation, and leads to conclusions that we might never have expected just going by *common sense*. Actually, what is referred to as *common sense* in this paragraph is nothing more than proportional reasoning. It is only the idea that doubling one variable should double the other variable that is flawed, not common sense in general. Perhaps it would be more correct to say that *unschooled* common sense can lead to errors. In any case, as this example illustrates, in quadratic growth models proportional reasoning is not appropriate. This point will be discussed further a little later in the chapter.

The questions we have answered so far involve setting a value for n and predicting the result for t_n. There is a second kind of question in which t_n is given but we don't know n. As an example, suppose we only have 2,000 telephone lines available to us, and we want to know how many computers can be connected. That means we know t_n (2,000) and want to find n. That leads to the equation

$$2,000 = (n - 1)n/2$$

This equation cannot be solved using the methods of the previous chapter; it is not a *linear* equation. It is called a *quadratic* equation, and will be studied in detail in the next chapter. We will return to this kind of question later. For now, the focus is on how quadratic growth models emerge in network problems.

The next example is a modified version of the computer network problem. As before, we wish to connect computers with phone lines. However, this time, in addition to the computers of the network users, there will be two special computers on the network to provide some network-wide services, such as email directories, a shared clock, and an electronic bulletin board. To avoid confusion, the two special computers will be referred to as *service computers,* and the rest will be called *user computers*. The two service computers will be connected together by a single phone line. Each user computer will be connected to each service computer by one phone line as well. Now suppose that in building the computer network, it is decided to have three phone lines connecting each pair of user computers. This will provide a greater capacity for message traffic, and will protect against interruptions in service. If one of the three phone lines is somehow damaged or disconnected, the other two will continue to function. So the overall plan is for two special service computers and a large number of user computers, all connected together. The service computers will be connected by just one phone line to other computers, but there will be three lines connecting each pair of user computers. In this design, if there are 500 user computers, how many phone lines would be needed?

When the number of computers is small, we can make a diagram that shows all of the required phone lines. This is shown in Fig. 5.3 for three examples: no user computers, 1 user computer, and 2 user computers. Just by counting the connecting lines, we can see that one phone line is needed when there are no user computers, three are needed when there is just one user computer, and 8 are needed for 2 user computers.

We want to know how many phone lines will be needed for 500 user computers. It would not be practical to draw a diagram and count the lines for so many computers. Instead, we will again use a difference equation approach. That means rather than looking directly at how many lines are needed for a given number of computers, we will focus on what happens when a new computer is added to the network. If there are 100 user computers already on the network, how many new phone lines are needed when one more is added? Try to figure out the answer, and write it in on the blank. _____

We need the 2 lines connecting the new computer to the two service computers. We also need three lines from the new computer to each of the original 100. That gives a total of $2 + 3 \cdot 100$.

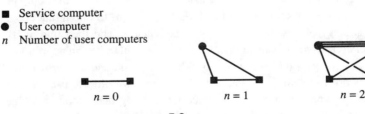

■ Service computer
● User computer
n Number of user computers

$n = 0$ $n = 1$ $n = 2$

FIGURE 5.3
Modified Network Model

A similar line of reasoning works no matter how many computers are originally on the network. If you have 400 computers and add one, that requires $2 + 3 \cdot 400$ new computer lines, 2 to connect the new computer to the service computers, and 3 lines for each of the existing 400 user computers. If there are 28 computers on the network, and a new computer is added, how many new lines would be needed then? _____

Once again, we can formulate this as a difference equation model. To keep this situation separate from the previous problem, let us use the letter s. Define s_n as the number of computer lines needed for a network with n user computers and 2 service computers, with each service computer connected by 1 line to every other computer, and each user computer connected by three lines to every other user computer. Reasoning as above, suppose you have n user computers, and they are connected by s_n telephone lines. If you add one computer, that will require $2 + 3n$ new phone lines, for a total of $s_n + 2 + 3n$. And that is the total for $n + 1$ user computers, or s_{n+1}. This gives the equation

$$s_{n+1} = s_n + 2 + 3n \tag{6}$$

This is another example of a quadratic growth difference equation. We can identify $d = 2$ and $e = 3$. The functional equation for this model is

$$s_n = s_0 + 2n + 3(n - 1)n/2$$

For the initial value s_0, we need to figure out how many phone lines are needed when there are no user computers. Remember that the two service computers need to be connected by a single phone line, so even with no user computers, one line is needed. Then, with $s_0 = 1$, we have

$$s_n = 1 + 2n + 3(n - 1)n/2$$

Now we can easily see that

$$s_{500} = 1 + 2 \cdot 500 + 3(499)500/2 = 375{,}251$$

The two examples in this section illustrate an important facet of difference equation methods. In many problems it is an effective strategy to study directly how one data value is related to the next, even if you don't know exactly what the data values are. It is the process of transition from one to the next that is considered. In the examples we looked at it was possible to formulate an exact equation for that transition. In other cases it is necessary to formulate a model that only gives an approximate description for the transition. In either case, the strategy is to formulate a difference equation directly. When that difference equation has a recognizable form, as it does in the computer network examples, we can use our knowledge of difference equations to find a functional equation. This overall strategy of first finding a difference equation and then finding a functional equation is very effective in many applied areas. In particular, it is an appropriate approach to a number of network problems, many of which lead to quadratic growth models.

For the network problems we have considered here, we did not use quadratic growth as a matter of choice. By studying the process of adding one computer to the network, we were forced to use a quadratic growth difference equation. That is simply the way it worked out. In the next section we will consider another situation in which quadratic growth models are dictated by the problem. In this new situation, we will use a model for one aspect of a problem to find a model for a related aspect. This time, it will be the decision to use an arithmetic growth model for the first aspect that will force us to use a quadratic model for the second.

Sums of Arithmetic Growth Models.   In most of the difference equation models we have studied so far, the focus has been on a single sequence of numbers, a_1, a_2, etc, typically representing regular periods of time. For example, the numbers might be daily, monthly, or yearly figures for some phenomenon. In some situations it is of interest to calculate a running total for these figures. Here is a specific example. A development plan for a city provides a certain number of lots for single family houses. Let us say for concreteness that there are 60,000 lots included in the plan. Over several years, the planning commission keeps track of the number of new homes constructed. The data might appear as in Table 5.3.

Year	1989	1990	1991	1992	1993	1994
Houses Built	3,000	3,800	4,600	5,400	6,200	7,000
Running Total	3,000	6,800	11,400	16,800	23,000	30,000

TABLE 5.3
New Home Construction Figures

The data in the second row of the table will be referred to as h_0, h_1, h_2, and so on. That is, h_n is the number of new houses built in year n, where 1989 is year 0. The table also shows a running total for the annual construction figures. This keeps track of the total houses built since the start of 1989. For 1989, the running total is the same as the number of houses built, 3,000. In 1990, there were 3,800 new homes built. That means the total for 1989 and 1990 combined was $3,000 + 3,800 = 6,800$. That figure appears in the bottom row of the table for 1990. In a similar way, each number in the bottom row is the total from 1989 up to the current year. So, for 1993 the entry in the running total row is 23,000, and that indicates that 23,000 homes were built from 1989 through 1993. Notice that the running total for each year can be found by adding that year's number of new houses to the running total through the previous year. For instance, the running total through 1992 was 16,800. In 1993 an additional 6,200 new homes were built. That brought the running total of homes up to $16,800 + 6,200 = 23,000$.

The running total figures are referred to as the *sums* for the original data sequence. We will denote these running totals or sums in this example by s_0, s_1, s_2, etc. In general terms, s_n is the sum of housing figures for years 0 through n. Put another way, s_n is the total of h_0, h_1, h_2, and so on up to h_n.

It is easy to understand why someone would be interested in s_n. The total number of houses built over the years can be used to figure out how many lots are still available, or how long it will be before all of the planned lots are used. But what does all this have to do with quadratic growth? The answer is this: if the original data follow an arithmetic growth model, then the running totals follow a quadratic growth model. We will see that next.

We can formulate a difference equation for s_n in two steps. First, we will relate s_n and h_n. Second, we will develop a *functional* equation for h_n. Combining these two steps will give the desired difference equation for s_n by itself.

For the first step, we use the observation made earlier: the total for any year is the number of houses built that year plus the total up to the previous year. Let us look at that in terms of s's and h's. The total for years 0 through 3 is s_3. It can be found by adding the previous year's total, s_2, to the number of houses built in year 3. That is expressed in the form of an equation as

$$s_3 = s_2 + h_3$$

Similarly, the total for years 0 through 2, s_2, is the total for the previous year s_1 plus the number of houses built in year 2, h_2. That is,

$$s_2 = s_1 + h_2$$

The examples reveal a pattern that can be expressed by the single equation

$$s_{n+1} = s_n + h_{n+1}$$

Next, just look at the h_n data. You should recognize that these figures follow an arithmetic growth pattern. The first year the figure is 3,000, and the figures increase by 800 each year. You should also remember how to express that as both a difference equation, $h_{n+1} = h_n + 800$, and as a functional equation, $h_n = 3,000 + 800n$. Combining both of these, we obtain a functional equation for h_{n+1}:

$$h_{n+1} = h_n + 800 = 3,000 + 800n + 800 = 3,800 + 800n$$

Now we will use this equation for h_{n+1} in the earlier difference equation for s_{n+1}, as follows

$$s_{n+1} = s_n + h_{n+1}$$

$$= s_n + 3,800 + 800n$$

By now, you should recognize this as another instance of quadratic growth.[7] This time, the parameters are $d = 3,800$ and $e = 800$. Observe that the starting value for s_0 is the same as h_0, namely, 3,000. Accordingly, we can immediately write out a functional equation for s_n.

$$s_n = s_0 + dn + e(n - 1)n/2$$

$$= 3,000 + 3,800n + 800(n - 1)n/2$$

---

[7] Since the sequence s_n is defined as the sums for the sequence h_n, it should come as no surprise that the h_n values are the differences of the s_n sequence. Similarly, the differences for h_n are the same as the second differences for s_n. If h_n is an arithmetic growth model, its differences are constant. That means that the second differences for s_n are also constant, and that shows in another way that s_n is given by a quadratic growth model.

Just to double check this result, use the equation to compute

$$s_5 = 3{,}000 + 3{,}800 \cdot 5 + 800(4)5/2 = 30{,}000.$$

This agrees with the number in the table for 1994, as it should.

Although we found the functional equation in this example by using a formula that works for all quadratic growth models, it is also easy to use patterns to find the formula. We know that s_n can be found by adding up the h values from h_0 through h_n. We can list out the first several h values using the functional equation $h_n = 3{,}000 + 800n$, and then add them up. Here is one way to organize the calculations:

$$
\begin{aligned}
h_0 &= 3{,}000 \\
h_1 &= 3{,}000 + 800 \cdot 1 \\
h_2 &= 3{,}000 + 800 \cdot 2 \\
h_3 &= 3{,}000 + 800 \cdot 3 \\
+ \; h_4 &= 3{,}000 + 800 \cdot 4 \\
\hline
h_0 + h_1 + \cdots + h_4 &= 5 \cdot 3{,}000 + 800(1 + 2 + 3 + 4) \\
&= 5 \cdot 3{,}000 + 800(4)(5)/2
\end{aligned}
$$

This calculation shows that $s_4 = 5 \cdot 3{,}000 + 800(4)(5)/2$. How would this change if you computed s_7 instead of s_4? On a piece of scratch paper, write down the equations for h_0 through h_7 in the same format shown above, and add them all up. You should end up with the final result $s_7 = 8 \cdot 3{,}000 + 800(7)(8)/2$. As these results suggest, there is a general pattern for these examples, leading to the equation

$$s_n = (n + 1)3{,}000 + 800(n)(n + 1)/2$$

This equation is not in the same form as the one we found earlier, but it gives the same results. As an illustration, here is the way s_5 is computed with our new equation.

$$s_5 = 6 \cdot 3{,}000 + 800(5)(6)/2 = 30{,}000$$

That is the same result we found earlier.

This example introduces the idea of the sums of a sequence, and shows that for an arithmetic growth model, the sums follow a quadratic growth model.

Sums of sequences are often of interest. Now we know that adopting an arithmetic growth model for a problem automatically leads to a quadratic growth model for the sums. This is the third of the three situations that leads to quadratic growth. As a final example in this section, we will revisit the oil reserves model discussed earlier.

On p. 46 the annual oil consumption model is used to develop a model for oil reserves. Remember that c_n is the average daily amount of oil consumed in year n, where 1991 is year 0, 1992 is year 1, and so on. Also recall that r_n is the amount of oil available at the start of year n. It should be clear that the reserves available at the start of any year will equal the reserves that existed at the start of year 0, minus all the oil used up in the years from 0 up to the current year. So this problem is again closely connected with the idea of summing the terms of a sequence, although this time we have to subtract that sum from the starting figure for the reserves.

Based on these ideas, the model presented on p. 46 involves the equations

$$c_{n+1} = c_n + .3$$

$$c_0 = 66.6$$

$$r_{n+1} = r_n - 365c_n$$

$$r_0 = 999{,}100$$

Like the discussion of home construction above, this problem involves two sequences of numbers, c_n for daily oil consumption figures, and r_n for the reserves available at the start of each year. Each sequence has a difference equation. We recognize the first difference equation, $c_{n+1} = c_n + .3$, as an instance of arithmetic growth, so the functional equation for c_n can be found using the methods for arithmetic growth. That gives $c_n = 66.6 + .3n$.

Now that equation ($c_n = 66.6 + .3n$.) can be combined with the difference equation for r. Simply replace c_n by $66.6 + .3n$. Then we have $r_{n+1} = r_n - 365(66.6 + .3n)$, or, removing the parentheses, $r_{n+1} = r_n - 24,309 - 109.5n$. With what has been presented so far in this chapter, at this point you should realize that r follows a quadratic growth model. The functional equation can be worked out as described earlier. Indeed, with $r_0 = 999{,}100$, $d = -24{,}309$, and $e = -109.5$, the parametric form of the quadratic growth functional equation (page 89) becomes

$$r_n = 999{,}100 - 24{,}309n - 109.5(n - 1)n/2$$

This equation can be rewritten in a little more compact form using algebra. Looking first at $109.5(n-1)n/2$, since $109.5/2 = 54.75$, we have $109.5(n-1)n/2 = 54.75(n-1)n = 54.75(n^2 - n) = 54.75n^2 - 54.75n$. Now combine this with the original equation. That leads to

$$r_n = 999{,}100 - 24{,}309n - 54.75n^2 + 54.75n$$

Finally, combining the two terms involving n produces

$$r_n = 999{,}100 - 24{,}254.25n - 54.75n^2$$

This is the equation that was given in Chapter 3.

This concludes the discussion of situations in which quadratic growth arises in applications. We have seen that a quadratic model can be adopted because a data set has constant (or nearly constant) second differences. Also, in many problems involving networks, quadratic growth difference equations arise from a direct study of the transition from n to $n + 1$. Finally, we considered the idea of summing up the terms of a sequence, and observed that when the original sequence follows an arithmetic growth pattern, the sums follow a quadratic growth pattern. In all of these cases, we only touched briefly on the kinds of questions that can be answered using quadratic growth models. As for other models, there are two fundamental kinds of questions that arise in quadratic models: what will the model predict for a given value of n, and for what value of n will a specified data value be observed? On a more general level, it may simply be of interest to derive a picture of the future evolution of the model. In order to answer these questions, it is helpful to have a better understanding of some additional properties of quadratic growth models. That is the topic of the next section.

Properties of Quadratic Growth Models

We have seen some of the ways that quadratic growth models arise in applications. All of these quadratic models share several common features. For one thing, they have the same form of difference equation and the same form of functional equation. There are a few additional aspects of quadratic growth that contribute to our understanding of these models.

Nonlinearity. We saw earlier that arithmetic growth models are closely related to linear functions, straight lines, and proportional reasoning. It is important to be aware that quadratic models do *not* behave as linear models do. For one thing, proportional reasoning is not correct for quadratic growth models. Remember that we saw this for the computer network model. Using Eq. (5), we can calculate that a 100-computer network will require $t_{100} = 4{,}950$ phone lines, according to the t_n model. That is nearly 5,000 phone lines. Based on that figure, it is natural to think that a 200-computer network will need around 10,000 phone lines. With twice as many computers shouldn't we need twice as many phone lines? The answer is no. We know how to compute the exact number of lines needed for 200 computers: $t_{200} = 199 \cdot 200/2 = 19{,}900$. This is almost 20,000, and is nearly four times the number of lines needed for 100 computers.[8] The problem here is that expecting the number of phone lines to double is using proportional reasoning. But that only works with arithmetic growth models, which are described by linear functional equations. The quadratic growth models have nonlinear functional equations. It is an error to apply proportional reasoning in this context, even for a rough estimate; applying proportional reasoning here will lead to incorrect conclusions. So, one aspect that is shared by all quadratic growth models is the trouble you can get into using proportional reasoning.

Graphs. As for any difference equation model, it is easy to create line graphs for quadratic growth models. Graphs for both t_n and s_n are shown in Fig. 5.4. At this point we merely observe that these graphs are definitely not straight lines. This is another indication that proportional reasoning doesn't apply to quadratic models. Proportional reasoning only makes sense in a model with a straight line graph. The farther the graph of a model departs from a straight line, the greater the error that will be committed by using proportional reasoning.

---

[8] Here you can see a connection with the idea of areas of rectangles and squares, the problems that give quadratic growth its name. Does a square of side 40 have twice the area of a square of side 20? No. The smaller square has an area of $20 \cdot 20 = 400$; the larger has an area of $40 \cdot 40 = 1{,}600$—four times greater. Doubling the sides of the square makes the area four times as great. This is what occurs in the computer network model: four times as many lines are needed when the number of computers is doubled.

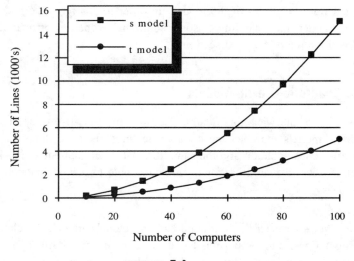

FIGURE 5.4
Graphs for t_n and s_n

There is a great deal that can be said about the graphs of quadratic growth models. The shapes revealed in Fig. 5.4 are characteristic for quadratic growth. Each curve in the graph can be recognized as part of a special shape called a parabola. We will take up parabolas and many interesting features of quadratic graphs in the next chapter.

Comparing Arithmetic and Quadratic Growth. Another interesting aspect of quadratic models is how they compare with arithmetic growth models. When we look at the difference equations and functional equations for these two kinds of models, a pattern emerges. This is shown in Table 5.4. The difference equation for quadratic growth has the same general form as the one for arithmetic growth, but there is one additional term, a multiple of n. Similarly, the functional equation for quadratic growth starts out like the one for arithmetic growth, but there is one additional term, a multiple of $(n-1)n$. For an arithmetic growth model, the differences of the data values are all equal. For a quadratic growth model, the *second* differences are all equal. It is interesting to think about extending this pattern. What would be the next kind of difference equation after arithmetic and quadratic growth? What kind of functional equation would you expect

Growth Type	Sample Difference Equation	Sample Functional Equation
Arithmetic	$a_{n+1} = a_n + 5$	$a_n = 7 + 5n$
Quadratic	$a_{n+1} = a_n + 5 + 4n$	$a_n = 7 + 5n + 4(n-1)n/2$

TABLE 5.4
Model Family Comparison

it to have, based on the pattern in the table? What would you expect to be true of the differences? Investigating patterns of this type is an important aspect of mathematics, and often leads to interesting new results.

Quadratic Equations. The functional equation for a model can always be used in two ways. First, if we specify a value of n, we can directly compute a_n. For example, in the school enrollment model, with the functional equation $a_n = 20,000+4,000n+1,000(n-1)(n)/2$, we can predict that $a_{10} = 105,000$. This is called function evaluation. The equation gives a_n as a function of n, and we are evaluating the function for $n = 10$. Function evaluation allows us to predict the enrollment for specific years. With 1991 as year 0, we know that year 10 will be 2001, and in that year the model predicts an enrollment of 105,000 students.

The second way to use the functional equation reverses the process of function evaluation. Instead of setting a value for n and computing a_n, we may wish to find an unknown n given the value of a_n. As an example, we return to the first computer network model. The functional equation $t_n = (n-1)n/2$ describes how many telephone lines are needed for a network of n computers. But dedicated telephone lines are expensive, and there are a limited number of lines available. For this reason, the designers of a network might know ahead of time the maximum number of phone lines that can be used. Let us say that only 2,000 lines can be used. How many computers can be put on the network? Here, the n is unknown, and t_n, the number of telephone lines is given as 2,000.

This kind of problem is referred to as inverting the functional equation, because it works in reverse. The easy way to use the equation is to plug in a value of n and then work out t_n as the answer. Here we are doing the reverse: we are starting with the answer (2,000 telephone lines) and trying to work backwards to figure out what n leads to that answer. When the answer to a functional equation is given, and we want to know what value of the variable produced that answer, that is called inverting the function.

Inverting a function typically involves solving an equation. For the computer model, the functional equation is $t_n = (n-1)n/2$. We want t_n to be 2,000, so make that substitution in the equation. Then we obtain

$$2,000 = (n-1)n/2$$

Here, n is the unknown, and the goal is to figure out what n must be to make the equation true. This equation cannot be solved using the methods of the previous chapter; it is not a *linear* equation. It is called a *quadratic* equation, and will be studied in detail in the next chapter.

Although we do not yet have algebraic methods to solve the equation above, we can apply graphic and numeric methods to find a solution. First let us look at the graph of t_n. In Fig. 5.4, look to see where the curve for t_n crosses the 2,000 line. (Remember that the vertical axis is in units of 1,000's, so on that axis 2 indicates the line for 2,000.) It appears that this crossing point is right above 60 on the horizontal axis. So, as a first guess, we expect that when n equals 60, t_n is pretty close to 2,000. That is a graphic method.

We can check that numerically using the functional equation: $t_{60} = (59)60/2 = 1,770$. This is close to 2,000. By trying some values close to 60 we can try to improve the answer. Since 60 computers use up only 1,770 lines, let us try a larger number, say 65: $t_{65} = (64)65/2 = 2,080$. That is just a little over. Next try 64: $t_{64} = (63)64/2 = 2,016$. That is even closer. Continuing in this way, it is possible to get closer and closer to a solution for the equation. If you try 63 you will find that t_{63} is less than 2,000. The conclusion is that 2,000 lines will accommodate up to 63 computers on the network. There are not enough lines for any more than that.

Does this answer seem unreasonably low? It might seem that with 2,000 phone lines it ought to be possible to connect many more than 63 computers. That perception is based on linear thinking, something most people do without even realizing it. This is another aspect of the fact that quadratic growth is not linear. In many instances, quadratic growth occurs much more rapidly than in a linear model. When you see a quadratic model, be wary of this effect. Things may grow more rapidly than you expect.

Let us look at another example of function inversion. Remember the housing construction model? There we found the functional equation

$$s_n = 3,000 + 3,800n + 800(n - 1)n/2$$

where s_n represents the total number of houses built in years 0 through n. An obvious question to ask about this model is, when will all the planned lots be used up? Originally, we said that there were 60,000 lots planned. When these are all gone, that means a total of 60,000 homes have been built, so $s_n = 60,000$. Make that substitution in the functional equation, and the result is

$$60,000 = 3,000 + 3,800n + 800(n - 1)n/2$$

The year, n, is unknown, so that remains a variable in the equation. Now you have to solve this equation to figure out when all the lots will be used up. Once again, this equation is not linear, and cannot be solved by the methods of the previous chapter. A little experimentation with a numerical approach will let you find an answer. That is, use trial and error with some guesses for n. Is $n = 10$ right? Is the answer too high or too low? This problem will be left as an exercise for you to work out.

This just about concludes the discussion of quadratic growth. You have read about what quadratic growth is, how to recognize quadratic growth difference equations and formulate the corresponding functional equations, and situations in which quadratic growth models are used. You have also seen some of the properties of quadratic growth models, and how they are used to answer questions about real problems. In all the applications of quadratic growth, the functional equation is of key importance. That equation, in turn, rests on the shortcut for adding up whole numbers. We conclude the chapter with a supplementary discussion of this shortcut, and a brief introduction to sigma notation. Neither of these topics is essential for the material in later chapters.

Whole Number Sums and Σ Notation

The patterns that permitted us to find functional equations for quadratic growth models were based on the shortcut for adding consecutive whole numbers. The shortcut can be

expressed as the equation

$$1 + 2 + \cdots + n = n(n+1)/2$$

Now we will take a closer look at this important idea.

At first, the shortcut was introduced without explanation. A few examples for small numbers were shown, and for those we saw that the shortcut does give correct answers. But why does it? How can we be sure it will be correct for $n = 100$, or $n = 1,000$, or for any n at all? One way to understand the pattern is to rearrange the terms of the sum. To illustrate this idea, here is the sum for $n = 10$ rearranged: $1+2+3+4+5+6+7+8+9+10 = (1+10)+(2+9)+(3+8)+(4+7)+(5+6)$. There are 5 groups of 11. Note that the number of groups is $n/2$ because each group has two terms in it. The sum of each group is $n+1$. So the sum is equal to $(n/2)(n+1) = 5 \cdot 11 = 55$. This agrees with the shortcut formula $n(n+1)/2$, and shows why that formula works when n is even. For an odd number of terms, things are a little bit different. For $n = 9$ group $1 + 2 + 3 + 4 + 5 + 6 + 7 + 8 + 9$ as $(1 + 8) + (2 + 7) + (3 + 6) + (4 + 5) + 9$. This time there are 5 groups of 9. The number of groups is $1 + (n-1)/2$. The first $n - 1$ terms (all except the last) are added in pairs, and there are $(n-1)/2$ such pairs. Each pair has a total of 9, or n. Then there is the one remaining term of n, for a total of $1 + (n-1)/2$ groups, each equal to n. When these are added together, the result is $n[1 + (n-1)/2]$. This can be rewritten as $n(n+1)/2$ using algebra. So for the odd numbers as well, the shortcut gives the right answer. This shows that for n even or odd, the sum $1 + 2 + 3 + \cdots + n$ is equal to $n(n+1)/2$.

The methods used in this discussion, and the formula $1+2+3+\cdots+n = n(n+1)/2$ belong to the mathematical subject called combinatorics. Another familiar pattern that belongs to combinatorics is Pascal's triangle, the first few lines of which are shown in Fig. 5.5. If the subject of combinatorics appeals to you, try to find patterns for these problems:

1. $1 + 3 + 5 + 7 + \cdots + n$ for odd n
2. $1 + 4 + 9 + 16 + \cdots + n^2$
3. $1 + 8 + 27 + 64 + \cdots + n^3$
4. $1 + 2 + 4 + 8 + 16 + \cdots + 2^n$

FIGURE 5.5
Pascal's Triangle

Surprisingly, the pattern for problem 2 is harder to find than the pattern for problem 3. Your instructor can help you find more information on the subject of combinatorics.

Throughout the chapter, we have used the three-dot notation to express sums. Although this is a big improvement over actually listing all the terms of the sum, there is an even more concise notation that is preferred in mathematics. It is called sigma notation because it involves the Greek letter Σ.

With the three-dot notation, the pattern of a sum is illustrated by listing the first several terms. In the sigma notation, the pattern is revealed using a variable, and the starting and ending points for the pattern are written below and above the Σ symbol. For example, the notation

$$\sum_{k=1}^{6} a_k$$

is read *the sum from k equals* 1 *to* 6 *of* a_k, and it simply means to add up all the $a's$ from a_1 to a_6. That is, it means $a_1 + a_2 + a_3 + a_4 + a_5 + a_6$. Similarly, $\sum_{k=0}^{100} c_k$ means to add up c_0, c_1, c_2, and keep going until you have added c_{100}.

This sigma notation is used throughout mathematics, and is found especially in statistics. Sometimes it is used with subscripts as in the two examples above, but in other instances it is used with an algebraic expression. For example, this notation

$$\sum_{k=2}^{5} 3k + 1$$

means the sum

$$
\begin{aligned}
&3 \cdot 2 + 1 \\
&3 \cdot 3 + 1 \\
&3 \cdot 4 + 1 \\
+\ &3 \cdot 5 + 1 \\
\hline
\end{aligned}
$$

The $k = 2$ below the Σ and the 5 above the Σ indicate a sum with terms for k equal to 2, 3, 4, and 5. The $3k + 1$ gives the general form for the term. That is, setting $k = 2$, we change $3k + 1$ into $3 \cdot 2 + 1$, and that gives the first line of the addition problem above. This process is repeated for k equals 3, then 4, then 5.

In this example, there are only 4 terms. But if we change the sigma notation to

$$\sum_{k=2}^{100} 3k + 1$$

that would indicate an addition problem similar to the one above, but having 99 lines! As you can see, the sigma notation packs a huge amount of information into a small number of symbols.

The sigma notation can be used to express the shortcut for adding up whole numbers in a very compact form:

$$\sum_{k=1}^{n} k = \frac{n(n + 1)}{2}$$

In this expression the k that comes after the Σ is similar to $3k + 1$ in the preceding example. As before, we think of k as becoming first 1, then 2, then 3, and so on. This time, however, rather than substitute that value of k into a more complicated expression like $3k + 2$, we just use the value of k itself. That means we will add first 1, then 2, then 3, and so on, all the way up to n. And for that sum, the shortcut gives the answer as $n(n + 1)/2$.

The sample combinatorics patterns posed above can be expressed in Σ notation as follows:

$$1. \quad \sum_{k=1}^{n} 2k - 1 \qquad 2. \quad \sum_{k=1}^{n} k^2 \qquad 3. \quad \sum_{k=1}^{n} k^3 \qquad 4. \quad \sum_{k=0}^{n} 2^k$$

As you can see, when expressed in this way, the problems all look very much alike. This ability to show the similarities of many different kinds of sums is one of the advantages of the Σ notation. Do not be alarmed if it is not immediately obvious to you that these really are the same problems listed earlier. The Σ notation definitely takes some getting used to. One thing that will help you become more familiar with it is to convert a sum from the Σ form to the three-dot form. To do this, take the first few values of k in the sum, substitute them into the formula that follows the Σ, and see what pattern of terms emerges. It is recommended that you do this for a few of the combinatorics problems listed above.

Summary

This chapter explores a new family of models, the quadratic growth models. An initial example was used to introduce these models. Quadratic growth models can be defined in two ways. First, a quadratic growth model is one in which the differences follow an arithmetic growth pattern. Equivalently, quadratic growth occurs when the second differences are constant.

All quadratic growth models share a difference equation with a common form: $a_{n+1} = a_n + d + en$, where d and e stand for constants that change from one application to another. This difference equation corresponds to the functional equation $a_n = a_0 + dn + en(n - 1)/2$. In general, a good first step for an application problem is to formulate a difference equation. In this task it is often useful to know that e is the constant second difference for the model, and that d is the difference between a_0 and a_1. Once the difference equation has been found, you will know the numerical constants d and e. These are then substituted in the general form for the functional equation to obtain the specific equation for the problem at hand. That equation, in turn, can be used to answer questions about the future performance of the model. Predictions in which n is given are computed by direct substitution in the functional equation. Questions in which n is the unknown lead to quadratic equations. These can be studied using graphical and numerical methods, and a theoretical method will be introduced in the next chapter.

Quadratic models can arise either out of choice, or because they are dictated by the nature of the model. The first possibility occurs when data are observed to have approximately equal second differences, and you simply choose to use a quadratic model. Two cases where quadratic growth is dictated include certain network problems and problems concerning sums of sequences. In the first of these, we saw that a direct analysis of the transition from one sequence value to the next simply worked out to be a quadratic growth difference equation. There was no choice involved. In the second, we observed that, when an arithmetic growth model is used for a particular sequence, the sums of the sequence have to follow a quadratic model. In this case, it might be a matter of choice to use an arithmetic model for the original problem, but, once that choice is made, you are forced to use a quadratic model for the sums. Examples of this kind of problem include a model for annual housing construction and a model for annual oil consumption. For the housing construction model, the sums tell us the total number of houses built over several years, and help us understand how many lots remain for future building. For the oil consumption model, the sums are closely related to a model for oil reserves, which can be used to predict how long the world petroleum supply will last.

In addition to difference and functional equations, we studied some other common properties of quadratic growth models. Quadratic models are nonlinear and, as one consequence, are not an appropriate context for proportional reasoning. The graphs of quadratic growth models are not straight lines. They follow a characteristic shape called a parabola. This will be studied in detail in the next chapter. We looked at an interesting pattern that emerges when arithmetic growth and quadratic growth models are compared. We also looked at several examples in which quadratic equations came up in problems with quadratic growth models. In this chapter we used graphical and numerical methods to solve these equations. A theoretical method will be developed in the next chapter.

The functional equation for quadratic growth models was formulated using a shortcut for adding whole numbers. This shortcut was presented without explanation at first. A discussion at the end of the chapter showed why the shortcut works. Two notations for sums were also presented. The three-dot notation shows enough terms in a sum to establish a pattern, and also specifies where the sum ends. The sigma notation uses an algebraic expression with a variable to describe the pattern of terms and specifies the starting and ending values of the variable.

Exercises

Reading Comprehension.   Answer each of the following with a short paragraph. Give examples, where appropriate, but also answer the questions in written sentences.

1. What is a quadratic growth model?

2. Is proportional reasoning appropriate with quadratic growth models?   Give an example.

3. In the equation $a_{n+1} = a_n + d + en$, d and e are called parameters. What does that mean?

4. What form does the functional equation take in a typical quadratic growth model? Don't give the parametric form of the equation, but do describe the appearance of the equation in general terms.

5. Describe three differences between quadratic growth models and arithmetic growth models.

6. What is meant by *nonlinearity*?

7. Explain the concept of a sum of an arithmetic growth model. Give an example illustrating why this is of interest, and how it can be useful.

8. Describe two kinds of shorthand notation useful when many terms are added together.

9. What is meant by *second differences*? Give an example. How are second differences related to quadratic growth models?

Mathematical Skills

1. In a quadratic growth model, the initial term is $b_0 = 6$, and the difference equation is $b_{n+1} = b_n + .5 + .2n$. Use the difference equation to figure out b_1, b_2, b_3, and b_4.

2. In a quadratic growth model, the initial term is $z_0 = 10$, and the difference equation is $z_{n+1} = z_n + 2 - n$. Use the difference equation to figure out z_1, z_2, z_3, and z_4.

3. Make a graph for the b_n's in problem 1. Include at least 6 points on the graph. (You may do this by hand, or use a graphing calculator or computer.)

4. Make a graph for the z_n's in problem 2. Include at least 6 points on the graph. (You may do this by hand, or use a graphing calculator or computer.)

5. What are the parameters d and e in problem 1?

6. What are the parameters d and e in problem 2?

7. Find the functional equation for b_n in problem 1. Use it to check the answer you got for b_4 in problem 1, and also to find b_{10}.

8. Find the functional equation for z_n in problem 2. Use it to check the answer you got for z_4 in problem 2, and also to find z_{14}.

9. Find the difference equation and functional equation for this sequence of numbers: 3, 5, 8, 12, 17, 23, $\cdots$.

10. Find the difference equation and functional equation for this sequence of numbers: 3, 13, 21, 27, 31, 33, $\cdots$.

11. The functional equation for the first computer network model is $t_n = (n-1)n/2$. Find the lowest n for which t_n is above 3,000. Use graphical and numerical methods. (See Fig. 5.4).

12. The functional equation for the second computer network model is $s_n = 1 + 2n + 3(n-1)n/2$. If s_n can not go above $1,000$, what is the largest possible choice for n? Use graphical and numerical methods to answer this question. (See Fig. 5.4).

13. The functional equation for the oil reserves model is $r_n = 999{,}100 - 24{,}254.25n - 54.75n^2$ at the start of year n, where year 0 is 1991, year 1 is 1992, etc. Use graphical and numerical methods to find the first r_n that is below 850,000. (This gives the first year the reserves fall below 850,000.)

14. If you add up all the whole numbers from 1 to 50, what is the total?

15. If you add up all the whole numbers from 1 to 1,000, what is the total?

16. Compute $1 + 2 + 3 + \cdots + 25$

17. An arithmetic growth model starts with $a_0 = 5$ and has the difference equation $a_{n+1} = a_n + 3$. What is $a_1 + a_2 + \cdots + a_{10}$? What is $\sum_{k=0}^{20} a_k$?

18. What is the sum of the first n odd whole numbers?

Problems in Context

1. Here is another modified version of the computer network problem. This time, there are 5 resource computers, each connected by one line to the other 4. Every user computer is connected to each resource computer by one line, and is connected to each of the other user computers by two lines. Let v_n be the total number of lines required when there are n user computers. Find the difference equation and the functional equation for v_n.

2. In a round-robin soccer tournament, each team must play every other team once. How many games must be scheduled if there are 10 teams? If there are 20 teams? If there are n teams? [Hint: Use the idea of difference equations. Try to figure out how the number of games increases when one new team is added to the tournament.]

3. A lumber company owns 50,000 acres of forest land. In the first year of operation they log 1,200 acres. The next year they log 1,400 acres. The next year they log 1,600 acres. Assuming that the number of acres logged each year follows an arithmetic growth law, how many acres total will have been logged after 10 years? How many years will it take to log the entire 50,000 acres?

4. A wise person once solved a problem for a wealthy merchant. The merchant told the wise person to choose a reward. The wise person pointed to the merchant's chess board and asked for one gold piece to be put on the first square, 2 gold pieces to be put on the second square, 3 on the third, and so on until all 64 squares on the checkerboard were filled. How many pieces of gold were then on the checkerboard?

5. In the problems at the end of the preceding chapter you developed a linear model for the number of autos in the US. Assume that each auto generates about 6 tons of carbon dioxide per year. According to your linear model, how many tons of carbon dioxide will be added to the atmosphere between 1995 and 2050 by automobile exhaust? If you did not do the problem at the end of the preceding chapter, use the equation $N = 2.45 \cdot \text{year} - 4,734$ where N is the number of automobiles in the US, in millions, in any year. [Hint: figure out how many tons will be added in 1995, in 1996, in 1997 and so on. This is an arithmetic growth model.]

6. Galileo discovered the following pattern about the motion of falling bodies. Say you check the distance traveled by the body over a series of equal periods of time, every second, or every two seconds, or every 5 seconds, or some other similar arrangement. However far the body falls in the first time period, it will fall 3 times that initial distance in the second time period, 5 times the initial distance in the third time period, 7 times in the fourth time period, and so on. Now consider a diver about to jump off a high platform. In the first one-quarter of a second the diver falls 1 foot. How long

will it take for the diver to reach the water if the platform is 16 feet high? How far does the diver fall in the last quarter-second before hitting the water? What is the diver's speed in feet per second for the last quarter-second? (Note, since you know how many feet the diver covered in a quarter-second, four times that amount would be covered in a full second, traveling at the same speed. So the speed in feet per second is four times the number of feet traveled in the last quarter-second.) Now repeat the same calculations for a platform 64 feet high. That is four times higher than the platform in the first part of the problem. Does the diver fall four times longer? Is the diver going four times as fast in the last quarter-second?

Solutions to Selected Exercises

Mathematical Skills

1. $b_1 = b_0 + .5 + .2(0) = 6.5.$ $b_2 = b_1 + .5 + .2(1) = 6.5 + .5 + .2 = 7.2.$ $b_3 = 7.2 + .5 + .2(2) = 8.1.$ $b_4 = 9.2$

4. See Fig. 5.6.

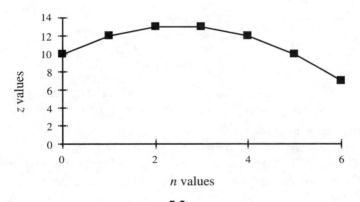

FIGURE **5.6**
Graph for problem 4.

5. $d = .5$ and $e = .2$

7. $b_n = 6 + .5n + .2(n - 1)n/2.$ For $n = 4$, $b_4 = 6 + .5(4) + .2(3)(4)/2 = 9.2.$ This is what we got before. $b_{10} = 6 + .5(10) + .2(9)(10)/2 = 20.$

9. First, look at the differences for the sequence: 2, 3, 4, 5, 6. The second differences are all 1. So this is quadratic growth. The differences start at 2 and increase by 1 each time, showing that $d = 2$ and $e = 1$. Then the difference equation is $a_{n+1} = a_n + 2 + n$, and the functional equation is $a_n = 3 + 2n + n(n - 1)/2.$

11. According to the graph, it looks as though t_n goes above 3,000 somewhere between $n = 70$ and $n = 80$. So for a first guess, try $n = 75$. That gives $t_{75} = 74(75)/2 =$

2,775. This is too low, so try $t_{78} = 77(78)/2 = 3{,}003$. This is so close to 3,0004, it must be the first t that is over 3,000. To be complete, check that t_{77} is below 3,000.

14. $1 + 2 + \cdots + 50 = 50 \cdot 51/2 = 1{,}275$

15. $1 + 2 + \cdots + 1{,}000 = 1{,}000 \cdot 1{,}001/2 = 500{,}500$

16. $1 + 2 + \cdots + 25 = 25 \cdot 26/2 = 325$

17. Solution 1. This is about the sums associated with an arithmetic growth model, and therefore can be solved using a quadratic growth model. Introduce the new variable $s_n = a_0 + a_1 + \cdots + a_n$. Then $s_0 = a_0 = 5$. Now a difference equation for s is given by $s_{n+1} = s_n + a_{n+1}$. To use this we need an equation for a_{n+1}. Since a_n is from an arithmetic growth model, we know $a_n = a_0 + dn = 5 + 3n$. Now we have $s_{n+1} = s_n + 5 + 3(n + 1) = s_n + 8 + 3n$. That allows us to write down the functional equation $s_n = 5 + 8n + 3(n - 1)n/2$. So $s_{10} = 5 + 80 + 135 = 220$. But wait: this is $a_0 + a_1 + \cdots + a_{10}$, and the problem only asked for the sum from a_1 up to a_{10}. So the final answer is found by subtracting a_0 from what we already found: $220 - 5 = 215$. For the second question, $\sum_{k=0}^{20} a_k$ is just exactly s_{20}. We find this as $s_{20} = 5 + 160 + 570 = 735$.

Solution 2: Use patterns:

$$
\begin{aligned}
a_1 &= 5 + 3 \\
a_2 &= 5 + 3 \cdot 2 \\
a_3 &= 5 + 3 \cdot 3 \\
a_4 &= 5 + 3 \cdot 4 \\
+ \ a_5 &= 5 + 3 \cdot 5 \\
\hline
a_1 + a_2 + a_3 + a_4 + a_5 &= 5 \cdot 5 + 3(1 + 2 + 3 + 4 + 5)
\end{aligned}
$$

Extending the same pattern, $a_1 + \cdots + a_{10} = 10 \cdot 5 + 3(1 + 2 + \cdots + 10) = 50 + 3 \cdot 10 \cdot 11/2 = 215$. Similarly, $a_1 + \cdots + a_{20} = 20 \cdot 5 + 3(1 + 2 + \cdots + 20) = 100 + 3 \cdot 20 \cdot 21/2 = 730$. However, we are supposed to find the total from a_0 to a_{20}, and 730 only has the total from a_1 to a_{20}. To get the final answer, add in $a_0 = 5$, to get 735.

18. The odd numbers are 1, 3, 5, etc. This is an arithmetic progression with $a_0 = 1$ and $d = 2$. We want to find the sum from a_0 through a_{n-1} (remember, the first n terms will be a_0 through a_{n-1}). Using either of the methods described above, the final answer is $n + (n - 1)n = n + n^2 - n = n^2$.

Problems in Context

1. Suppose there are no user computers. There are 5 resource computers, and they all have to be joined together. This is just like the first computer network model. We will need $t_5 = 5(4)/2 = 10$ lines. This shows that $v_0 = 10$. Now if we add one user computer, we just need one new line to each service computer, so $v_1 = v_0 + 5 = 15$. When we put in the second user computer, we need another 5 lines to service computers, plus 2 lines to the first user computer, for a total of $v_1 + 5 + 2 = 22$. These calculations will help you understand the difference equation

for this problem: $v_{n+1} = v_n + 5 + 2n$. This says that, when we already have n user computers, and we add one more, we need 5 lines to service computers plus 2 lines to each of the n existing user computers, for a total of $5 + 2n$ new lines. In all, that gives $v_{n+1} = v_n + 5 + 2n$, as stated. This is a quadratic growth model. The difference equation gives v_{n+1} as v_n plus a linear function of n. We can write down the functional equation as $v_n = v_0 + 5n + 2(n-1)n/2$. But we already decided $v_0 = 10$. So $v_n = 10 + 5n + (n-1)n$.

3. The total number of acres logged over several years can be found by adding up the amounts logged each year, so let us start with a model for that. Let a_n be the number of acres logged in the nth year. That is, $a_1 = 1,200$, $a_2 = 1,400$, and so on. The difference equation for a_n is $a_n = a_{n-1} + 200$. That is arithmetic growth. Therefore, we can solve this problem by using a quadratic growth model for the number of acres logged over n years. Alternatively, we can work directly with patterns. We want to find the total logged in 10 years, and that will be $a_1 + a_2 + \cdots + a_{10}$. The patterns look like this::

$$
\begin{aligned}
a_1 &= 1,200 \\
a_2 &= 1,200 + 200 \\
a_3 &= 1,200 + 200 \cdot 2 \\
a_4 &= 1,200 + 200 \cdot 3 \\
&\vdots \\
+ \ a_{10} &= 1,200 + 200 \cdot 9 \\
\hline
a_1 + a_2 + \cdots + a_{10} &= 10 \cdot 1,200 + 200(1 + 2 + \cdots + 9) \\
&= 10 \cdot 1,200 + 200(9)(10)/2
\end{aligned}
$$

Now we can use this last equation to compute the total for 10 years as $12,000 + 9,000 = 21,000$. Or, we can extend the pattern to any number of years, with the equation

$$
a_1 + a_2 + \cdots + a_n = n \cdot 1,200 + 200(n-1)(n)/2
$$

$$
= 1,200n + 100(n-1)n
$$

This is a functional equation for the total acres logged in n years. Let that be t_n, so

$$
t_n = 1,200n + 100(n-1)(n)
$$

We want to know how many years it takes to log the entire 50,000 acres. That is, find the n so that $t_n = 50,000$. Using the functional equation for t_n, we can find the answer by solving the equation $1,200n + 100(n-1)n = 50,000$. This is a quadratic equation, and you have not yet learned in this course how to solve it (although you might know a method from an earlier course). So let us take a numerical approach. Just guess different values of n and see what you find out. For example, we know that in 10 years 21,000 acres will be logged. That is about half of the total of 50,000, so try guessing $n = 20$. That leads to $t_{20} = 1,200 \cdot 20 + 100 \cdot 19 \cdot 20 = 62,000$. If they keep logging as described in the model, in 20 years they would consume 62,000 acres, far more than they own. So we should try a smaller n, say 15. Then

$t_{15} = 1{,}200 \cdot 15 + 100 \cdot 14 \cdot 15 = 39{,}000$. In 15 years they only log 39,000 of the 50,000 acres. We need a larger n. Continuing in this fashion, you can figure out that $t_{17} = 47{,}600$ and $t_{18} = 52{,}200$. This shows that they can log for 17 full years, but will run out of timber in the 18th year.

Incidentally, notice again that proportional reasoning does not work here. Given that they log 21,000 acres in the first 10 years, you might expect them to cut another 21,000 acres in the next 10 years. That is proportional reasoning. But it is wrong for this model. We know the total for 20 years is 62,000 acres, far more than the 42,000 acres suggested by proportional reasoning. Of course, we have already observed that the equation for t_n is not a linear equation. That should be a tipoff that proportional reasoning will not work.

4. The answer is $1 + 2 + 3 + \cdots + 64 = 64 \cdot 65/2 = 2{,}080$.

5. Let t_n be the number of tons of carbon dioxide added to the atmosphere by automobiles in the year n years after 1995. The given approximate equation for the number of autos in the US is $N = 2.45 \cdot \text{year} - 4{,}734$ (in millions of autos). This equation gives the number of autos in 1995 as $2.45 \cdot 1995 - 4{,}734 = 153.75$ millions. The number of tons of carbon dioxide generated by these cars is found by multiplying by 6: this gives $6(2.45 \cdot 1995 - 4{,}734) = 922.5$ million tons. This is t_0. Using similar methods you can compute $t_1 = 937.2$ million tons, $t_2 = 951.9$ million tons, $t_3 = 966.6$ million tons, and so on. If you examine these numbers carefully, you will see they go up by 14.7 million tons each year. That gives the difference equation $t_n = t_{n-1} + 14.7$, and we know $t_0 = 922.5$, all in units of millions of tons. We would like to add t_0 through t_{55}. As in earlier related problems, we can find this total as $56 \cdot 922.5 + 14.7 \cdot 55 \cdot 56/2 = 74{,}298$ millions of tons. That is 74.298 billion tons of carbon dioxide. This problem illustrates how linear growth models and their sums can be used to make an estimate of the effects of automobiles on atmospheric carbon dioxide. Such an estimate would then be used in further calculations to predict how that would effect the global weather patterns. As always, the assumptions must be understood before we can interpret the results. Assuming that automobile use in the US will continue to grow for the next 50 years at about the rate it grew for the past 50 years, and assuming that the amount of carbon dioxide added to the atmosphere each year by a typical auto remains unchanged for the next 50 years, we can predict that 74.3 billion tons of carbon dioxide will be added to the atmosphere by cars in the US.

6. Let d_n be the distance fallen in the nth quarter-second. The diver falls one foot in the first quarter-second, so $d_1 = 1$. According to Galileo's pattern, the diver will fall 3 feet in the second quarter-second, 5 feet in the next quarter-second, 7 in the next, and so on. So we have $d_2 = 3$, $d_3 = 5$, $d_4 = 7$, and so on. The total distance traveled after any number of quarter-seconds can be found by adding these together. In the first second, for example, the total distance is $d_1 + d_2 + d_3 + d_4 = 1 + 3 + 5 + 7 = 16$. So it takes one second to fall 16 feet. The diver falls 7 feet in the last quarter-second. At that speed, 7 feet per quarter-second, the diver would cover 28 feet per second. That is the speed in feet per second for the final quarter-second of the dive.

Now consider the 64 foot platform. As before, we could just start adding up more of the d's. In the first two seconds, for example, the total distance traveled is $1 + 3 + 5 + 7 + 9 + 11 + 13 + 15 = 64$. So it takes 2 seconds to fall 64 feet. This time the diver covers 15 feet in the last quarter-second, so the speed is 60 feet per second for the last quarter-second. So, even though the second jump is four times as high as the first, the dive only takes twice as long and the diver is only going twice as fast at the time of entry into the water.

This problem can be worked out using quadratic growth. Let s_n be the total distance traveled in the first n quarter-seconds. This is found by adding up the individual distances for each quarter-second. So $s_1 = 1$, $s_2 = 1 + 3 = 4$, $s_3 = 1 + 3 + 5 = 9$, and so on. Also, s_0 is clearly 0, because no distance is traveled within 0 quarter-seconds. Combining these ideas, the first several s values are 0, 1, 4, 9, 16, 25, 36. These follow a quadratic growth model, because the differences are 1, 3, 5, 7, 9, 11, and the second differences all equal 2. This shows that $d = 1$ and $e = 2$, leading to the functional equation $s_n = 0 + 1n + 2(n)(n-1)/2 = n + n(n-1) = n + n^2 - n = n^2$. Now to find out how long it takes to fall 64 feet, we need the total distance or s to be 64, and we leave n as an unknown. That gives the equation $64 = n^2$. The solution to this equation is $n = 8$, and, since we are counting in quarter-seconds, that means it takes 2 seconds to fall 64 feet. This is the same thing figured out earlier. However, the equation makes it easy to look at related problems. For example, if the platform is 100 feet high, we can now easily solve $100 = n^2$ to find out that it takes $n = 10$ quarter-seconds or 2 and a half seconds to reach the water. After 9 quarter-seconds, the diver has traveled $s_9 = 9^2 = 81$ feet. So in the last quarter-second, the diver falls 19 feet. That makes the speed 76 feet per second for the last quarter-second. So increasing the height of the jump from 64 feet to 100 feet only increases the speed of impact from 60 to 76 feet per second. The height nearly doubled but the speed only went up by about a third. This is another example of the way proportional reasoning fails for quadratic models.

6

Quadratic Graphs, Functions, and Equations

The arithmetic growth models of Chapter 3 lead to the linear functions and equations of Chapter 4. In the same way, the quadratic growth models of Chapter 5 lead to quadratic functions and equations, the subject of this chapter. Now we will take a systematic look at important properties of quadratic functions, including algebraic form and methods, graphical features, solving equations, and applications. As an introduction to this subject, we review two examples from previous chapters and introduce one new example.

Oil Reserves Model.   The first example is the oil reserves model. Recall that this model uses the estimated world oil supply in 1991, and a linear model for annual world oil consumption, to determine how much oil will remain each year after 1991. A functional equation was derived for the model:

$$r_n = 999,100 - 24,254.25n - 54.75n^2$$

This equation defines the oil reserves after n years as a function of n, meaning that the reserves can be immediately calculated as soon as a particular year is chosen. It is not a linear function because n^2 appears as part of the formula. This is an example of a quadratic function.

   As you have already seen several times, one of the ways a functional equation is used is to find out when something will occur. In this case, we might ask how many years it will take to use up half of the oil, or $999,100/2 = 499,550$ million barrels. When that point is reached, the oil reserves will also be $499,550$, so we would like to find the n for which $r_n = 499,550$. This leads us to the equation

$$999,100 - 24,254.25n - 54.75n^2 = 499,550$$

Finding the correct answer for n amounts to solving this quadratic equation. It is partly for problems like this that the information in this chapter is important.

Computer Network Model. The second example is the computer network problem presented at the beginning of Chapter 5. In that example, t_n is the number of telephone lines needed to create a network of n computers, where each computer has a direct line to every other. The functional equation for this model is

$$t_n = (n-1)n/2$$

which is again a quadratic function. One of the things we observed before is that proportional reasoning is not an accurate approach for this model. For example, we saw that doubling the number of computers on the network requires far more than twice as many phone lines. We also saw that a surprisingly small number of computers can be connected by what appears to be a large number of telephone lines. The information about quadratic functions in this chapter will add to your insight about these aspects of the computer network model.

Sales Revenue Model. For the third example, we will extend a linear model presented in Chapter 4.[1] This example concerns the relationship between the price p charged for a can of soda and the number of sales s that can be made at that price. The linear model has the equation

$$s - 200 = -4(p - 80)$$

which we can solve for s

$$s = -4(p - 80) + 200$$

Now we will extend this model by including a new variable, R, the revenue or income that we get by selling the sodas. For a concrete example, suppose we charge 50 cents per can of soda. The equation for s predicts that we can sell $-4(50 - 80) + 200 = 320$ cans. For each can we get 50 cents, so the revenue is $R = 320 \cdot 50 = 16,000$ cents, or 160 dollars. In a similar way, if we charge 90 cents per can, we will sell $s = -4(90 - 80) + 200 = 160$ cans. At 90 cents per can that is a revenue of $R = 160 \cdot 90 = 14,400$ cents, or 144 dollars. These calculations show that the revenue is a function of the price charged per can of soda. As soon as we pick the price, we can directly compute the revenue. In fact, we can express these calculations in a functional equation. The correct equation emerges as part of a pattern shown below:

Price	Number Sold	Revenue
50	$-4(50 - 80) + 200 = 320$	$320 \cdot 50$
70	$-4(70 - 80) + 200 = 240$	$240 \cdot 70$
90	$-4(90 - 80) + 200 = 160$	$160 \cdot 90$
p	$-4(p - 80) + 200$	$[-4(p - 80) + 200] \cdot p$

In the first three lines of the table, three sample calculations are shown. The pattern of these lines shows how the price enters into the determination of revenue. As part of the pattern notice that the entry in the revenue column is always found by multiplying

---

[1] See page 66.

together the entries from the other two columns. In the final line, the price is left as a variable, and the table is completed using the same steps that were used for the preceding lines. This leads to the equation

$$R = [-4(p - 80) + 200] \cdot p.$$

which expresses the revenue R (in cents) as a function of the price per can p. This is another example of a quadratic function. If we would rather have the revenue expressed in dollars, the equation can be modified to

$$R = [-4(p - 80) + 200] \cdot p/100.$$

In this application, a natural question to ask is *What price leads to the highest possible income?* If you look at a graph of the revenue function, the highest possible income corresponds to the highest point on the graph. As we will see, there are properties of quadratic functions that allow us to find high and low points, and so to answer questions like the one above.

Let us look again at the three functional equations presented in the examples above.

$$r_n = 999{,}100 - 66.45n - .15n^2$$

$$t_n = (n - 1)n/2$$

$$R = [-4(p - 80) + 200] \cdot p$$

These are all quadratic equations, but they do not look much alike. Just what is a quadratic function, and how do you recognize one? That is the topic for the next section.

Quadratic Functions and Equations

In addition to the examples above, we encountered several quadratic functions in the preceding chapter. A typical example is $1 - 2n + 3(n - 1)n/2$. This can be rewritten, using algebra, in the the form: $1 - 3.5n + 1.5n^2$. Similarly, the expression for R above can be rewritten in this form: $(520 - p)p$; or in this form: $520p - 4p^2$. An algebraic expression for a quadratic function can be put into so many different forms, it is difficult to come up with a description that will apply to them all. Instead, we will describe one standard form. If you can put an example into that form, it is a quadratic function. But you may have to do some work to see whether it can be put into the required form.

In order to describe the required form, we need to look at the various parts that go into an example like $1 - 3.5n + 1.5n^2$. First, there are the pieces that are separated by the $+$ and $-$ signs. These are 1, $3.5n$, and $1.5n^2$; they are called *terms*.[2] Each term is either a number, a number times a variable, or a number times a variable with an exponent. But remember that the exponent just means repeated multiplication. We can think of $1.5n^2$

---

[2] The word *term* was used earlier to mean one of the numbers in a sequence. If you have data a_1, a_2, a_3, and so on, each of the a's is called a term. Now we have a second meaning: parts of an expression separated by $+$ and $-$ signs. When the word *term* comes up in the writing, you will have to be alert to which is the correct meaning.

as $1.5 \cdot n \cdot n$. So the terms are made up by multiplying numbers and variables. For a quadratic function we permit at most two variables in a term. They can be different variables, as in $3xy$, or two appearances of the same variable, as in $1.5 \cdot n \cdot n$, although this will almost always be written using an exponent: $1.5n^2$. With this terminology, we can formulate the following definition:

> **A Quadratic Function** is any function that can be written in a form made up of one or more terms, each having at most two variables. **A Quadratic Equation** is any equation that can be written with a quadratic function on one side of the equal sign and a constant on the other.

The following are all examples of quadratic functions:

$$3x^2 + 4x - 5 \qquad (3x - 2)(4x - 5) \qquad (3x - 2)(4y - 5)$$
$$2xy + 3yz - 6xz \qquad (3x + 4)^2 \qquad \frac{x}{2} + \frac{3x^2}{4}$$

Each example can be rewritten using algebra in a form that reveals it to be a quadratic. For instance, $(3x-2)(4x-5)$ can be rewritten as $12x^2 - 23x + 10$. Similarly, since $3x^2/4$ is the same as $(3/4)x^2 = .75x^2$, the last example above can be expressed as $.5x + .75x^2$.

As these examples show, it sometimes requires some algebra just to recognize whether or not a function is a quadratic. Very often, quadratic functions appear in an application in some form other than the one in the definition. The functional equations from the previous chapter were generally in a form like $1 - 2n + 3(n - 1)n/2$ which doesn't immediately match the definition of a quadratic function. Also, as we shall see later, different forms of a quadratic function are especially useful for different purposes. One form is most convenient if you are trying to solve a quadratic equation, while another is more helpful if you are trying to graph the equation. For all of these reasons, algebra plays a fundamental role in much of what we will cover in this chapter. With that in mind, the next section will focus on some of the algebraic properties of quadratic functions.

Algebraic Considerations

In many earlier discussions, the idea of changing one algebraic form into another one has been used. Let us take a little closer look at this idea.

It is possible to write down two expressions that are different in appearance, but which always lead to the same numerical results. A very simple example is provided by the expressions $3x + 4x$ and $7x$. No matter what number is used for x, these two expressions produce the same result. For this reason both of these two expressions define a single function. Consider the equation $y = 3x + 4x$. This defines y as a function of x. As soon as a value is defined for x, y can be immediately computed from the equation. In the same way, $y = 7x$ defines y as a function of x. Both equations define the *same* function. When x is assigned a value, the two equations produce the same result for y. The steps that are taken to reach that y differ for the two equations, but the result is the same. In the case of the first computer network example of Chapter 5, $(n - 1)n/2$ and $.5n^2 - .5n$ are equally valid ways to express the number of telephone lines as a function of the number of computers.

How can you tell whether two expressions define the same function? One approach is to substitute some numerical values for the variables. If the two expressions produce different results, they can't define the same function. Unfortunately, if the results agree, this test is not conclusive. Consider $x^3 + 6x + 2$ and $7x + 2$. If you replace x by 1, each expression gives a result of 9. If you replace x by -1, the result is -5 in both cases. And if x is replaced by 0, both of the results equal 2. All of these might lead you to suspect that the two expressions define the same function. But these three cases of agreement do not prove equality. As mentioned, it only takes one case of disagreement to show that the two expressions define different functions. We can find just such a case in this example by taking $x = 2$. That leads to $2^3 + 6 \cdot 2 + 2 = 22$ from the first expression and $7 \cdot 2 + 2 = 16$ from the second. These are different results, so the expressions define different functions.

There is another method for determining whether two expressions define the same function: use rules of algebra to transform one into the other. For example, one rule of algebra states that a pattern of the form $A(B+C)$ can always be replaced by $AB + AC$. This is based on the distributive law for arithmetic: $4(3+2) = 4 \cdot 3 + 4 \cdot 2$. If this pattern is applied to $p(500 - 4p)$, the result is $p \cdot 500 - p \cdot 4p$ or, more simply, $500p - 4p^2$. The same pattern can be used in the reverse direction to combine like terms. For example $3x + 4x$ can be changed into $(3 + 4)x$ and so into $7x$. Combining both of these ideas, the expression $(2x - 5)(3x + 1)$ can be rewritten as follows:

$$(2x - 5)(3x + 1) = 2x(3x + 1) - 5(3x + 1)$$
$$= 2x \cdot 3x + 2x \cdot 1 - 5 \cdot 3x - 5 \cdot 1$$
$$= 6x^2 + 2x - 15x - 5$$
$$= 6x^2 - 13x - 5$$

Checking for Errors. Although in general no number of examples can be relied on to prove that two functions are equal, quadratic functions of a single variable are a special case. For these functions, it is only necessary to check three values of the variable. If two such functions have the same value for three different values of the variable, then they are actually the same function. This can be very useful in checking for algebra errors. Consider the example above, which starts with $(2x - 5)(3x + 1)$. After doing the algebra, a result of $6x^2 - 13x - 5$ is reached. If no errors have been made, $(2x - 5)(3x + 1)$ and $6x^2 - 13x - 5$ should produce the same result no matter what value is used for x. To check this, it is enough to try any three values of x, say 0, 1, and 2. With $x = 0$, each expression results in -5. With $x = 1$, we find $(2 \cdot 1 - 5)(3 \cdot 1 + 1) = (-3)(4)$ in the first expression and $6 \cdot 1^2 - 13 \cdot 1 - 5 = -12$ in the second. Similarly, using $x = 2$ in each expression produces the same result, -7. Because these are quadratic functions, three examples provide conclusive proof that the two expressions are equal, so no errors were made in the algebra. In practice, you will not want to check three examples for every algebra problem. It is a good idea to check at least one value though. That will usually reveal an error. It is not a conclusive test, as checking three examples is. It is possible for an error to go undetected when you test just one example, but you have to be pretty unlucky for that to happen.

Descending Order. In the example above, after several steps of algebra we arrived at the expression $6x^2 - 13x - 5$. This illustrates what is called writing a quadratic function in descending order. The idea is to write the function as a combination of terms, just as in the definition of quadratic functions, but to insist that the terms appear in a particular order: first the term with an exponent of 2, then the term with the variable and no exponent (if there is one), and then the term that is just a constant (if there is one.) The opposite order defines what is called ascending order. It is always possible to express a quadratic function in either ascending or descending order.

Most of the rest of this chapter will focus on quadratic functions involving a single variable. The descending order for such a polynomial can always be written in the form

$$ax^2 + bx + c$$

where a, b, and c are parameters. In any particular application the parameters will be replaced by specific numbers. These parameters are also called *coefficients*. They tell quite a bit about the graph of a quadratic function, in much the same way that the slope and y–intercept tell about the graph of a linear function.

Sometimes we abbreviate the form above a little. For example, if $a = 1$, $b = 0$, and $c = 2$ we would normally shorten $1 \cdot x^2 + 0 \cdot x + 2$ to $x^2 + 2$. In a similar spirit, a negative coefficient is indicated as a subtraction, rather than an addition, of the corresponding term. Illustrating this with $a = 2$, $b = -3$, and $c = -5$, the written form would be shortened from $2x^2 + -3x + -5$ to $2x^2 - 3x - 5$. Keep these points in mind when you identify the parameters for a particular example. In $x^2 - 2x$ the parameters are $a = 1$, $b = -2$, and $c = 0$. What are the parameters for $3x^2 - x + 5$? _____ For $3 - 4x^2$? _____.

Graphs

On a graph with two axes, you are already familiar with the idea of graphing ordered pairs such as $(2, 5)$. Find the first number (2) on the horizontal axis and then go up the amount specified by the second number (5). The graph of a function is thought of as being made up of many many points plotted in just this way. In each ordered pair, the first number is a value that can be substituted for the variable in the function, and the value that results is the second number in the ordered pair. As an example, consider the function $x^2 + 3x + 5$. If you make the variable 3, then you obtain a result of $3^2 + 3 \cdot 3 + 5 = 23$. That corresponds to one point on the graph, $(3, 23)$. If a point on the graph is labeled (x, y), then the y must be what you get by using x in the function. That means $y = x^2 + 3x + 5$. For this reason, the *equation* $y = x^2 + 3x + 5$ always goes with the graph of the *function* $x^2 + 3x + 5$.

If you graph a variety of quadratic functions, you will see some patterns emerge. The graphs are always in the shape of a ∪ or a ∩. The ∪ or ∩ is symmetric, there is a vertical line that divides it exactly in half, so that each side is the mirror image of the other. The dividing line intersects a ∪ at its lowest point, and it intersects a ∩ at the highest point. These qualitative properties are displayed in several graphs in Fig. 6.1.

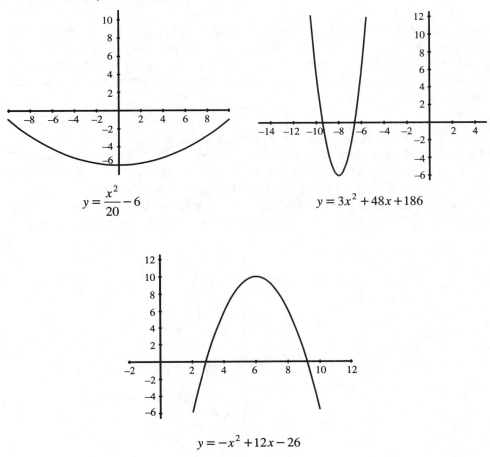

$$y = \frac{x^2}{20} - 6$$

$$y = 3x^2 + 48x + 186$$

$$y = -x^2 + 12x - 26$$

FIGURE 6.1
Sample Quadratic Function Graphs

The graph of a quadratic function makes a shape that is referred to as a *parabola*. Many other kinds of functions have graphs that are roughly in the shape of a ∪ or ∩, but only graphs of quadratic functions are called parabolas.

That characteristic parabolic shape provides a graphical explanation of the nonlinear behavior of quadratic functions. As you are aware, graphs of linear functions are straight lines, and using proportional reasoning amounts to assuming that some graph follows a straight line. In contrast, no part of a parabolic graph follows a straight line for very long. Even on the steepest parts of the sample graphs, if you draw a straight line along side the curve, the parabola will soon curve away from the line. This indicates that a quadratic function increases much more rapidly than any straight-line function. These ideas can be related to the computer network examples. There, after determining how many phone lines were needed for a small number of computers, we found that proportional reasoning badly underestimated the number of phone lines needed for a larger number of computers. That is because proportional reasoning assumes that the graph follows a straight line, while the true graph is a parabola, and curves up more steeply than a line. More generally, any quadratic function with a positive x^2 term increases much more rapidly than a linear function. Remember that we found with networks of 100 or so computers that the number of needed phone lines seemed surprisingly high. That is typical behavior for a quadratic function.

∪ Or ∩; y–Intercept.   Very precise information about the graph can be obtained using the coefficients. For example, consider the equation $y = 3x^2 - 12x + 6$ whose graph is shown in Fig. 6.2. The coefficients are $a = 3$, $b = -12$, and $c = 6$. What do these tell us? First, the coefficient a is positive. That means that the graph is a ∪, not a ∩. Second, the last coefficient, $c = 6$, gives the y–intercept of the curve.[3] Just looking at these two coefficients gives us the crossing point on the y axis and the general shape.

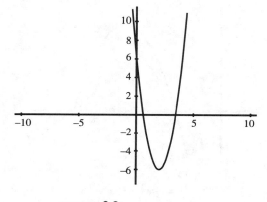

FIGURE 6.2
Graph of $y = 3x^2 - 12x + 6$

The Center Line and Vertex.   Next, look at the dividing line that cuts through the exact middle of the curve. Its location can also be determined from the coefficients. To find the exact location of the center line, simply draw a vertical line crossing the x axis at $-b/2a$. For the example under discussion, $-b/2a = -12/6 = 2$. The center line for this example crosses the x axis at $x = 2$.

One important aspect of the location of the center line is that it determines the highest or lowest point on the curve, depending on whether it is shaped like a ∪ or a ∩. In either case, that extreme point on the curve is called the *vertex* of the parabola. The x for the vertex is the same as the x for the dividing line: $-b/2a$. By substituting this for x in the equation of the curve, we can find the y, and hence the exact location of the vertex. In the example we have been working with, $-b/2a = 2$. Therefore

---

[3] Notice that the y–intercept appears in the same position in the $mx + b$ form for the equation of a line. In both cases, the constant coefficient is all that remains when the function variable is set to 0. And, when $x = 0$, you are at a point on the y axis.

$y = 3 \cdot 2^2 - 12 \cdot 2 + 6 = 12 - 24 + 6 = -6$. This shows that $(2, -6)$ is the lowest point on the curve.

The foregoing discussion shows how the parameters from the standard form $ax^2 + bx + c$ can be used to determine 4 characteristics of the graph. These are repeated below in summary form for emphasis.

On the graph of $y = ax^2 + bx + c$:

- The shape is $\cup$ for positive a; $\cap$ for negative a.

- The y–intercept is c.

- The vertical line that divides the graph into symmetric halves crosses the x axis at $-b/2a$.

- The high point for a $\cap$ or the low point for a $\cup$ is the point at which the vertical dividing line crosses the graph. This is called the *vertex*. The x at this point is $-b/2a$; the y can be found by substituting $-b/2a$ for x in the function.

Today it is easy to use a graphing calculator or a computer to obtain high quality graphs almost effortlessly. For this reason, the information about the graph given above may not seem very important. However, each of these characteristics is connected with the overall behavior of the function, and provides a bridge between its numerical and graphical properties. By studying both the graphs and the way aspects of the graphs can be found using the coefficients, you will develop a richer understanding of quadratic functions.

x–Intercepts. There is one more aspect of the graph of a quadratic function that can be determined from the coefficients — the x–intercepts. These are the points (if any) where the graph crosses the x axis. Whereas the y–intercept answers the question *What happens when $x = 0$?* the x intercepts answer the inverse question *Is there a point at which the function value is 0?* This distinction may make more sense in the context of an application. Later in this chapter we will see an example in which the function variable p is the price that is charged for a can of soda, and the function computes the profit P from selling as many cans as possible at that price. In this context, the horizontal axis is where the price is indicated, and the Profits are shown on the vertical axis. As we will see, the equation is

$$P = -231.5p^2 + 249.05p - 22.59$$

Now instead of a y–intercept we have a P intercept. That tells how much profit is made when the price is 0. The answer is -22.59, the constant coefficient c. (As expected, if you charge nothing for your product, you will have a negative profit.) On the other hand, in place of an x–intercept, this problem has a p–intercept. A p–intercept, if there is one, is a point at which the profit is 0.

Algebraically, the x intercepts for a quadratic function $ax^2 + bx + c$ are the solutions to the equation $ax^2 + bx + c = 0$. That is, they are the values of x which lead to a result of 0 when substituted in the function. These numbers are also referred to as the roots of the function. More generally, the solutions of any quadratic equation are called the roots of the equation.

The problem of finding the x–intercepts or roots of a quadratic function has a familiar form. The value that is supposed to be produced by the function (0) is given; it is the value entered into the function that we must find. We have seen this situation often; it is referred to as inverting the function. As usual, this requires us to solve an equation. Specifically in this chapter, we need to solve quadratic equations. That is the topic of the next section.

Solving Equations

As with functions, quadratic equations can be put into many forms. We will be interested in two particular forms. For the first, we will have a quadratic function in descending form on one side of the equation, and a number on the other. The second form has the additional restriction that the number is zero. So, the equation given earlier in the oil reserves example — $999{,}100 - 24{,}254.25n - 54.75n^2 = 499{,}550$ — can be put into the first standard form by writing

$$-54.75n^2 - 24{,}254.25n + 999{,}100 = 499{,}550$$

This equation will tell us when the oil reserves will reach 499,550. Similarly, to find out when the oil will all be gone, we must solve

$$-54.75n^2 - 24{,}254.25n + 999{,}100 = 0$$

This is an example of the second special form.

Graphical and Numerical Approach. One approach to solving any equation is to use graphical and numerical methods. We have seen this approach before for linear equations. Now, as an example with quadratic equations, consider again the equation $y = 3x^2 - 12x + 6$ whose graph is shown in Fig. 6.2. If the variable y is changed into a particular number, we obtain a quadratic equation that we can solve for x. For example, if we change the y to 2, the equation is $2 = 3x^2 - 12x + 6$, or, in a more common form, $3x^2 - 12x + 6 = 2$. Algebraically, solving this equation means finding a number for x that makes $3x^2 - 12x + 6$ work out to be 2. There is a graphical interpretation, too. Remember that the points on the graph are ordered pairs (x, y). When we replace y with 2 in the equation, that corresponds to looking for a point $(x, 2)$ on the graph. In other words, we are looking for a point on the graph which lines up with 2 on the y axis. Visually, we can locate this point by drawing a horizontal line crossing the y axis at 2, and seeing where it intersects the curve. This has been done in Fig. 6.3. We see that there are two different points where the line crosses the curve. At the first point, the x appears to be between 0 and 1, and for the other point, x is between 3 and 4. So let us say that the approximate solutions to the equation $3x^2 - 12x + 6 = 2$ are $x = .5$ and $x = 3.5$. Now these are not exactly correct. If you put $x = .5$ into the equation, the result is $3(.5)^2 - 12 \cdot .5 + 6 = .75$, which is not terribly close to 2. But the point of the graphical method is to get a first estimate.

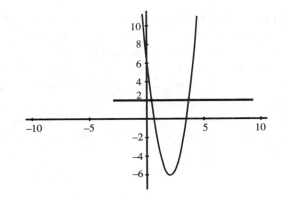

FIGURE 6.3
Points on Curve with $y = 2$

Next we use numerical methods to refine the approximation. Since .5 gives an answer that is too small (we got .75 and we wanted 2), the graph indicates we should pick a new guess that is further to the left. So, let's try .4 instead of .5. That gives $3(.4)^2 - 12 \cdot .4 + 6 = 1.68$. That is closer to 2. How does $x = .3$ work out? $3(.3)^2 - 12 \cdot .3 + 6 = 2.67$. That is too big, so we went too far. The correct answer must be between .3 and .4. A reasonable further guess would be .35. $3(.35)^2 - 12 \cdot .35 + 6 = 2.1675$, again too high. Continuing in this way you can slowly but surely zero in on better and better approximate solutions. That is the numerical method. Some graphing calculators can do this kind of calculation automatically, so that all you have to do is watch. But at the heart, what is going on is just what has been shown above, successively trying different values of x that get closer and closer to an answer.

There is a very important connection that you need to make here, between the *graph* of $y = 3x^2 - 12x + 6$ and an *equation* like $3x^2 - 12x + 6 = 2$. If we want to solve the equation for x, that is the same as finding a point on the graph where $y = 2$. If you set out to solve an equation with a single variable, say $x^2 + x - 5 = 8$, you will want to look at the graph of the related equation $y = x^2 + x - 5$. On this graph, try to find points where $y = 8$. That is, find the points where the graph crosses a horizontal line with a y–intercept of 8. The solutions of the original equation are the x's for these points, if any exist. This idea is applicable to any kind of function, not just quadratic functions, and is very useful.

There is also an important special case, when there is a 0 on one side of the equation. An example of that situation would be to solve $x^2 + x - 5 = 0$. As before, the idea is to look at the graph of $y = x^2 + x - 5$, but this time, we seek points where the graph crosses a horizontal line with $y = 0$. This is just the x axis. So to solve $x^2 + x - 5 = 0$ graphically, simply look for the x intercepts of $y = x^2 + x - 5$. Notice that we can always put an equation into the special form by moving all the terms to one side of the equation and leaving 0 on the other. The original equation $x^2 + x - 5 = 8$ can be rewritten in the form $x^2 + x - 13 = 0$. For quadratic equations, there is a theoretical approach to solving this kind of equation. We will look at that next.

A Theoretical Approach.   We will begin this topic by considering a very special kind of quadratic equation, exemplified by $x^2 = 7$. With a calculator, you can try various guesses for x. If you substitute 2.6 for x, the result is $x^2 = 6.76$, which is too low. Trying 2.7 leads to 7.29 which is too high. So the solution must be between 2.6 and 2.7. This is same kind of numerical approach we used earlier. Continuing with this approach we can get better and better approximations to an answer. For example, 2.6457 is too low and 2.6458 is too high. Clearly, there can be no exact answer in the form of a finite decimal between 2.6 and 2.7. Just imagine multiplying such a decimal by itself by hand. For example, consider the multiplication

$$2.64575$$
$$\times 2.64575$$

The very first thing you multiply is 5×5, and the result, 25, gives you the last decimal digit of the answer, namely 5. It is impossible to multiply two decimal expressions in this fashion and end up with all 0's following the decimal point. So there is no exact decimal solution for $x^2 = 7$. A similar argument shows that there is no exact fractional solution. If you replace x with a fraction in lowest terms, say $66/25$, then x^2 will still be a fraction. It can't work out to be exactly 7. Intuitively speaking, in $\frac{66 \cdot 66}{25 \cdot 25}$ there is nothing in the numerator to cancel the 25's in the denominator, so the result cannot be a whole number. Because $x^2 = 7$ has no exact solution as a fraction or finite decimal, the solution is said to be irrational. There are actually two solutions. One is approximately 2.6457; the other is the negative of the first. There is a mathematical notation for these solutions. The positive solution is written $\sqrt{7}$ (pronounced *the square root of* 7), and the negative solution is then $-\sqrt{7}$. In general, the $\sqrt{}$ of a number means the positive solution of the quadratic equation with the number on one side and x^2 on the other. So $\sqrt{13}$ means the positive solution of $x^2 = 13$. Put another way, $\sqrt{13}$ is the positive number which, when multiplied by itself, gives 13.

There are many algebraic properties of $\sqrt{}$. You have probably studied them in a previous course, and they will not be systematically described here. We will use the properties as needed. If you need to review the properties of $\sqrt{}$, consult any college algebra text. Your instructor can help you locate one.

There is one aspect of square roots that we will review, namely, that there are no real square roots of negative numbers[4]. Consider, for example, $\sqrt{-3}$. This is supposed to be a solution of the equation $x^2 = -3$. But there is no such x. Whatever number you use for x, multiplying it by itself to form x^2 produces a positive result, or zero. The result is never negative, and in particular, cannot equal -3. There is a graphical interpretation to this result. Rewrite the equation in the form $x^2 + 3 = 0$. The solutions to this equation would be x–intercepts on the graph of $y = x^2 + 3$, as discussed earlier. This graph, which is shown in Fig. 6.4, never crosses the x axis; there are no x–intercepts. This is another way of seeing that there are no solutions to the equation $x^2 + 3 = 0$.

---

[4]   In some courses, the real numbers are augmented by the addition of a new element, i, defined as the square root of -1. This results in a larger number system, the complex numbers, which contains the real numbers as a subset. We will restrict our attention to the real number system in this course.

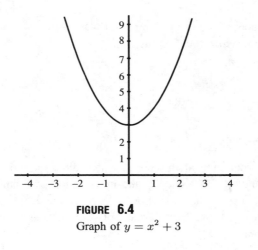

FIGURE 6.4

Graph of $y = x^2 + 3$

As discussed above, $\sqrt{13}$ can be approximated to any degree of accuracy using trial-and-error calculation. Of course, almost every calculator these days includes a button for performing this process. When you push the $\sqrt{}$ button, the calculator will compute as accurate an approximation to the true value as it can. In some cases this is actually exactly correct. For example, applying the $\sqrt{}$ button to 4 produces the correct result of 2. In other cases, the calculator will give an answer that is correct to as many decimal places as it can display.

Any quadratic equation can be transformed into a form that can be solved using the square root operation. This will be illustrated with the following example: $3x^2 - 5x - 8 = 0$. We make up a new variable, z defined by the equation $z = x - \frac{5}{6}$. The equation can also be used to express x as $x = z + \frac{5}{6}$. In the original equation, substitute $z + \frac{5}{6}$ in place of each x, and the result is

$$3\left(z + \frac{5}{6}\right)^2 - 5\left(z + \frac{5}{6}\right) - 8 = 0$$

This equation can be simplified using algebra, and written in the standard descending form. Without going into all the steps, the final result is

$$3z^2 - \frac{121}{12} = 0$$

The important feature of this equation is that there is no z term. So, a few more steps of algebra lead to

$$z^2 = \frac{121}{36}$$

which we can solve exactly using the $\sqrt{}$ operation. The final answer is

$$z = \pm\sqrt{\frac{121}{36}}$$

Now we are almost done. The original problem was to find x, not z. Since we know that $z = x - \frac{5}{6}$, the preceding equation turns into

$$x - \frac{5}{6} = \pm\sqrt{\frac{121}{36}}$$

which finally yields

$$x = \frac{5}{6} \pm \sqrt{\frac{121}{36}}$$

You are probably wondering at this point, where did $x = z + \frac{5}{6}$ come from? The $\frac{5}{6}$ is our familiar friend $\frac{-b}{2a}$. Graphically, the introduction of z amounts to shifting the parabola sideways by $b/2a$ units, just enough to bring the center line onto the y axis. The equation that goes with that kind of parabola has no z term, and so can be solved using the $\sqrt{}$ operation.

The Quadratic Formula. The method illustrated above can be applied to any quadratic equation. However, there are too many algebraic steps to make the method very attractive. Fortunately, what was done above with specific numerical coefficients can be carried out just as well with parameters a, b, and c. That is, start with $ax^2 + bx + c = 0$ and carry out the same steps that we followed above. After also using some of the properties of the square root operation mentioned earlier, the result is

$$x = \frac{-b \pm \sqrt{b^2 - 4ac}}{2a}$$

This is called the *quadratic formula*. It can be used to immediately determine the solutions to any quadratic equation. Or, if $b^2 - 4ac$ comes out to be a negative number, the formula does not give any solutions, since there are no square roots of negative numbers. In this case there are no solutions to the quadratic equation.

To use the quadratic formula, a quadratic equation must be expressed in the standard decreasing order. For example, given the equation $3x - x^2 = 4 + x$, we must first rearrange the terms to obtain $x^2 - 2x + 4 = 0$. In this form we can identify $a = 1$, $b = -2$, and $c = 4$. The quadratic formula then gives the solutions as

$$x = \frac{-b \pm \sqrt{b^2 - 4ac}}{2a}$$
$$= \frac{2 \pm \sqrt{4 - 4 \cdot 1 \cdot 4}}{2 \cdot 1}$$
$$= \frac{2 \pm \sqrt{4 - 16}}{2}$$
$$= \frac{2 \pm \sqrt{-12}}{2}$$

Of course, there is no square root of -12. This shows that the original equation has no solutions.

The quadratic formula completely clears up the problem of solving quadratic equations. It can be used with numerical values to find a solution to a specific quadratic equation, and the square root operation on a calculator can then be used to obtain a highly accurate numerical approximation to the solution. But the formula is also useful for theoretical purposes. For one thing it gives a definite indication when an equation has no solutions. It can also be used to invert a quadratic function. Let us apply this in the example involving soda sales. There the equation $r = 500p - 4p^2$ gives the income from sales (r) as a function of the price (p) per soda. Inverting this equation would allow us to determine the p given r. That is, we would be able to answer questions like *What price should be charged if we want to make* 200 *dollars of income?* To perform the inversion, write the quadratic equation in the standard form.

$$4p^2 - 500p + r = 0$$

In this equation, $a = 4$, $b = -500$, and c is the variable r. The quadratic formula then shows that there are two solutions for p:

$$p = \frac{500 + \sqrt{250,000 - 16r}}{8} \quad \text{and} \quad p = \frac{500 - \sqrt{250,000 - 16r}}{8}$$

These formulas give us two ways to compute a price, given the amount of income. Suppose we wish to obtain an income of \$120. Because the prices in the model are expressed in cents, we must express this income as $r = 12,000$ cents. The results are 93 and 32, rounded off to the nearest cent. That is, if we set the price of a soda to either 32 cents or 93 cents, the model predicts that we will have an income of \$120.

We saw earlier how the parameters a, b, and c can be used to obtain information about the graph of the function $ax^2 + bx + c$. Now we have seen that the quadratic formula uses these same parameters to compute the roots of the function. Thinking of the roots as the x–intercepts, the quadratic formula is just one more way to use the a, b, and c to find out something interesting about the graph. The list presented before can therefore be expanded to include the x–intercepts.

On the graph of $y = ax^2 + bx + c$:

- The shape is $\cup$ for positive a; $\cap$ for negative a.

- The y–intercept is c.

- The vertical line that divides the graph into symmetric halves crosses the x axis at $-b/2a$.

- The high point for a $\cap$ or the low point for a $\cup$ is the point at which the vertical dividing line crosses the graph. This is called the *vertex*. The x at this point is $-b/2a$; the y can be found by substituting $-b/2a$ for x in the function.

- If $b^2 - 4ac$ is negative, the function has no x–intercepts. If $b^2 - 4ac$ is positive or 0, the quadratic formula gives the x–intercepts as $\dfrac{-b \pm \sqrt{b^2 - 4ac}}{2a}$.

Factored Form. There is one more aspect of solving quadratic equations that should be mentioned. As already discussed, a quadratic function can be written in many different algebraic forms. One form in particular is handy for solving equations. Consider this equation

$$(x - 2)(x - 3) = 0$$

The solutions to the equation are 2 and 3. Check this by substituting 2 for x in the equation. Do you see why 2 and 3 are the solutions? For another example, consider

$$(3x - 5)(2x + 1) = 0$$

Can you guess the solutions this time? As in the previous example, one solution for x makes $3x - 5$ turn into 0, and the other makes $2x + 1 = 0$. Therefore, the solutions are 5/3 and −1/2.

    The examples above involve what is called the factored form of a quadratic polynomial. The parts enclosed in parentheses are called *factors*, indicating that they are items to be multiplied by other items. You may recall learning how to break a composite number down into its factors, for example $12 = 2 \cdot 2 \cdot 3$. The numbers on the right side of this equation are called factors of 12 because they are all multiplied together to give 12. In the same way, $(x - 2)$ and $(x - 3)$ are called factors of the function $(x - 2)(x - 3)$ because they are multiplied together to give that function.

    Using factors to find solutions to an equation only works when the opposite side of the equation is 0. If the equation is

$$(x - 4)(x - 7) = 1$$

it is no good to make x equal to 4. That produces 0 on the left side of the equation, not 1 as required. It also is no good making $x - 4$ equal to 1. Even if $(x - 4)$ does equal 1, there is still the $(x - 7)$ to worry about. The point is that 0 is very special. When $(x - 4)$ is equal to 0, we can forget about $(x - 7)$, because 0 times anything is still 0. It is this special property of 0 that makes the method of factors work, but only for equations with 0 on one side. In fact, the idea works for any number of factors. The equation

$$(x - 1)(x - 4)(x - 2.5)(x - 9)(x - 17) = 0$$

can be solved just by looking at it. The solutions are 1, 4, 2.5, 9, and 17, because each of these numbers makes one factor equal to 0. However, we are especially interested in problems with just two factors, because those are quadratic equations.

    Look again at the first example,

$$(x - 2)(x - 3) = 0$$

We can rewrite the expression on the left using algebra, just as we did on page 121. That will produce $x^2 - 5x + 6 = 0$. This is immediately recognizable as a quadratic equation. In general, if there are two factors of the form $(ax - b)$, then the result is a quadratic.

    It is easy to go from a factored form to the usual descending form. The reverse problem is not so easy. It usually can only be performed by trial and error. For example, to put $x^2 - x - 20$ into a factored form, you might try something like $(x - 2)(x - 10)$. Using algebra, you can find $(x - 2)(x - 10) = x^2 - 12x + 20$. This is similar to the expression we want, but we need a −20 not a 20, and we need a $-x$ not a $-12x$. By trying different combinations, you can eventually find that the factored form for $x^2 - x - 20$ is

$(x - 5)(x + 4)$. But the process is not direct and obvious. The point of these comments is to compare two algebra problems: (1) going from a factored form to the descending order form; and (2) going from descending order to a factored form. Generally, the first is easy, the second is not.

There are situations in which the factored form is preferred. Because the factored form is so closely connected with solving equations, you can use one to help with the other. We already saw how easy it is to find solutions to an equation if it is given in factored form. But it is also possible to find the factored form by finding the solutions. Here is an example. Suppose you want to express $2x^2 - 4x - 5$ in factored form. You can be certain that the answer will be of the form, $2(x-?)(x-?)$, and that the numbers that must go in place of the parentheses are the solutions to the equation $2x^2 - 4x - 5 = 0$. Solve that with the quadratic formula. The solutions are $(4 \pm \sqrt{16 + 40})/4$. On a calculator, we can approximate these solutions as 2.8708 and $-.8708$. Now put these numbers in place of the questionmarks to obtain

$$2x^2 - 4x - 5 \approx 2(x - 2.8708)(x + .8708)$$

(where the $\approx$ symbol emphasizes that the two sides are *approximately* equal).

Throughout this chapter we have seen that quadratic functions possess a number of well understood properties. In the final topic for the chapter we present an extended example illustrating how the properties of quadratic polynomials can be used.

An Example

This example uses data collected by some American University students in 1994. They surveyed their fellow students about soda prices. Each student was asked whether he or she would buy a can of soda between classes, given various prices for the soda. The results provided an idea of how many sodas could be sold at various prices.

The data from the survey are depicted in Fig. 6.5. Each data point consists of a price (p) in dollars for a can of soda and the number (s) of cans of soda that can be sold at that price. For example, one data point shows that, if the price is $.45, 133 sodas can be sold. This means that of all the students surveyed, 133 of them were willing to pay $.45 or more for a soda. The figure also includes a straight line that was drawn as near as possible to all the points. Although the line does not fit the data perfectly, it does give a good first approximation to the way the soda sales will decrease as the price is raised.

Using the methods of Chapter 4, the equation of the line in Fig. 6.5 can be found. It turns out to be

$$s = -231.5p + 225.9 \tag{1}$$

This equation expresses s as a function of p. That is, for any price p, we can immediately compute s, the number of cans of soda that can be sold at that price. The variable s is called the *demand* for sodas because it represents the number of sodas that consumers will want to buy. Eq. (1) gives a linear model for demand as a function of price. The model will be used in this section to study the way prices determine income and profit.

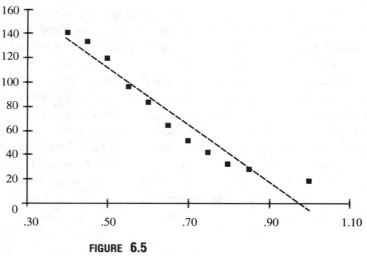

FIGURE 6.5
Soda Demand as a Function of Price

One problem with this model is that it predicts that demand will be a negative number for prices greater than 97 cents per can. This is wrong on two counts. First, the survey showed that there would be a positive demand at a price of a dollar. That is, you can still sell some cans of soda at that price. Second, in no case can the demand be negative. What meaning can there be to the statement that a price of one dollar causes you to sell a negative number of sodas? Perhaps someone who sees you selling soda at that price will come up and offer to sell you a few additional cans. In any case, in using this model, we should keep in mind that it only makes sense for a certain set of prices. This is a limitation of the model. We will see that the analysis will lead us to set the price far less than a dollar. For this reason, we need not be overly concerned with the applicability of the demand model for very high prices.

Note that when the survey was taken, the class was discussing the idea of selling soda on campus as a fund raiser. The demand model was interpreted as an indication of the number of sodas that could be sold at each price on any given day. We will see how this information leads to a model for the profits that can be obtained each day of the fund raiser.

The profits of a venture are determined by two things: income and expense. The income is generally referred to as *revenue*, while the expenses are referred to as *costs*. We will take up each of these in turn, beginning with revenue.

Revenue can be defined as in the example at the start of the chapter. Once a price is set, we can predict the number of cans that will be sold using Eq. (1): $s = -231.5 \cdot p + 225.9$. If the price is set at .60, for example, the number of sales will be $s = -231.5 \cdot .60 + 225.9 = 87$. Each can sells for $.60, so the total income is $R = .60 \cdot 87 = 52.20$. In a similar way, given any choice for p we can calculate the revenue as $R = ps = p(-231.5p + 225.9)$. This gives revenue as a function of price. Notice that it is a quadratic function. We can use the properties of quadratic functions to gain a good deal of understanding about the relation of revenue to price. For one thing, the equation can be graphed, as shown in

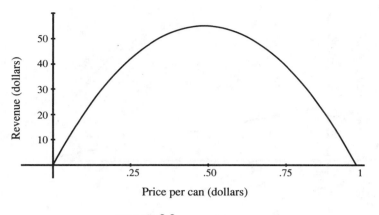

FIGURE 6.6
Revenue as a Function of Price

Fig. 6.6. The ∩ shape is just what should have been expected from the earlier discussion of quadratic polynomials. In this context, though, the high point on the curve has a special significance. It indicates the maximum revenue. That is, finding the vertex in this model tells us what price to charge per can of soda to obtain the highest possible income, and also predicts what the highest income will be. What is the vertex? To answer that question most easily, we will put the equation for revenue into the standard form of descending order:

$$R = -231.5p^2 + 225.9p \qquad (2)$$

The coefficients are $a = -231.5$, $b = 225.9$, and, since no constant term appears, $c = 0$. To find the center line of the parabola, we compute $-b/2a = .488$. This tells us where the center line crosses the horizontal axis. That is, it tells the price that corresponds to the highest point on the graph. At that price, the revenue is $R = -231.5 \cdot .488^2 + 225.9 \cdot .488 = 55.11$. This is the highest possible revenue. Of course, it is not possible to actually charge a price of 48.8 cents. Instead, we would set the price at 50 cents and expect a revenue then of 55.08. This is one of the limitations of using a continuous model for the price variable. In fact, the possible prices are restricted to a discrete set of values, namely whole numbers of cents. As we proceed with the analysis of this problem, we will continue to pretend that the price variable is continuous. It will not be difficult to modify the conclusions we reach so that the true discrete nature of the price variable can be observed.

At first glance it seems reasonable to set the price so that we obtain the maximum revenue. But profits depend on costs as well as revenue. How do the costs C vary as a function of price p? To keep matters as simple as possible, let us say that the only expense involved is the cost of buying the sodas. If we buy the soda at a big volume retail store, such as the Price Club, perhaps we can get it for $.10 per can. We will ignore any costs for driving to the store, ice to keep the soda cool, and so on. The total costs will be modeled as the price for the sodas at $.10 per can. Of course, the number of cans we buy will depend on how many we expect to sell, and that in turn depends on the price we will

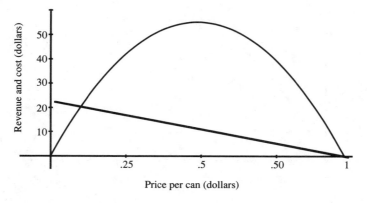

Price per can (dollars)

FIGURE 6.7

Cost and Revenue as Functions of Price

charge. Reasoning as before, if the price we charge is set at $.60 per can, then we need 87 cans. This will be a cost of $87 \cdot .10 = 8.70$. Similarly, for any price p we will sell $-231.5p + 225.9$ cans, and the costs for these cans will be $C = .10(-231.5p + 225.9)$. This is a linear model. Put in standard form

$$C = -23.15p + 22.59 \qquad (3)$$

we can read off the slope -23.15 and the intercept 22.59, and so we can easily graph this equation. In Fig. 6.7 both this cost function and the revenue function are shown. There are some interesting features on this graph. For any price, we can find both the costs and the revenue. Anywhere that the cost line is *above* the revenue curve, there will not be enough income to cover expenses. In this case the venture will suffer a loss. On the other hand, where the expense line appears below the revenue curve, the costs will be less than the income. This will produce a profit. The points where the line and the curve cross are called break-even points. They tell us the prices that produce just enough income to meet expenses. Of course, what we are more interested in is the profit, which is the amount of income left after subtracting the expenses. Visually, the profit will be a maximum at the price for which the cost line is farthest below the revenue curve. This is not easy to determine on the graph. Applying a numerical method, you can simply compute both the cost and revenue for a variety of prices and see where the greatest profit occurs. Even better, let us use a theoretical approach. We know that the profit is found by subtracting the costs from the revenue. That gives us the equation $P = R - C$ where P stands for profit. We will use the upper case (capital) letter here because the lower case letter p is already being used for the price we charge per soda. Now, applying Eq. (2) and Eq. (3), we can express R and C as functions of p. This leads to $P = (-231.5p^2 + 225.9p) - (-23.15p + 22.59)$, which expresses the profit P as a function of the price p. The graph is shown in Fig. 6.8. Observe that profit is a quadratic function of price, and the graph again has the characteristic $\cap$ shape. As before, we put

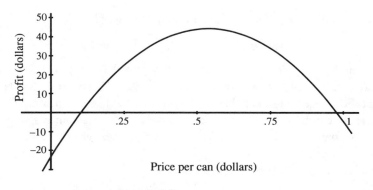

FIGURE **6.8**
Profit as a Function of Price

the equation into the standard descending order:

$$P = -231.5p^2 + 249.05p - 22.59 \qquad (4)$$

Now we can use this equation to find both the maximum profit and the break-even points. The center line of the parabola occurs for the price $p = -b/2a = 249.05/(2 \cdot 231.5) = .538$. If we charge this price for the sodas, then the highest possible profit will be obtained. Notice that this is different from the price that produces the maximum revenue. It may not be clear why the greatest profit does not go with the greatest revenue. Here is one way to think about it. At the point of greatest profit, we are bringing in less than the maximum revenue, but more of each dollar that is brought in goes to profit. A numerical approach should convince you that the analysis is correct. At the point of greatest revenue, the price is .488. At the center of the profit curve, the price is .538. Using Eq. (4) we can compute the profit for each of these prices:

$$P(.488) = 43.82$$
$$P(.538) = 44.39$$

If you calculate the profit for a few more prices, you will see that .538 really does give the highest possible profit. There is not much difference in this example between the profit at the point of maximum revenue and the maximum profit. In fact, for convenience in making change we would probably just decide to make the price 55 cents. However, this analysis gives us confidence that a price of 55 cents per can will result in just about the best profit possible. Moreover, with a profit of about \$44 each day of the fund raiser, it appears that selling sodas is a feasible project.

The break-even points are prices that result in a profit of 0. We can find these by solving the equation

$$-231.5p^2 + 249.05p - 22.59 = 0$$

using the quadratic formula. Alternatively, recalling that the profit is given by $R - C$, and that $R = ps$ while $C = .10s$, we can substitute to see that

$$P = ps - .10s = (p - .10)s$$

This is a factored form, and it shows immediately that the profit will be 0 if we set $p = .1$. That is reasonable. If we sell the sodas for just what we pay for them, there will be no profit. We can also see that, if $s = 0$, there will be no profit. That is also reasonable. There can be no profit if we sell no sodas. These comments show that the two break-even points are for $p = .10$ and at the price for which the demand model predicts zero demand.

This example illustrates several important points:

A Quadratic Model.   Here, starting from a linear model for demand, we are led naturally to a quadratic model for revenue. Combined with a linear model for costs, we are led to a quadratic model for profits as well. This illustrates also how one model is built on top of another.

Properties of Parabolas.   The example illustrates how a knowledge of graphs of quadratic functions can be useful. From the vertex of the parabola we determine the maximum revenue and profit. From the quadratic formula we can find the break-even points. Even the general shape of the profit and revenue curves is immediately understood in terms of the general results on graphs of parabolas.

Using Algebra.   At each step, our understanding of soda costs and income leads us to combine separate equations: revenue is found by multiplying the price per can by the number of cans we can sell; profit is found by subtracting costs from revenue. The algebraic forms that these combinations produced were not immediately useful. Instead, we had to use rules of algebra to transform the equations into familiar standard patterns. This shows one of the reasons that algebra is so useful. It allows us to change the form of an equation or a function. In some cases one form is easier to work with; in other cases another form is preferred. And often the original form taken by an equation is not the one we want to use. Algebra is the tool that allows us to translate from one form to another.

The Theoretical Approach.   Most of the analysis in this example uses the theoretical approach. The ultimate result is an equation that gives the profit as a function of price. This is a powerful result. It allows us to easily study the way profit increases or decreases. We can also generalize the results to determine the effect of changing basic assumptions. Suppose for example that we have to pay 20 cents per can of soda, rather than 10 cents, as assumed. What will the effect be on the profits? Would this still be a feasible fund raising activity? In more sophisticated models, the profit might depend on a large number of variables. But the fundamental idea of making predictions of the profit based on assumed values of the variables is identical to what we did in the example.

Limitations.   This example also gives us a chance to remember the limitations of the model. For one thing, we have been acting as if the price per can is a continuous variable. This is not true. It is not possible to choose a price to be a fraction of a cent, and for convenience we would probably want to set a price that is a multiple of 5 cents. So even after we carefully determine that the best price to charge is .538, we end up rounding that price off to .55. Another limitation is the dependence on the linear demand model. We know that the demand line only approximates the survey data collected, and even the survey data is only an approximation. When concluding that the maximum profit will occur at a price of .538, we must remember that this is only approximately true because it is based on other models that are only approximately true. One important way to judge the significance of these errors is to recompute the model based on modified demand or cost equations, and see how much the optimal price and maximum profit change.

Summary

This chapter is all about quadratic functions. It began with several examples from earlier work. There followed a definition of quadratic functions, as well as discussions of algebraic and graphical properties of these functions. In particular, every quadratic function can be put into a general form $ax^2 + bx + c$. The parameters or coefficients a, b, and c provide a lot of information about the graph of the function.

A closely related topic involves solving quadratic equations. Graphical and numerical solution methods were reviewed. For the graphical method, it was observed that solving an equation can be related to finding where a graph is crossed by a horizontal line. A special case occurs when that horizontal line is the x axis. Then the solutions to the equation give the x intercepts of the graph.

There is a theoretical approach to solving quadratic equations expressed in the standard form $ax^2 + bx + c = 0$. It culminates in a formula for the solutions, called the quadratic formula. This can also be viewed as a way of finding the x intercepts for the graph of a quadratic function.

A quadratic function can sometimes be expressed in a factored form, which can be useful at times. It is easy to change the factored form into the standard descending order, but it is not so easy to start with the descending order and figure out a factored form. If you have the factored form, you can find the roots by dealing with the factors individually. Conversely, if the roots are known, they can be used to find the factored form.

Finally, an extended example was presented applying many of the results about quadratic functions. The example showed how linear demand and cost models lead to quadratic models for revenue and profit. These were analyzed graphically, numerically, and theoretically. Among the conclusions of the analysis were predictions about what price would produce the greatest profit, and what prices result in just breaking even with neither a profit nor a loss.

Exercises

Reading Comprehension

1. Write short definitions or explanations of the following terms and phrases used in the reading: quadratic function, quadratic equation, terms, descending order, coefficient,

x and y intercepts, roots, factors, factored form, square root, quadratic formula, vertex.

2. Explain what it means for two different algebraic expressions to define the same function. How can you tell whether two expressions do or do not define the same function?

3. Explain what is meant by the statement: $x^2 = -3$ has no solutions. Give both theoretical and graphical reasons in support of the statement.

4. Describe all that is known about the graphs of quadratic functions. Tell about the general shape and explain what information about the graph is revealed by the coefficients of the quadratic.

5. What is revenue? What is a break-even point?

Mathematical Skills

1. Put each polynomial into descending order:
   a. $2x - 3x^2 - 5x + 7$ b. $(7x - 2)x + 3$
   c. $(2x - 5)(x + 3)$ d. $(x - 2)(x + 2)$

2. Use graphical and numerical methods to solve the following polynomial equations approximately:
   a. $3x^2 - 10 = 0$ b. $x^2 - x - 1 = 0$
   c. $3x^2 + 1 = 3x + 2$ d. $x^2 + 2x + 5 = 20$

3. Use theoretical methods (the quadratic formula or the factored form) to solve these equations exactly:
   a. $(x - 2)(x + 3)(3x - 6) = 0$ b. $3x^2 = 15$
   c. $3x^2 + x - 2 = 0$ d. $(x - 2)(x + 3) = 3$

4. For each equation, find the following graphical features by using the coefficients: x and y intercepts; orientation ($\cup$ or $\cap$); center line; and vertex. Verify the results by using a graphing calculator or computer program.
   a. $y = 5x^2 - 20x + 15$ b. $y = -2x^2 + 3x - 7$

5. The equation $y = x^2 - 4x$ defines y as a function of x. Use the quadratic formula to invert this function, expressing x in terms of y.

Problems in Context

1. In the oil reserves model, the equation $r_n = 999{,}100 - 24{,}254.25n - 54.75n^2$ expresses the reserves after n years as a function of n. Here, $n = 0$ in 1991, and at that point 999,100 million barrels of oil were available. According to the model, how many years will it take to use up half of that supply? How many years to use up all of the oil?

2. According to one model we looked at for a computer network, the number of telephone lines needed to connect up n computers is given by $1 + 2n + 3(n - 1)n/2$. Suppose there are a total of 100,000 telephone lines available to the developers of

a network. How many computers can they include in the network, according to the model?

3. One of the exercises in the last chapter concerned a logging company. Using an arithmetic growth model for the number of acres logged each year, it was determined that the total acres logged over n years is given by $1{,}200n + 100(n-1)(n)$. Using this equation, determine how many years it will take to log 50,000 acres.

4. A small amusement park brings in about 60,000 dollars in income each year. When the park was new, it cost about 20,000 dollars to run and maintain, but as the park gets older, the maintenance costs increase. This is described by an arithmetic growth model: $c_n = 20 + 3n$ is the cost (in thousands of dollars) to operate and maintain the park for year n, where the year the park opened is considered to be year 0. Thus, the initial cost was $c_0 = 20$ but after 5 years the cost was $c_5 = 20 + 3 \cdot 5 = 35$.

Each year the park's owners put the profits in a savings account. Let b_n be the balance after n years, in thousands of dollars. Each year, the profit is found by subtracting the costs (given by c_n) from the income of 60 (thousand). So, assuming $b_0 = 0$, after one year the bank account will have $b_1 = 0 + 60 - c_0 = 60 - 20 = 40$. Similarly, $b_2 = 40 + 60 - c_1 = 100 - 23 = 77$. As these examples suggest, the difference equation for b is $b_{n+1} = b_n + 60 - c_n$ or $b_{n+1} = b_n + 40 - 3n$.

   a. Find the functional equation for b_n. (Hint: use the method of the previous chapter.)

   b. Make a graph for b_n showing what happens over 20 years. (Hint: use a computer or graphing calculator.)

   c. According to the functional equation, when is the bank account the largest? At that point, how much is in the bank?

   d. Based on this model, formulate a long-term plan for the owners of the park.

5. In Chapter 4, one of the homework exercises concerned a linear model for monthly operating costs of a company making backpacks. We will refer to this as the Chapter 4 Backpack Company.[5] Let b stand for the number of backpacks made in a month. The cost of each backpack is \$23, and the company has \$12,000 of monthly expenses that remain constant. So the monthly operating costs are given by $23b + 12{,}000$. A market survey is conducted to see how many backpacks will be purchased by retail stores each month for various prices. The results are shown in Table 6.1. For this

Price	30	35	40	45	50
Number Sold	10,000	7,200	6,2004	5,300	3,500

TABLE 6.1
Wholesale Demand for Backpacks

---

[5] The stockholders are very glad that the company did not first appear in Chapter 11!

problem you will develop a model for the profit of the company and use it to decide what price should be charged for each backpack.

a.   Use a computer module or paper and pencil to graph the data in Table 6.1. If you do it by hand, use graph paper and be very accurate. Draw the best line you can for the data, or use the computer module to choose the best line for the data. This should give you a linear equation for the number of backpacks that can be sold as a function of the price that is charged. If you draw the line by hand, find the equation of the line using the point–slope form.

b.   For several different prices, compute the number of backpacks that can be sold per month, based on the equation from part a, not on the original data. For each price, use the predicted number of backpacks sold to determine how much money the company will bring in from selling the backpacks. (This is called the revenue.) Also compute the company's monthly expenses, assuming that it makes just the number of backpacks that it can sell. How much profit does the company make for each price you examine?

c.   Express the profit as a function of the price. Let p stand for the price per backpack. Use your linear equation from part (a) to give you an expression for the number of backpacks that will sell as a function of the price. Use that to write down an expression for the income from sales of backpacks (revenue) as a function of price. Use the number of backpacks sold to determine the expenses, and subtract that from the revenue to find the profit.

d.   Your equation should give profit as a quadratic function of the price. Use a graphing calculator or computer module to examine the graph of this profit function. What price should be charged for each backpack in order to get the most possible profit?

e.   Put the profit function into the standard descending order form so that you can find the coefficients a, b, and c. Use these to determine the vertex of the parabola. This will give a theoretical derivation of the best price to charge, and of what the maximum profit will be. Note that it can also be used to determine how many backpacks to make per month.

f.   Suppose the Chapter 4 Backpack Company is competing with another maker of backpacks. In order to capture as large a share of the market as possible, Chapter 4 will sell as many backpacks as they can for a very low price. What is the lowest price they can charge without losing any money? That is, if they are willing to give up all profits for a while and just break even, so that expenses exactly match income, what price should be charged per backpack?

6. This problem concerns a model for managing a population of fish in a lake. The model uses a quadratic function to predict how many fish will be in the lake each year based on the number that were in the lake the preceding year. Let p be the number of fish in the lake (in thousands) at the start of this year. Then the number of fish in the lake at the start of next year is given in the model by

$$(2 - .01p)p - 16$$

In this exercise, you will explore some of the properties of this model and gain some insight about its significance. In a later chapter you will see how the model was formulated in the first place.[6]

a. Suppose there are 85,000 fish in the lake this year ($p = 85$). How many fish will be in the lake next year, according to the model? Is that an increase or decrease in the fish population?

b. Suppose there are 70,000 fish in the lake this year. How many will be in the lake next year, according to the model? Is that an increase or decrease in the fish population?

c. Starting at a population of 70,000, compute the population for the next several years. What do you notice?

d. Starting with a population of 85,000, compute the population for the next several years. What do you notice?

e. The preceding parts of the problem suggest that a population of 80,000 fish will remain unchanged year after year. Verify this by checking what the population will be next year if it is 80,000 this year. A population that remains constant over time is said to be *stable*.

f. The stable population of 80,000 was discovered numerically in this problem. However, we could look for it algebraically by setting next year's population equal to this year's. This gives the equation

$$(2 - .01p)p - 16 = p$$

Put this equation into the form $ap^2 + bp + c = 0$ and use the quadratic formula to solve for p. One solution should be $p = 80$, the stable population we already found. Is the other solution also a stable population?

g. In the model, the term 16 represents 16,000 fish that are harvested each year. That is, if we harvest no fish, the population will be expected to go from p one year to $(2 - .01p)p$ the next. Since we harvest 16,000 fish, we have to subtract 16 from our previous figure to find the next year's population. The preceding question shows that (according to the model) starting with a population of 80,000, we can harvest 16,000 fish year after year. That is a renewable resource. In the same way, starting with a population of 20,000 fish we can harvest 16,000 year after year. Which of these situations do you think would be better? Why?

h. Earlier, you found that when the population is a bit above or a bit below 80,000, after a few years the population approaches the stable size of 80,000. Does the same thing work with the stable population size of 20,000? That is, if you start with a population of 22,000, after a few years does it go down to 20,000 and stay there? What about if you start with a population of 18,000? 19,000? $19,900$?

---

[6] If p_0 is the population size at some starting year, and p_n stands for the population size n years later, then the model described above corresponds to the difference equation $p_{n+1} = (2 - .01p_n)p_n - 16$. In this problem, we will leave off the subscripts to make the notation simpler.

   i.   With a population of 80,000, harvesting 16,000 each year results in a stable population size. Suppose you now decide to increase the harvest. If you harvest 20,000 fish, instead of 16,000, the model predicts that a population of size p this year will become one of size $(2 - .01p)p - 20$ next year. In this case, what will the stable population be? [Hint: solve the equation $(2 - .01p)p - 20 = p$.]

   j.   If you harvest 22,000 fish each year what will the stable population be?

   k.   If you harvest 30,000 fish per year, there is no stable population. In this case, a population of p this year becomes $(2 - .01p)p - 30$ next year. Starting with 80,000 fish, what happens to the fish population after several years? What happens if you try to use algebra to find a population p that will remain stable? Is there a solution to $(2 - .01p)p - 30 = p$? Why or why not?

   l.   The questions leading up to this point show that in some cases you can harvest the same number of fish year after year. But if you try to harvest too many, the entire fish population can be killed off. That is the effect of over-harvesting. What is the maximum amount you can harvest each year and still have a stable population? If you try to harvest that many each year, and the population falls a little below the stable size, what will happen in later years?

   m.   Based on this model, if you were in charge of managing the lake and its resources, how many fish would you permit to be harvested each year?

Group Activities.   This project involves a model for the way a liquid leaks out of a container through a small hole. These models are useful in understanding how gasoline leaks out of buried tanks at filling stations, potentially contaminating underlying ground water[7]. As a simplified physical model, a plastic soda bottle is filled with water and a small hole is punched in the bottom. As the water leaks out, the height of the water in the bottle is recorded every 10 seconds. This leads to the data in Table 6.2. The variable n represents the number of each reading; H_n is the height of water in the bottle at reading n. The data for H_n are in units of centimeters.

n	0	1	2	3	4	5	6	7
H_n	10	8.7	7.5	6.4	5.4	4.4	3.5	3.0

TABLE 6.2
Water Level for Leaky Tank Model

1. Use a computer module (or do it on paper by hand) to graph the data points of the form (n, H_n). Fit a line to the H_n data. As in past exercises, find the equation for the line. For each n, let h_n be the value you obtain using n in the equation for the line, so that h_n is a model for the true data H_n. Make a table similar to Table 2.2. This gives an idea of how good the model is.

---

[7] This exercise is adapted from material in *Functioning in the Real World: A Precalculus Experience* by The Math Modeling/Precalculus Reform Project, Sheldon P. Gordon and B. A. Fusaro, Project Directors, 1993.

2. Compute the differences between successive values of H_n. Although these difference are pretty close together, they definitely seem to be decreasing in a regular way. This suggests looking at the second differences.

3. Compute the second differences. That is, list the differences from part (b) and then look at the differences between the differences. Note that these are nearly constant. There is some variability, but no definite trend as there was for the first differences. This suggests that a quadratic model might be appropriate.

4. Referring again to the graph of the points (n, H_n), try to find a quadratic function of the form $h_n = an^2 + bn + c$ that comes close to all the data points. That is, try to find a quadratic model h_n for the true data points H_n. [Note: the coefficient a should be half of the constant second difference. The c coefficient is supposed to be the y intercept. This should give you some idea of starting values for a, and c. Then try different choices of b.]

5. For each data point, use the n value to compute an estimated h_n using your best fitting formula $an^2 + bn + c$. How close are these estimates to the true data? Again make a table similar to Table 2.2. Is this fit better than the linear model from part (a.)? How much?

Solutions to Selected Exercises

Mathematical Skills

1. a. $-3x^2 - 3x + 7$ b. $7x^2 - 2x + 3$
   c. $2x^2 + x - 15$ d. $x^2 - 4$

3. a. $x = 2$, $x = -3$, and $x = 2$, are the solutions found by equating each factor to 0. Notice that $x = 2$ appears twice. The solutions are 2 and -3.

   b.   This equation can be simplified to $x^2 = 5$ which shows that solutions are $\sqrt{5}$ and $-\sqrt{5}$. The quadratic formula can also be used, after writing the equation in the form $3x^2 - 15 = 0$. Use $a = 3$, $b = 0$, and $c = -15$ to find $x = [0 \pm \sqrt{180}]/6$. Numerically, these are the same solutions already found.

   c.   Use the quadratic formula with $a = 3$, $b = 1$, and $c = -2$. The result is $x = [-1 \pm \sqrt{1 + 24}]/6 = (-1 \pm 5)/6$ which gives the two solutions -1 and $2/3$.

   d.   The factored form in this problem is not of much use because the other side of the equation is not 0. So put the left side into descending form, $x^2 + x - 6 = 3$, and then move the 3 to the left side, $x^2 + x - 9 = 0$. Now use the quadratic formula. Solutions are $(-1 \pm \sqrt{37})/2$.

4. a. The coefficient of x^2 is 5, which is positive. That means it is a $\cup$. The y–intercept is 15, the constant term. The center line is at $x = -20/10 = 2$. The vertex has $x = 2$ and $y = 5 \cdot 2^2 - 20 \cdot 2 + 15 = -5$. Finally, to find the x–intercepts we solve $5x^2 - 20x + 15 = 0$. As a first step, divide the entire equation by 5: $x^2 - 4x + 3 = 0$; then use the quadratic formula: $x = (4 \pm \sqrt{4})/2 = (4 \pm 2)/2$. This gives x intercepts at 1 and 3.

5. Given $x^2 - 4x - y = 0$ we can solve for x using the quadratic formula. $a = 1$, $b = -4$, and $c = -y$. Therefore, $x = (4 \pm \sqrt{16 + 4y})/2$. This only makes sense if y is at least as big as -4, in which case there are two different x's that can be used in the original function to obtain a result of y. Otherwise, if y is less than -4, we will get a negative number in the square root, which is not permitted. This also makes sense because the graph of $y = x^2 - 4x$ doesn't have any points below a y of -4. The algebraic properties of square roots permit the answer to be rewritten in the simpler form $x = 2 \pm \sqrt{4 + y}$.

Problems in Context

1. Half the original amount of oil is $499{,}550$. We want to solve for n in the equation $999{,}100 - 24{,}254.25n - 54.75n^2 = 499{,}550$. Put this into the standard form for a quadratic equation: $-54.75n^2 - 24{,}254.25n + 499{,}550 = 0$. The quadratic formula gives $n = (24{,}254.25 \pm \sqrt{24{,}254.25^2 - 4(-54.75)(499{,}550)})/(-108.9)$. This simplifies to $(24{,}254.25 \pm \sqrt{697{,}670{,}093})/(-108.9)$. This gives two answers, but only one of them is positive. It is 19.8273. So according to the model, half of the oil will be used up in about 20 years. Similarly, to find when all the oil will be gone (according to the model) we must solve $-54.75n^2 - 24{,}254.25n + 999{,}100 = 0$. The answer this time is 37.94.

2. We know how many lines there are, but n is unknown. The equation is $1 + 2n + 3(n-1)n/2 = 100{,}000$. First put this into standard form: $1.5n^2 + .5n - 99{,}999 = 0$. Now use the quadratic formula to find solutions 258.03 and -258.36. Only the positive solution is of interest. It shows that with 258 computers, 100,000 telephone lines will be just enough.

3. For this problem the equation is $1{,}200n + 100(n-1)(n) = 50{,}000$. In standard form that becomes $100n^2 + 1{,}100n - 50{,}000 = 0$. The solutions are 17.53 and -28.53. Again, we only care about the positive answer. The 100,000 acres will be used up in 17.53 years.

4. a. This is a quadratic growth model with $b_0 = 0$, so the functional equation is $b_n = 40n - 3(n-1)n/2$

   b. The graph is shown in Fig. 6.9.

   c. The functional equation is a quadratic. In the standard form it becomes $b_n = -1.5n^2 + 41.5n$. The center line for this quadratic occurs for $n = -41.5/(2 \cdot -1.5) = -41.5/-3 = 13.833$. If you use that figure for n we get $b(13.833) = 287.04$. Remember that is in thousands of dollars, so it is actually \$287,040. As the model is defined, it might make more sense to use only whole numbers for n, so let us compare b_{13} and b_{14}. $b_{13} = -1.5(13^2) + 41.5(13) = 286$. $b_{14} = -1.5(14^2) + 41.5(14) = 287$. This latter value is the highest b_n can get. After the 14th year, the bank balances decrease.

   d. The bank balance starts to decrease after 14 years because at that point more money is needed to operate the park than will be made in income. Based on the model, the owners should plan to sell the park or close it after 14 years. Of course,

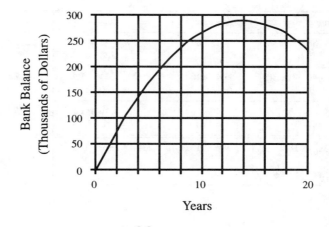

FIGURE 6.9
Bank Balance over 20 Years

this is a very simple model. In a real business setting many more factors would be included, such as effects of raising ticket prices over the years, interest on the profits, reinvestment in the park to increase income, and tax considerations.

5. a.   See Fig. 6.10. The equation of the best line is $s = -.298p + 18.36$ where s is the number of backpacks that can be sold (in thousands) and p is the price per backpack. Your line may be slightly different.

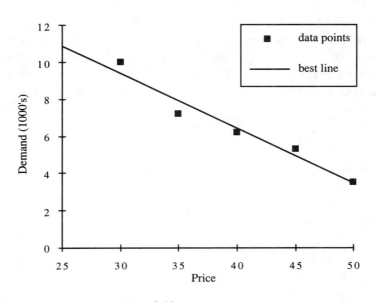

FIGURE 6.10
Demand Model for Backpacks

price	number sold	income	costs	profit
30	$-.298 \cdot 30 + 18.36$ $= 9.42$	$9.42 \cdot 30$ $= 282.6$	$23 \cdot 9.42 + 12$ $= 228.66$	$282.6 - 228.66$ $= 53.94$
40	$-.298 \cdot 40 + 18.36$ $= 6.44$	$6.44 \cdot 40$ $= 257.6$	$23 \cdot 6.44 + 12$ $= 160.12$	$257.6 - 160.12$ $= 97.48$
50	$-.298 \cdot 50 + 18.36$ $= 3.46$	$3.46 \cdot 50$ $= 173$	$23 \cdot 3.46 + 12$ $= 91.58$	$173 - 91.58$ $= 81.42$
p	$-2.98p + 18.36$	$(-2.98p + 18.36)p$	$23(-2.98p + 18.36)$ $+12$	$(-2.98p + 18.36)p$ $-23(-2.98p + 18.36)$ -12

TABLE 6.3
Profit Calculations

b.   Suppose the price is $p = 30$. The equation gives the number sold as $s = -.298 \cdot 30 + 18.36 = 9.42$ (in thousands). The income will be $9,420 \cdot 30 = 282,600$. Or, in units of thousands of dollars, it will be $9,420 \cdot 30/1,000 = 9.42 \cdot 30 = 282.6$. The expenses for the 9,420 backpacks will be (in thousands of dollars) $(23 \cdot 9,420 + 12,000)/1,000 = 23 \cdot 9.42 + 12 = 228.66$. The profit will be $282.6 - 228.66 = 53.94$. This same kind of computation has been made for several prices and entered in Table 6.3.

c.   As shown in the last line of the Table 6.3, the equation is $P = (-.298p + 18.36)p - 23(-2.98p + 18.36) - 12$.

e. $P = -.298p^2 + 25.214p - 434.28$. The center line is at $p = -25.214/(2 \cdot -.298) = 42.31$, and there the profit is 99.064 (in thousands).

f.   We must find the p that results in a profit of $P = 0$. That means solve the equation $-.298p^2 + 25.214p - 434.28 = 0$. Using the quadratic formula we find two solutions: 24.07 and 60.54. These are the break-even points. We will want to set the price to \$24.07 to capture as large a share of the market as possible, without losing money.

7
Polynomial and Rational Functions

In past chapters you have learned about linear functions and quadratic functions. One way to think about these functions emphasizes the operations they entail. Both linear and quadratic functions can be defined in terms of the usual operations of arithmetic: addition, subtraction, multiplication, and division. Linear functions can use any combination of addition and subtraction, but multiplication and division are subject to a restriction: only multiplication and division by constants are allowed. This restriction is relaxed somewhat for quadratic functions, so that multiplication of two variables is allowed, but not multiplication of three or more. These restrictions may be relaxed even further, giving rise to two more collections of functions. *Polynomial* functions can be defined by any combination of addition, subtraction, and multiplication, as well as division by constants. *Rational* functions can be defined by any combination of addition, subtraction, multiplication, and division, with the only restriction being that division by 0 may not appear. This chapter will provide a brief overview of some of the properties of these types of functions.

Although polynomials and rational functions can be defined in terms of operations, they can also be recognized by their appearance. A polynomial, like a quadratic function, can be thought of as a combination of terms. Each term is made up by multiplying together constants and/or variables, and the terms are then added or subtracted. The expressions 6, x^2, xyz^4, $2xy$, and $x^3/9$ are all terms; $6 - x^2 + xyz^4$ and $2xy + x^3/9 - x^2$ are polynomials. Polynomials can appear in many other forms, too. This is a polynomial: $(((3x + 5)x - 4)x + 2)x - 8$. The way to recognize this as a polynomial is to observe that it involves only addition, subtraction, and multiplication, with no division by variables. Any function of that type can be rewritten using only the addition and subtraction of terms, as described above.

Notice how similar this is to quadratic functions. They too can appear in many forms, but can always be recognized by rewriting them in a standard form. In fact, quadratic and linear functions fit the description of polynomials, so they are sometimes called quadratic and linear polynomials.

Rational functions can also be put into a standard form: a single fraction with a polynomial for the numerator (top) and denominator (bottom). For example $\dfrac{x^2 + xy + 2}{x + 3y^2 z}$. Rational functions can appear in many other forms as well. Basically any expression that can be calculated using only addition, subtraction, multiplication, and division is a rational function.

Both polynomials and rational functions are important in applications. Remember that a function is simply a recipe for computing something. Since we often compute things using the four operations of arithmetic, it is no surprise that polynomials and rational functions are so prevalent. In this book we are focusing on the kinds of functions that appear in simple difference equation models. In that context, the most common polynomials are the linear and quadratic ones, which you have already studied in detail, and rational functions are not often encountered. For this reason, we won't see much in the way of applications of polynomials and rational functions. They are presented primarily to round out your ideas of functions and graphs, and to make you aware of a terminology that often appears in mathematics and its applications. To simplify matters, we will restrict our attention to functions involving just one variable.

Algebraic Aspects of Polynomials

As already mentioned, any polynomial can be expressed as a combination of one or more terms. As for quadratic functions, it is a common practice to list the terms in descending order, meaning the term with the highest exponent comes first, then the next highest, and so on. Here, we are referring to the exponent on the variable; any exponents on constants are ignored for the purpose of ordering the terms. In each term, the exponent on the variable is referred to as the *degree* of the term. When there are two or more terms, the degree of the polynomial is the highest degree among its terms. As an example, $3x^5 + x^4 - 7x + 5$ is a fifth-degree polynomial written in descending order. The constant part of each term is called the coefficient of the term. For $3x^5$ the coefficient is 3; for the x^4 term the coefficient is 1 ($x^4 = 1x^4$); there is no x^3 term so that coefficient can be considered to equal 0; for the $7x$ term the coefficient is -7 because the term is subtracted.

Algebraic operations were discussed in the previous chapter for writing quadratic polynomials in different forms. These same operations are used with any kind of polynomials. As with quadratics, it is a good idea to check for algebra errors. If the expression you started with is truly equivalent to the one you found using algebra, they should both give the same answer when you replace the variable with a number. Do you remember the rule for quadratics? If two quadratic expressions give the same answer for three different x's, that is conclusive proof that the two expressions are equal for all x's. The number of different x's you need for this kind of test is one more than the degree of the polynomial. So, since a quadratic is of degree 2, we need to test 3 examples to be sure. In the same way, to test whether two fifth degree expressions are equivalent, it is enough to check 6 examples. Although it is not really a practical idea to check your algebra this thoroughly, the idea of the conclusive test is an interesting aspect of polynomials.

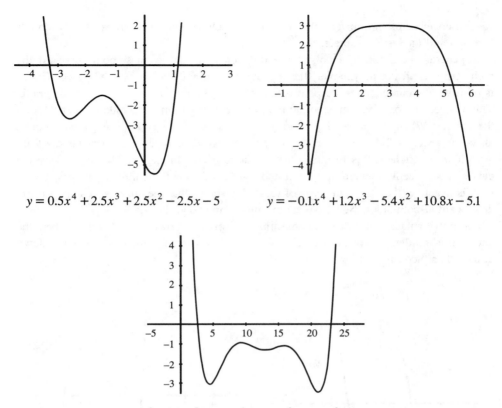

$$y = 0.5x^4 + 2.5x^3 + 2.5x^2 - 2.5x - 5$$

$$y = -0.1x^4 + 1.2x^3 - 5.4x^2 + 10.8x - 5.1$$

$$y = 22x^6 - 168x^5 + 508x^4 - 772x^3 + 613x^2 - 235x + 31$$

FIGURE 7.1
Sample Polynomials of Even Degree

Graphing Polynomials

As for linear and quadratic functions, a polynomial function is represented by a graph. For each point (x, y) on the graph, the y value is what is found by substituting the x value in the function. For example, if $(2, 13)$ is a point on the graph, that means that when 2 is substituted for the variable in the function, the result is 13. Usually, this idea is simply expressed as an equation in the form

$$y = \text{polynomial expression involving } x$$

for example, $y = x^3 + x^2 + x + 1$. As usual, we also say that this equation defines y as a function of x, because if a particular value for x is chosen, we can immediately compute the value of y. Take $x = 3$, for instance. That gives y as $3^3 + 3^2 + 3 + 1 = 40$, and produces one point on the graph: $(3, 40)$. By graphing many many points, we see that a curve of some sort appears. This is best done by a computer program. After studying many different examples of polynomial functions, a number of patterns are discovered. In this section we will describe a few of these patterns. There are many more patterns than we can cover here, and, in fact, the study of the behavior and properties of polynomials is a rich and interesting subject.

In general, the graph of a polynomial function is an unbroken smooth curve. If the highest power in the polynomial (the degree) is even and the corresponding coefficient is positive, the curve slopes steeply upward on both the left and right sides of the graph. This is because for large values of x, the highest-power term is so much larger than all the others. Whether x is a very large positive number or very large negative number, an even power such as x^2 or x^4 or x^{20} is positive and, of course, very large. With a positive coefficient, this results in a very large y coordinate. This explains why the curve slopes steeply upward at the left and right. If the coefficient of the highest power term is negative, then the same type of reasoning shows that the curve will slope steeply downward at both sides. Some typical examples are shown in Fig. 7.1.

For polynomials of odd degree, meaning the highest power is an odd number, the situation is similar, except in this case the curve slopes steeply up at one side and down at the other. See Fig. 7.2.

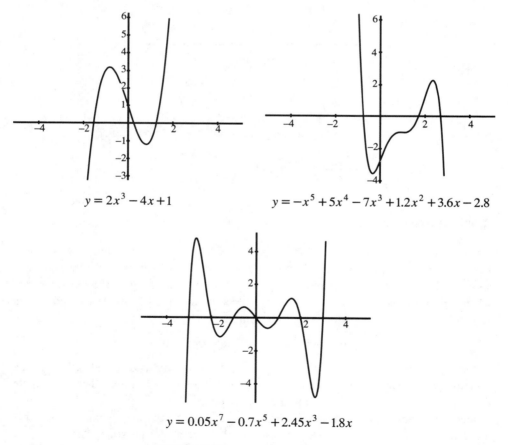

$$y = 2x^3 - 4x + 1$$

$$y = -x^5 + 5x^4 - 7x^3 + 1.2x^2 + 3.6x - 2.8$$

$$y = 0.05x^7 - 0.7x^5 + 2.45x^3 - 1.8x$$

FIGURE 7.2
Sample Polynomials of Odd Degree

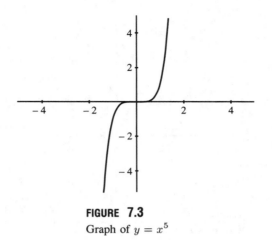

FIGURE 7.3
Graph of $y = x^5$

Polynomials tend to wiggle around a little bit somewhere in the center of the graph. The degree places a limit on the number of wiggles. For example, a polynomial of third degree (highest exponent is 3) can cross the x axis at most three times. In fact, it can cross any horizontal line at most three times. Similarly, for any polynomial, the degree indicates the maximum possible number of times the graph can cross any horizontal line. Not every polynomial wiggles to the full extent allowed by this law. The polynomial $y = x^5$ crosses each horizontal line just once (see Fig. 7.3).

These comments show how the degree (the highest exponent) and the corresponding coefficient give some qualitative information about the graph. The constant coefficient, the one that has no power of x, gives some very specific information: it is the y–intercept. This is exactly what was true of linear and quadratic functions, and it holds for the same reason. The y intercept is the point that comes from setting x to 0. If x is zero, every term that includes an x is also zero, leaving only the constant term. For example, if $y = 17x^3 - 2.5x^2 + .3x + 7$, then when $x = 0$, y is easily seen to equal 7. This gives the point $(0, 7)$ on the graph, which is the y intercept.

Solving Polynomial Equations

How would you find the places where the graph crosses the x axis? If $y = 23x^4 - 17x^3 - 14x^2 + 3x + 3$, graphing many points produces a graph as shown in Fig. 7.4. The approximate locations of the x axis crossings can be read off the graph. To be more accurate, a trial and error numerical approach can be followed. Using the example shown, substituting $x = 0$ and $x = 1$ we find y values of 3 and -2. When these points are graphed, the first is above the x axis and the second is below. Somewhere in between, the curve must cross the axis. By checking values of x that are closer and closer together, you can zero in on the value of x that produces a y of 0. An exact answer is rarely possible, because the x intercepts of a polynomial are almost always irrational numbers. But we can get a good approximation.

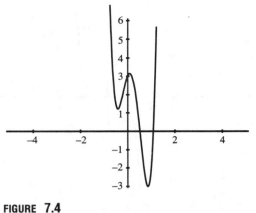

FIGURE 7.4
Graph of $y = 23x^4 - 17x^3 - 14x^2 + 3x + 3$

Notice that as for quadratic polynomials, the search for x intercepts is really a search for solutions to an equation: $23x^4 - 17x^3 - 14x^2 + 3x + 3 = 0$. We are looking for the x that results in y being 0. Solutions to an equation of this form are called *roots* of the polynomial that appears in the equation. Remember that for a fourth-degree polynomial there can be at most 4 solutions.

The comments above concern solutions to an equation where 0 appears on one side. Graphically, the solutions are the points where the graph of the function crosses the x axis. More generally, we may wish to see where the graph crosses any other horizontal line. This kind of situation arises in a model when we ask questions about inverting a polynomial function. If the polynomial above models the radiation level at various times inside a reactor, we might ask whether the radiation ever reaches the level of 5,000. That corresponds to solving the equation $23x^4 - 17x^3 - 14x^2 + 3x + 3 = 5,000$. The solutions are then given by the points where the graph crosses a horizontal line which hits the y axis at 5,000. As before, there can be at most 4 solutions, because a fourth-degree polynomial can cross any horizontal line at most 4 times.

The most general form of a polynomial equation consists of two polynomials with an equal sign between, such as this: $3x^5 - 4x^2 = 2x + 7$. The degree of this equation is the highest exponent on either side of the equation. It sets a limit on the number of possible solutions. Unless the two sides of the equation are actually the same function, the equation just shown can have up to 5 solutions. It might actually have fewer solutions, but can never have more than 5. This is connected with the idea of testing for algebra errors mentioned earlier. When we want to see whether two third-degree polynomials are truly the same, we test 4 points. If the polynomials were not the same, they could not be equal at more than 3 points. That is the rule for solutions to polynomial equations. So if they are equal at 4 points, they cannot be different polynomials. They have to be different guises of the same polynomial, and that is what we were trying to test.

Factored Form.   There is no theoretical procedure that can be counted on to determine the exact solutions to a polynomial equation. Often we must settle for numerical methods. An important exception to this case occurs when a polynomial equation is in factored form, as discussed for quadratic functions. An example is the equation $(2x-1)(x+3)(x-5) = 0$. Recall that this is called factored form because it is made up of several things multiplied together: $(2x - 1)$, $(x + 3)$, and $(x - 5)$, and that these are called *factors*. Just as for quadratics, in factored form we can obtain solutions by finding values of x that make each factor 0. In this form, it is easy to see at a glance that setting $x = 5$ gives one solution. Similarly, $x = -3$ and $x = 1/2$ are solutions of the equation.

In general, when a polynomial equation has zero on one side and a factored form on the other, each solution to the equation makes one of the factors become 0. It is rare for a polynomial equation to occur in this form, unfortunately. If the equation above is written in descending order, it becomes $2x^3 - 5x^2 - 28x + 15 = 0$. In this form it is not easy to see what the solutions are. Finding the exact solutions by a theoretical method is difficult. So, while the factored form is easy to rewrite in descending order, the reverse is not true. Given a polynomial in descending order, it is not a simple matter to write it in a factored form.

Other Clues. Although the theory of polynomials does not provide simple methods for finding exact roots in general, there are many theoretical results that can help provide *clues* about the roots. Here is one example. Suppose that all the coefficients of a polynomial are whole numbers, and that there is an exact root of the polynomial that is expressed as a fraction. Then the numerator (top) of the fraction must divide evenly into the constant term of the polynomial, and the denominator (bottom) of the fraction must divide evenly into the coefficient of the highest power. For the example $2x^3 - 5x^2 - 28x + 15 = 0$, the constant term is 15, and the coefficient of the highest power is 2. So if we want to look for exact fractional solutions, we should consider fractions such as $3/2$ or $5/1 = 5$. For each of these, the top of the fraction divides evenly into 15 and the bottom divides evenly into 2. With this rule, you can usually form a fairly short list of possible fractional solutions of a polynomial. If you are lucky, this will lead to one or more solutions. Usually, though, the polynomials that come up in real applications have solutions that are not fractions. Instead they are irrational numbers, and we must be content with approximate numerical solutions. In the special case of quadratic equations, exact solutions can be written in a form involving square roots, for example $3 + \sqrt{2}$ is an exact solution of $x^2 - 6x + 7 = 0$. However, for a numerical solution, we can only obtain approximate answers, because $\sqrt{2}$ can only be approximately expressed as a fraction or decimal.

Graphical and Numerical Methods. If exact solutions to a polynomial equation cannot be found, how can we find approximate numerical solutions? One convenient approach is to use graphical and numerical methods, just as for quadratic equations. If the polynomial equation is of the form $3x^5 - 6x^3 + 2 = 0$, the solutions are the x intercepts for the graph of $y = 3x^5 - 6x^3 + 2$. Looking at the graph gives a good first approximation to these intercepts. For more accuracy we can use a trial and error numerical approach. This is very easy to do using a graphing calculator or computer program. By looking at the graph, you can get an idea of where the x intercepts are. Then you can zoom in on these to get better and better accuracy. Many graphing calculators have built-in procedures to do this automatically.

Applications of Polynomials

We have already seen in Chapter 5 how quadratic polynomials can arise in difference equations. In particular, when a data set has second differences that are nearly constant, that is an indication that a quadratic model will be a pretty good approximation. This idea extends to higher degree polynomials as well. If the third differences[1] are nearly constant, a third degree polynomial will provide a good fit; if the fourth differences are nearly constant, then a fourth degree polynomial will be a good fit. The same pattern holds true for any number of differences, and it can often be used to find a polynomial that is a good approximation to a data set.

Polynomial models are also selected sometimes based on the shape of a graph. A $\cup$ or $\cap$ shape suggests a quadratic polynomial. A sideways S shape suggests a third degree polynomial. As you have seen, the graphs of polynomials have a recognizable shape. When such a shape shows up in a data set, that is a clue that a polynomial model might be of interest.

There is another way that polynomial models appear. Sometimes, the way we calculate something is just naturally expressed as a polynomial. That is what happened in the example in the last chapter involving profit and revenue. Once a linear function is adopted for the demand curve, we are led by simple arithmetic to a quadratic function for revenue.

Finally, polynomials can be used to obtain very good approximations to other operations. For example, it is possible to approximate square roots using polynomials. Methods of higher mathematics show that

$$\sqrt{1+x} \approx 1 + \frac{1}{2}x - \frac{1}{8}x^2 + \frac{1}{16}x^3$$

provides an excellent approximation for small values of x. Consider $x = .2$ The equation says

$$\sqrt{1.2} \approx 1 + \frac{1}{2}(.2) - \frac{1}{8}(.2)^2 + \frac{1}{16}(.2)^3$$

You can check this by performing the calculation on both sides of the equal sign using a calculator. You will see that the polynomial gives an approximation that is correct to 4 decimal places. Polynomials can be used in this way to find good approximations using only addition, subtraction, and multiplication. Of course, if you have a calculator, you don't need to make an approximation. But there were not always calculators. More importantly, being able to approximate a complicated operation using a polynomial is an important theoretical tool. It allows us to use some of the properties of polynomials and their graphs in studying the more complicated operation.

This concludes the discussion of general polynomials. We have seen some of the properties of graphs of polynomials, discussed solving polynomial equations, and described two ways that polynomials arise in models. Next we turn to the topic of rational functions.

---

[1] Just as the second differences of the data set are the differences of the differences, so the third differences are the differences of the differences of the differences, and similarly for fourth and higher differences.

Graphs of Rational Functions

Like polynomials, rational functions have graphs with certain characteristic shapes. A few representative examples are shown in Fig. 7.5. These graphs illustrate two features that never appear in the graphs of polynomials, *discontinuities* and *asymptotes*. A discontinuity is simply a break. You can see that none of the graphs in Fig. 7.5 is a single continuous curve. A point where there is a break from one part of the graph to the next is called a discontinuity. An asymptote is a straight line that is gradually approached by a graph. In the figure, asymptotes have been represented using dashed lines. If a graph is drawn on a large enough scale, parts of the curve will appear to lie directly on top of the dashed asymptote lines. In some sense, these lines show where the graph *straightens out*. Polynomial graphs do not have asymptotes. They do not ever straighten out. This difference between rational functions and polynomials can be understood in numerical terms. Very often, rational functions are very closely approximated by linear functions for extreme values of the variable. This is not true for polynomials of degree 2 or more.

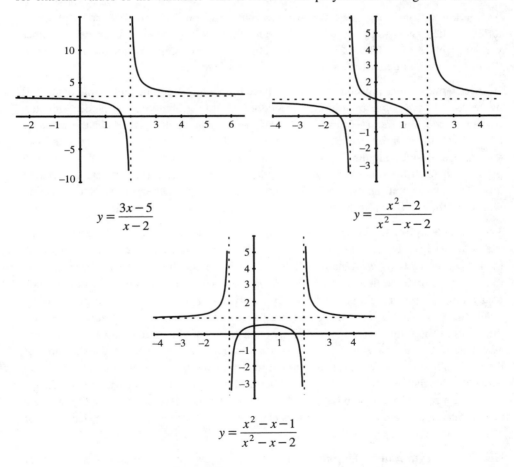

$$y = \frac{3x - 5}{x - 2}$$

$$y = \frac{x^2 - 2}{x^2 - x - 2}$$

$$y = \frac{x^2 - x - 1}{x^2 - x - 2}$$

FIGURE 7.5
Sample Graphs of Rational Functions

Discontinuities and asymptotes can be found by putting a rational function into a special form. As mentioned earlier, it is always possible to express a rational function as a fraction with one polynomial for the numerator and another for the denominator. If those polynomials are expressed in factored form, it is possible to identify and eliminate common factors. This is similar to reducing a fraction. For example, the expression

$$\frac{(2x - 1)(x + 5)}{(x + 6)(x + 5)}$$

can be simplified to

$$\frac{(2x - 1)}{(x + 6)}$$

by removing the $(x - 5)$ factor from both the numerator and denominator. To see why this makes sense, substitute 2 for x in the original fraction. The result is

$$\frac{3 \cdot 7}{8 \cdot 7} = \frac{3}{8}$$

As you know from your previous experience with fractions, the two 7's can be removed to reduce the answer to lowest terms. The same thing will happen no matter what number we put in place of x. That is why we can eliminate the common factor of $(x + 5)$ from the numerator and denominator of the original fraction.

x–Intercepts. Once any common factors have been eliminated, the remaining factors tell us the roots of the polynomial in the numerator and of the polynomial in the denominator. A root of the numerator corresponds to an x–intercept of the graph. In the example $(2x - 1)/(x + 6)$ the numerator is 0 for $x = 1/2$. If that is substituted for x in both the numerator and denominator, the result will be $0/6.5$. The result is 0. So, $x = 1/2$ produces a function value of 0, and that indicates an x–intercept. The same situation holds for any rational function in lowest terms. A root of the numerator leads to a fraction with 0 on the top (but not the bottom) and so makes the rational function 0.

Vertical Asymptotes. What is the significance of a root of the denominator? In the example of $(2x - 1)/(x + 6)$, -6 is a root of the denominator. If we try to put $x = -6$ in the entire fraction, we obtain $-13/0$. This is an invalid expression because it is impossible to divide a number by 0. It turns out this corresponds to a break in the graph of a rational function. This is not terribly surprising. How can you put a point on the graph for $x = -6$? In fact, the graph must have a break at $x = -6$ precisely because there is no point for that x. What is more, a break of this sort is always at a point where the curve is straightening out along a vertical line. In other words, the roots of the denominator identify the vertical asymptote lines for the rational function. You can verify this for the graphs in Fig. 7.5. Each vertical asymptote is for a particular x. That x should be a root of the denominator of the fraction. For the first graph in the figure, there is a vertical asymptote at $x = 2$, and it is clear that $x = 2$ produces 0 in the denominator of $(3x - 5)/(x - 2)$.

Horizontal Asymptotes. The preceding discussion shows that many important aspects of a rational function can be determined if the numerator and denominator appear in factored form. Indeed, that is one of the contexts in which the factored form of a polynomial is important. However, for locating horizontal asymptotes, it is more convenient to express the numerator and denominator in descending order. Consider the example

$$\frac{3x^2 - 5x + 1}{8x^2 + 4}$$

and imagine using a very large number for x, say on the order of 1,000. Then x^2 will be about 1,000,000. It is so much larger than x and the constants in this example that the numerator will be closely approximated by $3x^2$ and the denominator by $8x^2$; the other terms are too small to have much effect. So, if you use a large number for x, the final result will be very close to $(3x^2)/(8x^2) = 3/8$. This reveals a horizontal asymptote. If you graph points for many large x values, they will all produce y values of just about 3/8. Those points will all line up on a horizontal line, and that shows the location of the asymptote. In general, if there is a horizontal asymptote, it is easily found using descending form for both the numerator and denominator of a rational function.

It is interesting that both of the different forms for polynomials, factored and descending order, are useful for understanding rational functions. The factored form in both the numerator and denominator reveals common factors, x–intercepts, and vertical asymptotes. The descending order for both the numerator and denominator helps you find any horizontal asymptotes. However, in applications, rational functions may not show up in either form. In fact, rational functions might not even appear as a single fraction. Remember that a rational function is simply a recipe for combining variables and constants using addition, subtraction, multiplication, and division. There are many different ways these operations can be combined in an application. For example, suppose you are planning a trip. The first 300 miles will be traveled on the highway, traveling at H miles per hour. But then you have to cover 100 miles on back roads, traveling at B miles per hour. How long will the trip take? The first part takes $300/H$ hours, and the second takes $100/B$ hours, so the total is $\frac{300}{H} + \frac{100}{B}$. That is a rational function, but it doesn't look like a single fraction. This shows again how important algebra can be. Rational functions show up in applications because we have to compute something. The form of that computation can be any combination of addition, subtraction, multiplication, and division. In order to analyze these kinds of functions, we need to use algebra to put them into the form of a single fraction, and then to express the numerator and denominator using both factored form and descending order.

Choosing A Rational Function Model. Although rational functions often appear in applications as a result of some computation, sometimes they are used as a matter of choice. If a set of data has a shape that resembles a rational function, we might just decide to use a rational function to approximate the data. This is particularly true when the data resemble one of the parts in the first graph of Fig. 7.5. Look at the part of the graph in the upper right corner. It has a sort of L shape that appears quite often in data. That shape can not be approximated very well by a straight line, or by a polynomial function. For that kind of shape, you should think of a rational function. More specifically, consider a function of the form

$$\frac{ax + b}{x - c}$$

for appropriately chosen constants a, b, and c. Based on the remarks made earlier, the constants a and c relate directly to the graph. The value of a gives the horizontal asymptote, showing where the curve levels off. The value of c gives the vertical asymptote, where the curve straightens out going straight up or down. The value of b can be adjusted to make the curve appear more or less rounded.

To illustrate this idea, we will take another look at the soda demand data first shown in Fig. 6.5. Look back at that figure now, and observe that the data seem to follow a mildly curved L shape, especially if you disregard the first few points. Using a computer graphing tool, I experimented with various values of the a, b, and c mentioned above, until I obtained the graph in Fig. 7.6. The curve shown in the figure has the equation

$$s = \frac{90 - 75p}{p} \tag{1}$$

where p is expressed in units of dollars.

This provides an alternative demand model for the soda problem analyzed in the preceding chapter. Using Eq. (1) in place of the linear model we used before, we can find the revenue as

$$R = \left(\frac{90 - 75p}{p}\right) p$$

the costs as

$$C = 0.1 \left(\frac{90 - 75p}{p}\right)$$

and the profit as

$$P = \left(\frac{90 - 75p}{p}\right) p - 0.1 \left(\frac{90 - 75p}{p}\right)$$

The point of this example is to illustrate how a rational function can be chosen based on the shape of a graph. It also illustrates the idea that rational functions can appear in a problem in some other form than as a single fraction. If you are interested in comparing

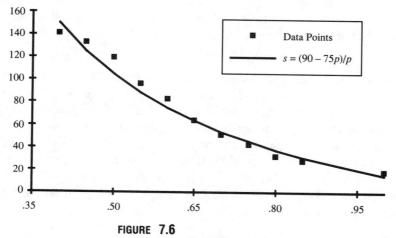

FIGURE 7.6
Rational Function Demand Model

the profits predicted by the rational function model with what we found earlier, a good starting point would be to look at the graphs of the two profit functions. Do both models suggest charging about the same price to obtain the greatest profit? And how big a profit does each model predict? These questions will be left to the interested reader.

Summary

This chapter has discussed some of the general properties of polynomial and rational functions. Polynomials are an extension of the idea of quadratic functions; in fact, a quadratic function is a polynomial. The idea of degree was introduced, and it was observed that quadratic functions are polynomials of degree 2. A polynomial function can also be described as any combination of variables using addition, subtraction, and multiplication, as well as division by constants. Rational functions are similarly defined, but without any restriction on how division is used, except for the inevitable prohibition of division by 0. A rational function can always be expressed as a single fraction with polynomials for numerator and denominator, although there are many other forms that can be used as well.

Graphs of polynomials share some common characteristics. A rough idea of the shape can be determined from the term of highest degree. In general, the graphs curve steeply up or down at the sides of the graph, and wiggle around in the middle. The number of wiggles is limited by the degree of the polynomial. Specifically, if the degree is n, a horizontal line can cross the graph of a polynomial at most n times. In particular there can be at most n x–intercepts. These are also called the roots of the polynomial.

There was a discussion of solving polynomial equations. There are no general theoretical methods for solving these equations, so graphical and numerical methods must often be used. However, there are interesting theoretical results that give partial information about the solutions. Solutions of a polynomial equation can be found by a graphical method. If the equation is of the form *polynomial = b,* where b is a number, the solutions are found as intersections of the graph of the polynomial and a horizontal line which crosses the y axis at b. In the special case that $b = 0$, the intersections are just the x intercepts. If the polynomial is in factored form, the roots can be found by finding the roots of the individual factors. Applications of polynomials were also discussed.

For rational functions there are also some common features that appear in graphs. Two features that were discussed are discontinuities and asymptotes. When a rational function is expressed as a single fraction with polynomials in the numerator and denominator, and when those polynomials appear in factored form, it is possible to eliminate any common factors. Then the roots of the numerator give the x–intercepts and the roots of the denominator give the vertical asymptotes of the graph of the rational function. If the numerator and denominator are expressed in descending order, any horizontal asymptotes can be found easily.

In applications, rational functions may appear in some other form than as a single fractional expression. Algebra can be used to express the function as a single fraction, and to express the numerator and denominator in factored form. These ideas illustrate the importance of algebra.

At times, data graphs follow shapes that are conveniently modeled using rational functions. Particular shapes of this kind were described, as well as the kind of rational functions that might prove applicable. An example using the soda demand data from Chapter 6 illustrated these ideas.

Exercises

Reading Comprehension

1. Write short definitions or explanations of the following terms and phrases used in the reading: polynomial, terms, descending order, coefficient, degree, x and y intercepts, roots, factors, factored form, asymptote, discontinuity.

2. Describe some general patterns of graphs of polynomials. Tell how the graphs of even and odd degree polynomials differ. What does the degree tell you about the number of wiggles?

3. Describe some general patterns of graphs of rational functions. Explain how to locate the horizontal and vertical asymptotes, if any, using the formula for a rational function.

4. In Fig. 7.7 there are several graphs shown. Decide for each graph whether it is a polynomial or a rational function.

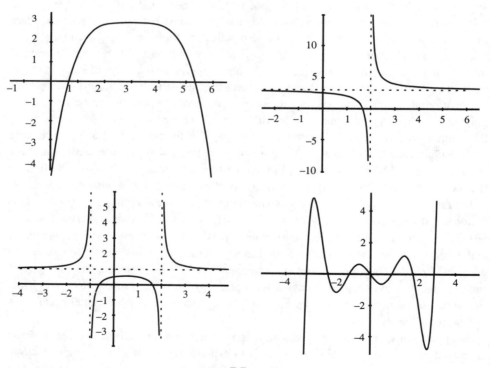

FIGURE 7.7
Graphs for Problem 4

Mathematical Skills

1. Put each polynomial into descending order:

   a. $3x^2 + x^3 - 5x + 2$ b. $x^2(3x + 4) - 3x$

   c. $(2x - 5)(x + 3)$ d. $(x - 2)(x^3 + 2x^2 + 4x + 8)$

2. Use graphical and numerical methods to solve the following polynomial equations approximately:

   a. $3x^5 - 10 = 0$ b. $x^3 - x - 1 = 0$

   c. $2x^3 + 1 = 3x + 2$ d. $x^4 + 2x^2 + 5 = 0$

3. For the polynomial $5x^3 - 6x^2 + 31x - 6$ make a list of all possible fractional roots. Use a graphical method to see whether any of these appear to actually be roots. Then check numerically if they are exact roots.

4. For the rational function $\frac{(x-1)(3x-27)}{(x+5)(2-x)}$, determine the x–intercepts, y–intercept, vertical asymptotes, and horizontal asymptote. Verify your answers by using a computer or graphing calculator to obtain a plot of this function.

Solutions to Selected Exercises

Mathematical Skills

1. a. $x^3 + 3x^2 - 5x + 2$ b. $3x^3 + 4x^2 - 3x$

   c. $2x^2 + x - 15$ d. $x^4 - 16$

2. a. The solutions to this equation are the x intercepts of the graph of $y = 3x^5 - 10$. So look at the graph to get a first estimate for where there might be solutions (see Fig. 7.8). From the graph, there seems to be just one solution at about $x = 1.25$. Now

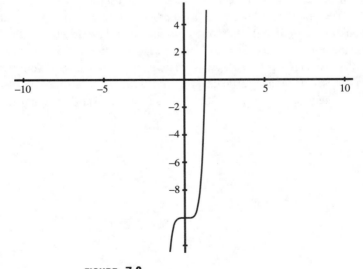

FIGURE 7.8

Graph of $y = 3x^5 - 10$ for problem 2a.

let us switch to a numerical method. When $x = 1.25$, we find $3x^5 - 10 = -.845$. On the other hand, when $x = 1.28$, we find $3x^5 - 10 = .308$. This shows that the true solution is between 1.25 and 1.28. So try $x = 1.27$. That gives $3x^5 - 10 = -.0885$. So now we know the solution is between 1.27 and 1.28. Trying 1.285 we get a result of $3x^5 - 10 = .108$. Continuing in this fashion you can find that the correct value for x is between 1.272 and 1.273. There is also a theoretical solution to this problem. Rewrite the original equation in the form $3x^5 = 10$ and divide by 3 to get $x^5 = 10/3$. Now raise each side of this equation to the $1/5 = .2$ power: $(x^5)^{.2} = (10/3)^{.2}$ or $x^1 = (10/3)^{.2}$. This gives x in a form that can be directly computed on the calculator: 1.272260. Substituting that into the original equation gives the correct result to 12 decimal places.

3. If there is a fraction that is a root of the polynomial, the numerator (top) of the fraction must divide evenly into 6 (the last coefficient in the polynomial), and the denominator (bottom) of the fraction must divide evenly into 5 (the first coefficient). So the possible numerators are 1, 2, 3, and 6, and the possible denominators are 1 and 5. Also, each possible fraction can be either positive or negative. Combining all of this gives the following list of possible roots: $1, 2, 3, 6, 1/5, 2/5, 3/5, 6/5$ and $-1, -2, -3, -6, -1/5, -2/5, -3/5, -6/5$. If you look at a graph of this polynomial, for x between -6 and 6, the only place there is an x intercept is very close to 0. So $1/5$ is the only reasonable possibility from the ones we listed earlier. Substituting that into the equation, we find $5x^3 - 6x^2 + 31x - 6 = 0$. That shows that $1/5$ is a root of the polynomial. There are no other whole numbers or fractions that are roots.

4. For the x intercepts, we look for where the numerator of the fraction is 0. That is, we solve $(x - 1)(3x - 27) = 0$, finding x to be 1 or 9. Similarly, for the vertical asymptotes we look for where the denominator is 0. That gives $(x + 5)(2 - x) = 0$ so x is -5 or 2. The y intercept is found by setting x equal to 0: the result is $y = 27/10 = 2.7$. The horizontal asymptote is found by imagining that x is a humongously large number. For this question, it is best to change the numerator and denominator of the rational expression into polynomials in descending order: $(3x^2 - 30x + 27)/(-x^2 - 3x + 10)$. Now for a very large x only the highest powers of x will have a significant effect on the answer, so we can estimate y to be approximately $3x^2/(-x^2) = -3$. That shows that the horizontal asymptote is $y = -3$.

8

Fitting a Line to Data

In several previous chapters we have discussed the idea of fitting a line to a set of data points. The idea is that the data points are graphed using x and y coordinates, and a line is drawn that comes as close as possible to all the points. In this section, we will explain some of the ideas that go into this process. These ideas can be used to work out formulas for the slope and intercept of the best-fit line, but that is not our purpose here. Rather, the emphasis is on understanding how the procedure works. In practice, you should use a computer program or a statistics calculator to find the best-fit line. But be aware that the methods these devices use are based on the ideas presented in this chapter. Also, you will see that what you learned about quadratic polynomials contributes to the solution of the best-fit line problem.

Defining "Best"

The first issue to consider is what is meant by *best*. That is, if we draw two different lines, and each one comes close to some of the data points, how do we decide which line is better? To make the discussion concrete, we will consider a particular example. In Chapter 1, data concerning atmospheric carbon dioxide levels were presented. The data are repeated in Table 8.1. The data can be represented on a graph with year (Y) on

Year	Carbon Dioxide Concentration
1965	319.9
1970	325.3
1980	338.5
1988	351.3

TABLE 8.1
Carbon Dioxide Concentration in Parts per Million

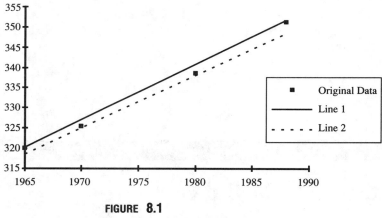

FIGURE 8.1
Two Lines and Carbon Dioxide Data

the horizontal axis and carbon dioxide concentration (C) on the vertical axis. There are four data points, $(1965, 319.9)$, $(1970, 325.3)$, $(1980, 338.5)$, $(1988, 351.3)$, as shown in Fig. 8.1. The figure also shows two lines. These were simply drawn with a straight edge to go near the points. Using the methods of Chapter 4, the equations of these lines are found to be

$$C = 1.37Y - 2372 \tag{1}$$

$$C = 1.30Y - 2236 \tag{2}$$

Neither line hits all the data points. Which one is better?

Our answer emphasizes the idea that each line can be used to define C as a function of Y. That is, given a numerical choice for the Y, each equation allows us to immediately compute a value of C. Taking $Y = 1965$, Eq. (1) produces $C = 1.37 \times 1965 - 2372 = 320.05$, while Eq. (2) produces $C = 1.30 \times 1965 - 2236 = 318.5$. Neither equation gives us the correct data value of 319.9. Instead, each equation gives an approximate value. From this viewpoint, each equation defines a function that is supposed to approximate the true data values. The errors in these approximations can be used to compare how well the lines fit the data. Looking just at the data point for 1965, the error for Eq. (1) is $320.05 - 319.9 = .15$. For Eq. (2) the error is $318.5 - 319.9 = -1.4$. The error is negative in this case because the estimate is too low. Comparing these errors, it is clear that Eq. (1) gives a better estimate of C for 1965. The same kind of calculation can be done for all the data points. The results are shown in Table 8.2. Graphically, each error tells how far one data point is above or below the approximating line. If the true data point is below the line, measure straight up from the data point to the line. That distance is the error as calculated in the table.

Which line is better? For some years one line gives a better estimate, and for other years the other line does. By combining the errors for all the years, we can get an overall measure of how well the line fits the data. Here there is a slight twist. Instead of adding up the errors for each line, we square the errors first and then add them up. There are statistical and mathematical reasons for doing this, but we cannot go into them here. One

Year	Actual C	Estimate 1	Estimate 2	Error 1	Error 2
1965	319.9	320.05	318.5	0.15	−1.4
1970	325.3	326.90	325.0	1.60	−0.3
1980	338.5	340.60	338.0	2.10	−0.5
1988	351.3	351.56	348.4	0.26	−2.9

TABLE 8.2
Estimates and Errors for Two Lines

consequence of squaring the errors is the elimination of negative signs. This prevents a large positive error in one year from balancing out a large negative error for another year. For each line, all of the errors are squared, then added together. This gives for the first line $.15^2 + 1.6^2 + 2.1^2 + .26^2 = 7.0601$ and for the second line $1.4^2 + .3^2 + .5^2 + 2.9^2 = 10.71$. Because the errors produce a smaller total for the first line, that line is a better overall fit to the data.

With this background, a definition for the best fitting line can now be given. For any line, a total error can be computed just as for the two lines considered above: find the error for each individual year, square all these errors, and add them all up. We seek the line for which this total error is the smallest possible.

An Error Function

Given a proposed line for the data (which really means an equation for C as a function of Y), we now know how to compute an overall error. The numerical approach would now be to pick lots of different lines, calculate the error for each one, and try to find the smallest error. The equation of any line we choose is specified by giving a slope and an intercept. Once we know those numbers, we use the procedure above to compute an error. This is really a new function: the calculations define the total error as a function of the slope and intercept of a chosen line. This means that once you know the slope and intercept for an approximating line, you can use them to calculate an overall error for that line. Our next step is to find an equation for this function.

To begin with, we will use the variables m and b for the slope and intercept of a chosen line. The equation of the line will be of the form $C = m \times Y + b$. Observe that Eq. (1) and Eq. (2) are of that form. Leaving m and b as variables, we must work out the error for each data point. For 1965, the estimated value for C will be $1965m + b$. The true data value is 319.9. The error is $319.9 - 1965m - b$. Now we are supposed to square this error. That is, we are to compute

$$(319.9 - 1965m - b)(319.9 - 1965m - b)$$

Using the methods of algebra, the result is

$$3,861,225m^2 + 3,930mb + b^2 - 1,257,207m - 639.8b + 102,336.01$$

This looks a little complicated because the coefficients have so many digits, but the general form is not too bad. It is a polynomial in two variables, m and b. There are terms for m^2, mb, b^2, m, b, each with a coefficient, and a constant at the end. In fact, this is a quadratic polynomial; the highest degree that appears is 2. We can substitute the value of m and b for any line we have in mind, and the result will be the error when that line is used to approximate the carbon dioxide level for 1965. For example, in Eq. (1), $m = 1.37$ and $b = -2372$. Substitute those in the preceding equation and work out the result on a calculator. Do you get .0225? This is the correct answer, and it agrees with the squared error for 1965 from Eq. (1).[1] In the same way, if you substitute the m and b from Eq. (2) you should get 1.96, the same squared error we found for Eq. (2) for 1965. More generally, for any m and b the equation above gives the squared error for the line $C = mY + b$ for 1965.

So far this is just the squared error for 1965. We need to compute a similar error for each of the other years. The same method is used, and the result has the same form: a quadratic polynomial with terms involving m^2, mb, b^2, m, b, and a constant. For each year the coefficients will be different, but the general form of the polynomial will be the same. The squared errors are tabulated in Table 8.3.

The total is the result of adding up the squared errors for all of the years. This gives the total error as a function of m and b. We repeat it here for future reference

$$\text{Total Error} = 15,614,669m^2 + 15,806mb + 4b^2 \tag{3}$$

$$-5,276,117.8m - 2,670b + 446,150.04$$

The total error has the same form as the individual squared errors, namely, a quadratic polynomial in two variables. By choosing different values of m and b, we should be able to quickly calculate the total error, and so use a numerical approach to hunt for the best line.

To illustrate how this process works, let us repeat the calculation of the total error for the lines defined by Eq. (1) and Eq. (2). For the first line, we have $m = 1.37$ and $b = -2372$. Substitute these values in Eq. (3) and we should obtain a result of 7.0601,

1965 :	$3,861,225m^2 +$	$3,930mb +$	$b^2 -$	$1,257,207m -$	$639.8b +$	$102,336.01$
1970 :	$3,880,900m^2 +$	$3,940mb +$	$b^2 -$	$281,682m -$	$650.6b +$	$105,820.09$
1980 :	$3,920,400m^2 +$	$3,960mb +$	$b^2 -$	$1,340,460m -$	$677b +$	$114,582.25$
1988 :	$3,952,144m^2 +$	$3,976mb +$	$b^2 -$	$1,396,768.8m -$	$702.6b +$	$123,411.69$
Total:	$15,614,669m^2 +$	$15,806mb +$	$4b^2 -$	$5,276,117.8m -$	$2,670b +$	$446,150.04$

TABLE 8.3
Errors for Each Data Point

[1] Your calculator might show a different result, due to the effects of roundoff error. Every calculator is limited to some preset number of decimal places of accuracy, and any digits that extend beyond this range are lost. For example, if you compute $1,000,000 + .0000001 - 1,000,000$ on paper, you will get the correct answer of .0000001. Many calculators will not compute this correctly, because the result of the first addition, $1,000,000.0000001$ is rounded off to $1,000,000$. This is called a roundoff error.

the total error we computed earlier. Similarly, for equation Eq. (2), the values of m and b are 1.30 and -2236, respectively. When these values are substituted in Eq. (3) the result is 10.71. As mentioned before, your calculator might not get these answers exactly because of roundoff errors.

Now for any other choice of m and b, we can compute the error using the same procedure. For example, looking at Fig. 8.1 it appears that a better approximation would be obtained by choosing a line in between the two we have already examined. Accordingly, we might try $m = 1.33$ and $b = -2300$. This leads to a total error of more than 147, not at all what you might expect. Taking a little more systematic approach, let us hold m at 1.33 and try several different values of b. Some results for this type of numerical approach are shown in Table 8.4. As this table shows, the numerical approach can be systematically applied to find a line that is better than either of the two we considered originally. With a slope of $m = 1.33$ and an intercept of $b = -2294$, the approximating line produces a total error of about 3.2, which is definitely less than the value of 7.2 we found earlier. Can an even better line be found? To find out, we could experiment with some other values of m and see what errors are produced. You can see that this numerical approach will require us to consider a large number of different lines. It is for this reason that Eq. (3) is so useful. You wouldn't want to have to go through all the work of building Table 8.3 for each and every line you considered. It is much less work to simply choose different values of m and b and substitute them into Eq. (3).

The data in Table 8.4 also show that how well the line fits is very sensitive to the value of the intercept. Making a very small change in the value of b can make a large difference in the total error. This is one of the pitfalls of a numerical approach: sometimes you have to be very careful and systematic in order to find the answer you are looking for. In this problem, the process is further complicated by the possibilities of roundoff errors. If you make the calculations shown in the table on your own calculator, you might get very different results.

Luckily, there are also a graphical approach and a theoretical approach to this problem. In fact, the graphical and theoretical properties of a quadratic polynomial involving two variables are very similar to the properties of quadratic polynomials in a single variable.

Slope m	Intercept b	Total Error
1.33	-2297	39.3001
1.33	-2296	19.2801
1.33	-2295	7.2601
1.33	-2294	3.2401
1.33	-2293	7.2201
1.33	-2292	19.2001
1.33	-2291	39.1801

TABLE **8.4**
Systematic Variation of b with $m = 1.33$

The graph must be visualized in three dimensions, rather than in two, but many of the same features occur. In particular, we will be able to apply our knowledge of parabolas to find the m and b that produce the lowest possible error.

Finding the Best Slope

The theoretical approach for finding the m and b that result in the lowest possible error is somewhat involved. To provide some insight into that process, we begin with a simpler problem. Suppose that we have a good idea what number we wish to choose for b, but want to find the best possible slope. This problem will be illustrated with an example. Returning again to the example considered earlier, let us agree to accept a y intercept of $b = -2350$. That is about halfway between the y intercepts of the two lines we considered earlier, given by Eq. (1) and Eq. (2). Once we choose $b = -2350$, the equation for total error becomes

$$\text{Total Error} = 15{,}614{,}669m^2 + 15{,}806mb + 4b^2$$

$$-5{,}276{,}117.8m - 2{,}670b + 446{,}150.04$$

$$= 15{,}614{,}669m^2 + 15{,}806m(-2350) + 4(2350)^2$$

$$-5{,}276{,}117.8m + 2{,}670(2350) + 446{,}150.04$$

$$= 15{,}614{,}669m^2 - 37{,}144{,}100m + 22{,}090{,}000$$

$$-5{,}276{,}117.8m + 6{,}274{,}500 + 446{,}150.04$$

$$= 15{,}614{,}669m^2 - 42{,}420{,}217.8m + 28{,}810{,}650.04$$

This total error is a quadratic function of m. The coefficients are huge numbers, but the general ideas of quadratic functions are valid for any kind of coefficients. For example, looking at the coefficient of m^2, which is positive, we know that the graph will be a parabola with the vertex at the bottom, a $\cup$ shape. At the vertex the total error function produces the smallest possible value. Do you remember how to find the vertex? For an equation in the form $y = ax^2 + bx + c$, the vertex is found by making x equal to $-b/2a$. Here, we are using m as the variable, rather than x. But the procedure is the same. Divide the negative of the coefficient of m by twice the coefficient of m^2. That leads to $m = 42{,}420{,}217.8/(2 \cdot 15{,}614{,}669) = 1.358345$. That is the best possible choice for the slope of a line, given that the intercept was already selected as -2350.

Now we can repeat this entire process for many different choices of b. For each choice, the error function will turn out to be a quadratic function of m, just as in the preceding paragraph. The least possible error will correspond to the vertex of this quadratic, and can be found just as in the preceding paragraph. In Table 8.5 you will find for several different choices of b the best choice of m, and the corresponding value of the total error function. Keep in mind that the error value shown for each b is the least possible error for that particular b.

In Fig. 8.2 these results are shown in the form of a graph. The different choices for b appear on the horizontal axis; the least value of the error is shown on the vertical axis.

y–Intercept (b)	Best Slope (m)	Total Error
−2475	1.422	3.81
−2450	1.409	3.42
−2425	1.396	3.12
−2400	1.384	2.93
−2375	1.371	2.84
−2350	1.358	2.85
−2325	1.346	2.96
−2300	1.333	3.17
−2275	1.320	3.49

TABLE 8.5
Best m and Total Error for Various b's

The shape of the graph appears to be a parabola. There is a lowest point that occurs for b somewhere around −2360. That error is the smallest of the small. Remember that each point in the graph is the smallest possible error for a particular choice of b. The lowest point on the graph is the smallest of these smallest possible errors. That gives us the best possible b. Then we can find the best possible m to go with that best possible b using the same process we followed earlier. In this way, a two-step process can be used to find the best possible m and b. In the first step, we choose a variety of values for b and compute the best possible m for each. In the second step we select among those choices the best possible b.

In fact, the shape of the graph in Fig. 8.2 is exactly a parabola. It is possible to find the equation for that graph and then compute the vertex exactly. That would give the best

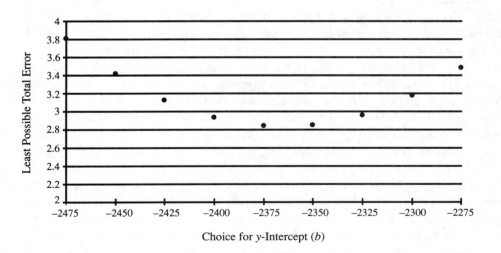

FIGURE **8.2**
Least Error for Various Values of b

possible choice of b. However, we will not delve into the details of that approach. The main point for you to learn from all of this is that the exact values of m and b that provide a line with the least possible total error for a data set can be found using properties of quadratic equations. In this way, the methods you learned earlier for finding the vertex of a parabola lead to a surefire procedure for finding the closest fitting straight line to a set of data. There is an alternate approach to this problem that is explored in the exercises. It too ends up using the properties of quadratic functions. We will close this chapter with a look at a third alternative: an approach using quadratic functions with two variables.

Three-Dimensional Parabolas

As described above, the properties of parabolas can be applied in a two-step approach. At the first step we choose the best possible m for many different b's. At the next step we choose the best possible b. Now we will look at an approach that zeros in on the best possible m and b all at the same time. In the new approach we extend our ideas for a quadratic function depending on a single variable to very similar properties for a quadratic function with two variables. The total error function is an example of such a function. The two variables, m and b, appear in the equation for the total error function (Eq. (3)) with degree at most 2. We shall see that there is a graph for this kind of function, and that the graph has a vertex. By locating the vertex we will find the m and b that produce the least possible total error, and we will do it in one step. To get started on this approach, we must consider how to graph a function which depends on two variables.

For a function of one variable, we generally draw a two-dimensional graph. For example, the equation $y = x^2$ defines y as a function of the single variable x. We use one axis to locate the value of x, and the other to locate the resulting y. Taking the same approach for a function of two variables, we will need a three-dimensional graph. For example, the equation $z = x^2 + y^2$ defines z as a function of the two variables x and y. Here we have to specify both an x and a y before we can compute z. In the three-dimensional graph, we will use x and y axes to locate the two variables that are used to compute the function, and a third axis for the result. Suppose we choose $x = 2$ and $y = 1$. We locate these values as the point $(2, 1)$ on a standard graph with x and y axes. The function then gives $z = 5$. To visualize how this is included in the graph, imagine the xy graph as lying flat on a table, with the point $(2, 1)$ marked. Float a little spot directly above that point, 5 units high. That spot is denoted $(2, 1, 5)$ meaning 5 units above the point $(2, 1)$. That is one point on the graph of $z = x^2 + y^2$ (see Fig. 8.3). Now we can repeat this process for many many (x, y) points. Say $(x, y) = (1, 1)$. Then $z = 1^2 + 1^2 = 2$, so we float a little spot 2 units above the point $(1, 1)$. The floating spot is called $(1, 1, 2)$. With enough points, the little floating spots blend together to form some kind of sheet. The sheet is not flat—it may have bends and wrinkles, hills and valleys. For example $z = x^2 + y^2$ looks like a cup with a round bottom, sitting on the xy plane at the point $(0, 0)$ (Fig. 8.4). For other functions it might float suspended above the xy plane like a magic carpet in hover mode, or it might sit partly above the plane and partly below the plane. Generally speaking, the graph of a function of two variables is some kind of wavy sheet. This type of object is referred to as a *surface*.

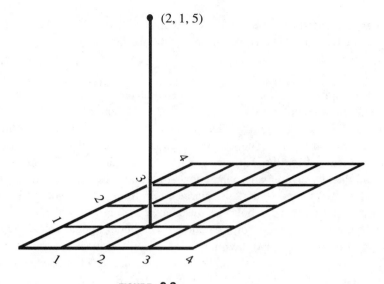

FIGURE 8.3
Graphing with Three Variables

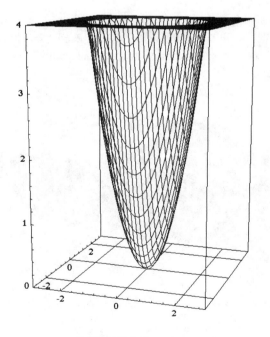

FIGURE 8.4
Graph of $z = x^2 + y^2$

Now we are interested in a special kind of function of 2 variables. Instead of x and y, our variables are m and b. The general form of the function is quadratic:

$$Am^2 + Bmb + Cb^2 - Dm - Eb + F \tag{4}$$

The uppercase (capital) letters stand for numbers. For the total error function in Eq. (3) the coefficients are as follows: $A = 15,614,669$, $B = 15,806$, $C = 4$, $D = 5,276,117.8$, $E = 2,670$, and $F = 446,150.04$. The graph of this function is a three-dimensional parabola—called a paraboloid. The graph in Fig. 8.4 is a particularly simple example of a paraboloid. Viewed from the side, the graph of a paraboloid has the appearance of a parabola. And this is true no matter where you are to the side. Imagine standing at some distance from the graph, and walking in a large circle around it. As you walk around the graph, it always appears as a parabola. From some viewpoints it is wider, and from others it is narrower, but the shape is always parabolic. The bottom point, called the vertex, is the lowest point of the graph and the minimum value of the function. For the total error function, the graph is shown in Fig. 8.5. For this function, the bottom curves so slowly in one direction that it looks more like a trough than a cup. But it truly is a paraboloid, and the vertex corresponds to the lowest possible value of the total error function. If we could read the m and b accurately off the graph, that would give us the slope and intercept of the best-fit line, because for that choice of m and b the total error is as small as possible. Reading information off the graph is difficult for this function because the paraboloid is so steep in one direction and flat in another. However, an expert with graphics software could, with some effort, use different points of view and different scales to get a pretty good idea of the location of the vertex.

A more practical alternative is to use a theoretical approach. For quadratic functions of one variable, we saw how to locate the vertex using the coefficients. A similar procedure can be used to find the vertex for a three-dimensional parabola, although it is not quite

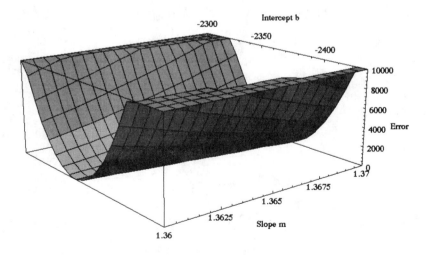

FIGURE 8.5
Graph of Total Error Function

as simple as in the two-dimensional case. Here it is: using the coefficients A, B, and C, in Eq. (4), compute $d = B^2 - 4AC$. Then the vertex is specified by

$$m = \frac{BE - 2CD}{d}$$

$$b = \frac{BD - 2AE}{d}$$

Not every two-variable quadratic function has a graph in the shape of a paraboloid. If the graph *is* a paraboloid, the equations for m and b give the location of the vertex. (The coefficients can also be used to determine whether the graph is a paraboloid or not, but we won't go into that here.) For the special case of our total error function Eq. (3), we can apply the equations to find the vertex, and hence the m and the b for the lowest possible error. The results are $m = 1.3659$ and $b = -2364.9$. These define the best line for the carbon dioxide data:

$$C = 1.3659Y - 2364.9$$

Compare these values of m and b with the data in Table 8.5 and with Fig. 8.2. You should see that $b = -2364.9$ is in about the right place to be the vertex for the curve in Fig. 8.2, and that $m = 1.3659$ fits in well with the other m values in the table. This provides some evidence that $m = 1.3659$ and $b = -2364.9$ really do give the best fitting line for the original data. Using these values of m and b in the error function Eq. (3), the result is about 2.8. This also fits in well with the table and the figure. The final conclusion is this: when the carbon dioxide data are plotted on a graph, the straight line that is nearest to the points is given by $C = 1.3659Y - 2364.9$. This is the best line in the sense that it gives the lowest possible total error. For this best line, the total error is about 2.8. This indicates that the distance from the straight line to the data points is about $\sqrt{2.8} = 1.7$ on the average.

All of the preceding discussion has been in the context of a specific example. However, the general flow of the analysis is applicable any time a line is fit to a set of data points. For each choice of such a line, we can compute the squared errors between the line and each data point. The total of these squared errors determines how good a fit the line is. The best line is the one with the lowest total error.

The problem of choosing the best line reduces to picking the slope m and the intercept b. The total error is a function of m and b, and, what is more, it is always a quadratic function in the form of Eq. (4) with a graph in the shape of a paraboloid. The equations for the vertex of this paraboloid then determine the best slope and intercept.

The whole process of finding the best-fit line can be boiled down to a series of routine computations. Many statistical and scientific calculators, as well as computer programs for data analysis, have built-in procedures for this process. All the user has to do is enter the data. Then the calculator or computer uses the data values to compute the equivalent of the coefficients in the error function, and the location of the vertex of the paraboloid. The results that are returned to the user generally consist of the slope and intercept of the best line for the data, together with the total error corresponding to that best line.

These procedures for fitting a line to a set of data constitute an example of linear regression. This is a subject that is covered in most introductory statistics courses. As we proceed in this text, we will often refer to the idea of fitting a line to data points. Now you know how it is done. What is more, you have seen that the process is not much more complicated, conceptually, than finding the vertex of a parabola.

Summary

In this chapter the main ideas of fitting a line to a set of data points were covered. As a first step, a method for comparing two lines was demonstrated. As a result of the comparison method, we developed the concept of the total error between a line and the points of the data set. The best-fit line is the line with the lowest total error. Moreover, it was shown that the total error is actually a function of the slope and intercept chosen for a line.

Once an equation for the total error function is known, the problem of finding the minimum possible total error can be attacked by numerical and graphical methods. The graph of the error function is a paraboloid, and the vertex is the location of the lowest error. A graphical method of completing the line-fitting problem is to find the vertex visually. The slope and intercept of the best line are the first two coordinates of the vertex.

A theoretical method is also possible. This is similar to the method used to find the vertex for a quadratic function of one variable. The coefficients of the total error function are used to compute the vertex values of m and b. This gives the best-fit line for the original data set.

These procedures can be applied to any set of data points. Many calculators and computer programs are designed to carry out the procedures. All the user has to do is enter the data points.

Exercises

Reading Comprehension

1. Write a short paragraph explaining what is meant by the *best* line for a set of data points.

2. The reading describes how to compute a total error for a line that is meant to fit a set of data points. Explain how that error is computed.

3. Explain what a paraboloid is. What kind of equation does it have? What is meant by the vertex of a paraboloid, and how can it be found?

4. In the reading, an equation is given for total error as a function of m and b. What are m and b in this context? If you use the error function to compute a value, how should that value be interpreted? How is the error function used to find the best line for the set of data?

Mathematical Skills

1. Make a graph showing the line $y = 3x + 1$ and these points: $(1, 3.9)$, $(2, 7.2)$, and $(3, 10.3)$. This line does not pass exactly through the 3 points, but it is close to them. Compute the total error, in the same manner as in the reading.

2. Repeat the previous problem, but use the line $y = 3.1x + .9$.

3. Treat the slope m and the y intercept b as unknowns. Considering just the first data point $(1, 3.9)$, express the error as a function of m and b.

4. This equation describes a three-dimensional graph: $z = 3x^2 + 4y^2 - 6x - 16y + 24$. The graph is a paraboloid. Use a computer graphics tool to view the graph of this paraboloid. Find the vertex using the equations

$$d = B^2 - 4AC$$
$$x = \frac{BE - 2CD}{d}$$
$$y = \frac{BD - 2AE}{d}$$
$$z = 3x^2 + 4y^2 - 6x - 16y + 24$$

Does the vertex that you find from the equations agree with the appearance of the graph?

5. A function of two variables is given by the equation

$$z = 3x^2 + 4y^2 - 6x - 16y + 24$$

If all possible choices for x and y are considered, what is the least possible value that can result from this function?

6. This chapter shows how the total error function is defined and how it can be used to find the best line for a data set. In this exercise you will learn about an alternate approach to this problem.

The main idea is to transform the data before graphing it, so that the x values and y values each have an average of 0. Once that is done, the error function is a little simpler than the one we looked at before. In fact, it is easy to see that the best line has a y intercept of $b = 0$, so that all we have to find is the slope.

a.   We will use the three points $(1, 3.8)$, $(2, 7.2)$, and $(3, 10.3)$. The average of x values is $(1+2+3)/3 = 2$. The average of the y values is $(3.8+7.2+10.3)/3 = 7.1$. Make up three new data points by subtracting 2 from each x and subtracting 7.1 from each y. We will refer to these new data points using the variables u and v. That is, $u = x - 2$ and $v = y - 7.1$.

b.   Draw a graph with axes labeled u and v and plot the three data points. Draw a line on the graph that is close to these data points.

c.   Assume that the line you drew has the equation $v = mu + b$. Work out the total error function from the line to the three points. You will see that the formula has no mb term and no b term. This is always the case when we make new data points by subtracting the average x and y values from the original data.

d.   The total error can be split into two parts. The first part is $3b^2$, and the second part is $2m^2 - 13m + 21.14$. Considering each part separately, what is the smallest that that part can be?

e.   Combine the two parts of your previous answer, what is the smallest the total error can be? What are the m and b for this answer?

f.   What is the equation for the best line for the drawing with u and v axes? The answer is an equation involving the variables u and v.

g.   Remember that for the original data points, we have x and y values with $u = x - 2$ and $v = y - 7.1$. Substitute these expressions for u and v in your equation to get an equation involving x and y.

The steps you followed in the preceding problem can be applied in any problem to find the best line for a data set. The first step is always to create a new data set by subtracting the average x and y values from the original data set. The total error function is then formulated for fitting a line to the new data set. That error function will always have the same form: a multiple of b^2 added to a quadratic function in m. The least possible total error comes from setting b equal to 0, and choosing the m at the vertex of the quadratic function. That gives a best line for the u and v data. The corresponding equation for the original x and y data is then found replacing u with $x - a$, where a is the average of the x data, and replacing v by $y - c$ where c is the average of the y data. Notice that this shows that the best-fit line always goes through the point (a, c) whose coordinates are the average of the x data and the average of the y data. This point is called the *centroid* of the data set.

Group Activities

1. Use a graph from the newspaper or a magazine that shows data points that are nearly in a line. (You may use a graph that was already used in a previous homework assignment). Find the equation of a line that is close to the data in three different ways: (1) draw the line by eye with a ruler, find two points on that line, and obtain the equation using the methods of Chapter 4; (2) use a computer or statistical calculator to find the best line; and (3) repeat the analysis that was given in this chapter. In the third step, you will have to formulate a total error function and put it into the standard algebraic form for a paraboloid, then find the vertex. To keep things simple, use only a few data points, no more than 4.

Solutions to Selected Exercises

Mathematical Skills

1. For the line with equation $y = 3x + 1$ we will compute y values for the three x values 1, 2, and 3. The results are shown in the table below

x	y from equation	y from data point	difference of y's	square of difference
1	4	3.9	.1	.01
2	7	7.2	.2	.04
3	10	10.3	.3	.09

The total of the squared differences is $.01 + .04 + .09 = .14$. This is the total error for that line.

2. Using the line $y = 3.1x + .9$ the table becomes

x	y from equation	y from data point	difference of y's	square of difference
1	4.0	3.9	.1	.01
2	7.1	7.2	.1	.01
3	10.2	10.3	.1	.01

The total error is $.03$.

3. The true y value for the first data point is 3.9. Using the equation $y = mx + b$, the y value would be estimated as $m \cdot 1 + b = m + b$. The difference between these two y values, squared, is $(m + b - 3.9)^2$. This gives the error for that one data point as a function of m and b.

4. The constants for the formula are $A = 3$, $B = 0$, $C = 4$, $D = 6$, and $E = 16$. This gives $d = -48$, so $x = (0 - 2 \cdot 24)/-48 = 1$, and $y = (0 - 2 \cdot 48)/-48 = 2$. Then we compute $z = 3 + 16 - 6 - 32 + 24 = 5$

5. This is the quadratic function considered in the preceding problem, and the least possible value for this function occurs at the vertex. The value of the function there is 5. That is the smallest the function can ever be.

6. a.   The new data points are $(-1, -3.3)$, $(0, .1)$, and $(1, 3.2)$.

   c.   For the first point, the data value for v is -3.3. Using the u value -1 in the equation of the line, we get an approximate v value of $m(-1) + b = -m + b$. The error for this point is $-3.3 + m - b$. Squaring the error gives $m^2 - 2mb + b^2 - 6.6m + 6.6b + 10.89$. In a similar way we find squared errors for the other points: for $(0, .1)$ the squared error is $(.1 - b)^2 = b^2 - .2b + .01$, and for $(1, 3.2)$ it is $(3.2 - m - b)^2 = m^2 + 2mb + b^2 - 6.4m - 6.4b + 10.24$. Adding the three errors together, the total error is $3b^2 + 2m^2 - 13m + 21.14$.

   d.   The smallest that $3b^2$ can be is 0. This happens when $b = 0$. For any other value of b, $3b^2$ will be positive.
   For the second part, $2m^2 - 13m + 21.14$ is a quadratic function of m. The smallest that can be is its value at the vertex. So set $m = 13/(2 \cdot 2) = 13/4$. Then $2m^2 - 13m + 21.14 = .015$.

   e.   Since the first part of the error is at least 0 and the second part is at least $.015$, the total is at least $.015$. This total error occurs when $b = 0$ and $m = 13/4 = 3.25$.

   f.   The equation is $v = 3.25u + 0$ or simply, $v = 3.25u$.

   g.   The equation is $y - 7.1 = 3.25(x - 2)$.

9

Geometric Growth

In Chapter 2 several different kinds of difference equations were displayed. This equation

$$s_{n+1} = .1s_n \tag{1}$$

was introduced in a discussion of salt contamination in a tank of fresh water. This difference equation provides an example of what is called *geometric growth*. In this chapter we will study geometric growth, and the difference equations that go with such growth. These are equations of the form

$$a_{n+1} = ra_n \tag{2}$$

where the parameter r stands for a numerical constant. In different applications, different values will be chosen for r, but the form of the difference equation will remain the same.

There is a close analogy between geometric growth and arithmetic growth. Recall that arithmetic growth arises from making a simple assumption about some phenomenon: in equal periods of time, the phenomenon will grow (or shrink) by equal amounts. When we have a sequence of data points taken at regularly spaced times (say every hour, or every day), the assumption of arithmetic growth leads to a difference equation describing how each data point depends on the preceding data point. There is also a functional equation. This tells how to determine each data point directly without knowing the preceding data point. In the case of arithmetic growth, the functional equation is linear, and it leads to the study of linear equations. Often, we can interpret the functional equation as a continuous model. This allows us to think about the data that would occur in between the times of the original data.

As you will see, the development of geometric growth follows a very similar pattern. We will again begin with a simple assumption, in equal periods of time, the phenomenon will grow (or shrink) by an equal *percentage*[1]. That will lead us to a difference equation

---

[1] It will be important throughout this chapter to keep in mind the difference between *amount* and *percentage*. Here is an example: Suppose a new stereo used to cost $500 but the price goes up to $550. The *amount* of this increase is $50. The *percentage* of the increase is 10%, because 50 is 10 percent of 500.

of the same form as Eq. (1). We will also find a functional equation. As for arithmetic growth, the functional equation leads to an important family of functions, with continuous model interpretations. As we proceed through this material, we will again see numerical, graphical, and theoretical aspects of the subject.

The Geometric Growth Assumption

As an illustration of the idea behind geometric growth, consider a new species of mouse that is introduced into an ecological system. When the mouse is first discovered, researchers estimate that about 100 of the mice are living in the area. A month later a follow-up study finds that the population of mice has grown to 200. How big would you expect the size of the population to be after one more month? At first glance, you might be tempted to say 300. After all, if the population increased by 100 in the first month, maybe the same will occur in the second. That kind of thinking leads to a model with arithmetic growth. It amounts to the following assumption, (labeled *Assumption A* for future reference).

     Assumption A: In any month, the mouse population will increase by 100 mice.

Although an arithmetic growth assumption of this type is very useful in many applications, there is good reason to doubt it here. To see why, consider the following even more restricted assumption.

     Assumption B: In any month, a population of 100 mice will grow to 200 mice.

In this assumption, we do not assume that the increase depends only on the amount of time, as in arithmetic growth. Rather, we recognize that the size of the population at the start of the month will have an effect on the number of new members added to the population during the month. We are not willing to say that *any* size population will increase by 100 mice in a month—only that a population of size 100 will increase by 100 in a month.

Using this assumption, what can we figure out about the future growth of the mouse population? It starts out at 100 mice. After 1 month it grows to 200 mice. Now think of those 200 mice as making up 2 populations each of size 100. This makes some sense, for as the mouse population grows, the mice will spread out and inhabit a larger territory. Maybe we can divide the territory into two regions, north and south, with 100 mice each. In any case, our assumption says that each of these separate groups of mice will grow from 100 to 200 during a month. Therefore, after the second month, there will be two populations of size 200, for a total of 400 mice. Now repeat this process. Consider the 400 mice as being divided into 4 separate groups of 100 mice each. In another month, each will grow in size to 200 mice, for a total of 800 mice. To summarize, Assumption B leads us to predict the following growth for the mouse population:

Months	0	1	2	3
Mice	100	200	400	800

Geometric Growth

This is an example of geometric growth. It is quite different from what we would predict using a linear growth model. Under linear growth the amount of increase would remain 100 mice month after month, leading to the following predicted populations:

Months	0	1	2	3
Mice	100	200	300	400

Arithmetic Growth

Given such different models, how do we know which to use? Of course, one important step in developing models is to collect data and determine empirically whether the model is accurate. But in this case, common sense seems to indicate that the linear model is not right. The other assumption, the one that depends on the population size, seems more consistent with what we know about how animals live. It recognizes that the growth of the mouse population in one month will depend on how many mice were alive at the start of the month. The more mice we have to start with, the more females are available to have young, and so the more new mice will be added to the population. This is one of the main features of geometric growth. The amount of growth in any period depends on the amount you have at the start of the period.

The main idea of geometric growth is this: under a geometric growth assumption, the population grows by the same fraction or percentage in each month. This is in contrast to the arithmetic growth assumption in which the population grows by the same absolute amount each month. In the example the mouse population increased by 100% in the first month, because the amount of increase was equal to 100% of the amount we started with. The geometric growth assumption states that the percentage of increase is an invariant, so we predict that the population will increase by 100% in *every* month. Accordingly, for the second month we start with 200 mice. A 100% increase adds as many mice as we already have—200 mice. That leads to a total of 400. In another month the population again grows by 100%, to a total of 800. This is the same pattern we found using Assumption B, by thinking about splitting the population into smaller sizes. In that earlier discussion, commonsense suggested a pattern of future growth. The point of this discussion is to show that the common sense approach gives the same results as assuming a fixed percentage of increase each month. That is geometric growth.

Let's work through another example using a different percentage. Suppose the mouse population increases by 50% in any month. If we start with 100 mice, then a 50% increase will add 50 mice to the population. Now we have 150 mice. The next month produces another 50% increase. But 50% of 150 is 75. So the increase in the second month is 75, to produce a total of 225 mice. What happens in another month? There will be another 50% increase. We have 225 already, and 50% of that will be 112.5. This is clearly nonsense. The population can't have half a mouse. Remember that this is just a model, and it is only supposed to give an approximation of the true situation. So maybe there are really going to be 113 or 112 mice added to the population. This kind of situation is typical in geometric growth models. It is one of the limitations that must be accepted when geometric growth is assumed. Geometric growth models seem to fit so well in so many situations, that they are used extensively. Remembering to round off from a fraction to a whole number is at worst a minor inconvenience.

Notice that in the previous paragraph, the *amount* of increase in the mouse population goes up each month: 50 in the first month, 75 in the second, 112.5 in the third and so on. The actual amount of increase is not the same month after month as it would be for arithmetic growth. Rather, the population grows by the same *percentage* each month. As already stated, this is the main idea of geometric growth.

> **First Definition of Geometric Growth:** *Growth of a variable concerns the way the variable changes over time. Under the assumption of Geometric Growth, the variable increases by a fixed percentage in any period of a given length.*

In this definition, the fixed percentage of increase might occur each month, or each week, or each day, or each period of any other length. That is what is meant by the words *any period of a given length.* To illustrate this idea, here is another variation on the mouse model. Suppose that the mouse population is found to be 100 on the first measurement, but that the next measurement is made after 6 weeks, instead of after a month. At that time the population is found to be 280. That is an increase of 180 mice, which is 180% of the original population of 100. A geometric growth assumption then states that there will be a 180% increase in *any* six-week period. So 12 weeks after the first measurement, the predicted size of the population is 180% higher than 280. That gives us an increase of $1.80 \times 280 = 504$ and a total of $280 + 504 = 784$.

There is another way to think of geometric growth. For the original mouse population example, notice that the population doubled each month. That is, the population at the start of each month was twice what it was at the start of the preceding month. In terms of a difference equation, we could write

$$p_{n+1} = p_n \times 2$$

Compare that to an arithmetic growth model that says the population will grow by 100 each month:

$$p_{n+1} = p_n + 100$$

See how similar these equations are. In geometric growth we *multiply* each data value by a fixed amount to get the next one. In arithmetic growth we *add* to each data value a fixed amount to get the next value. To emphasize the idea of multiplying (instead of adding) we refer to 2 as a *growth factor*. A growth factor is a number that is multiplied by one data value to get the next data value. Under a geometric growth assumption, the growth factor remains constant (or invariant) over time.

What is the growth factor for the second example, where the mouse population was assumed to grow by 50% each month? The initial population size was 100. A month later it grew to 150. That is a growth factor of 1.5, because $100 \times 1.5 = 150$. In general, to find a growth factor between two data values, simply divide the second by the first. In this example we wanted to grow from 100 to 150, so we divided $150/100 = 1.5$ to find the growth factor. Now the same growth factor can be applied again. With 150 mice at the start of the second month, we expect $150 \times 1.5 = 225$ at the start of the third month. This is the same result we got before. However, now we are getting it by multiplying by a constant, rather than by adding a constant percentage. Mathematically, the result is the same. This gives an alternate definition of geometric growth,

> **Second Definition of Geometric Growth:** *Growth of a variable concerns the way the variable changes over time. Under the assumption of Geometric Growth, the variable increases by a fixed growth factor in any period of a given length.*

The connection between increasing by a fixed percentage and increasing by a fixed growth factor can be made a little clearer. Suppose we are going to increase the population by 50%. We start with 100 mice. Then we calculate 50% of that. This calculation involves multiplication. In fact, there is a direct translation from the phrase 50% *of* 100 to the formula $.50 \times 100$. This pattern is generally valid. To calculate a 15% tip on a bill of \$37.80, translate 15% *of* 37.80 into $.15 \times 37.80$. Now back to the mice. We have the 100 mice we started with, plus half that amount again. The total is one and a half times what we started with. That is a growth factor of 1.5.

Here is another example. Suppose the mouse population grows by 20% each month. If we start with 370 mice, at the end of one month we will have the original 370 plus 20% more. Calculate the additional mice as 20% *of* $370 = .20 \times 370$. After adding that to what we already had, the total is $370 + .20 \times 370 = 1.20 \times 370$. Increasing the population by 20% is identical to multiplying it by 1.20. All of these examples show the same pattern. In each case, the growth factor is 1 plus the decimal form of the percentage of increase: A 20-percent increase means a growth factor of 1.20; a 50-percent increase means a growth factor of 1.50; a 13.38-percent increase means a growth factor of 1.1338.

The preceding comments show that a geometric growth assumption may be expressed as increasing by a constant percentage each month, or as increasing by a constant growth factor each month. These ideas both produce the same result. What is more, the growth factor can be found by adding 1 to the decimal form of the constant percentage.

Each of these approaches can be expressed in a difference equation. First consider the statement *the mouse population at the start of any month is* 20% *higher than the population at the start of the preceding month.* If p_n is the population at the start of this month, then p_{n-1} was the population at the start of the preceding month. To find an amount that is 20% higher, we have to start with p_{n-1} and add 20%. Now translate *20% of* p_{n-1} to $.20 \times p_{n-1}$. Adding that to the amount we started with gives the equation

$$p_n = p_{n-1} + .20 \times p_{n-1}$$

On the other hand, if we say there is a growth factor of 1.20 each month, then the population size this month is 1.20 times what it was last month. That gives the equation

$$p_n = 1.20 \times p_{n-1}$$

The rules of algebra show that these two equations have just the same meaning.

The growth factor idea is a useful way to recognize patterns of geometric growth in data. Divide each data value by the preceding value. For true geometric growth, the results should be same for each pair of data points. In real data, exact equality will rarely be seen. However, if the divisions all produce nearly equal results, then the data may be closely approximated by a geometric growth model.

As an example of this idea, consider again the oil consumption data in Table 2.1. Although the oil consumption figures for the last two years shown are projections, we will again pretend that these are true data values. Do these numbers seem to follow a geometric growth model? We divide each number by the preceding number in the table:

Year	1991	1992	1993	1994	1995
Daily Oil Use	66.6	66.9	66.9	67.6	68.4
Growth Factors	—	1.005	1.000	1.011	1.012

Here, we compute a growth factor for each year. For example, from 1991 to 1992 the growth factor is $66.9/66.6 = 1.005$. These growth factors correspond to increases of .5%, 0%, 1.1% and 1.2%. It is not possible to decide in the abstract if that is a close enough fit to justify an assumption of geometric growth. As discussed in Chapter 1, an accurate model for oil consumption would be built out of many factors, some of which might involve geometric growth. The point here is to illustrate how the idea of growth factors can be used to investigate whether data appear to exhibit geometric growth.

Here is another example of the same idea. Recall the Fibonacci sequence from Chapter 2: 1, 2, 3, 5, 8, 13, 21, etc. In this pattern, each number is found by adding the two preceding numbers. Does this pattern exhibit geometric growth? Let's calculate the growth factor from each number to the next. From 1 to 2 the growth factor is $2/1 = 2$. From 2 to 3 the factor is $3/2 = 1.5$. We can continue to calculate growth factors in this fashion as long as we wish. Here is a list of the first 10 results:

$$2 \quad\quad 1.5 \quad\quad 1.667 \quad 1.600 \quad 1.625$$
$$1.615 \quad 1.619 \quad 1.618 \quad 1.618 \quad 1.618$$

The first few growth factors are not very similar, but after a while, the growth factors become nearly identical. This shows that the long term behavior of the Fibonacci pattern is very nearly geometric growth. This suggests that we can approximate the Fibonacci numbers by a geometric growth pattern. More advanced mathematical techniques then lead to a surprising discovery: the Fibonacci numbers can be obtained *exactly* by adding together *two* patterns of geometric growth. This idea will be further explored in an exercise.

We have seen so far that there are two ways to think of geometric growth. One focuses on the idea of percentage of increase, and the other on the idea of a growth factor. In either case we will end up with the same difference equation. The general form of this difference equation is

$$a_{n+1} = r a_n$$

where r is a parameter, or numerical constant, that remains fixed in any particular model. In some contexts, we are led to this kind of difference equation by simple reasoning. This is what happened in the discussion of the water tank situation at the end of Chapter 2. In other cases, we may simply choose to use a geometric growth model. This often occurs when it seems reasonable to assume that growth occurs by a constant percentage. A third possibility is that we have data that appear to approximate geometric growth, as for the

Fibonacci numbers. In any of these cases, we are led to a difference equation of the same form. As we shall see shortly, this type of difference equation always leads to the same general patterns of growth, and the same kinds of mathematical properties. In the next section, we will discuss some of these properties. At the end of the chapter, there will be several more examples of geometric growth models.

The Geometric Growth Difference Equation

In any geometric growth model, there is a difference equation of the form

$$a_{n+1} = ra_n$$

The parameter r will be a particular numerical constant. We can examine the behavior of the model both numerically and graphically. Consider again the sample equation

$$p_{n+1} = 1.5p_n$$

with the initial value $p_0 = 100$. We can systematically use the difference equation to compute successive values:

$$
\begin{array}{rcccccr}
p_1 & = & 1.5 \times p_0 & = & 1.5 \times 100 & = & 150 \\
p_2 & = & 1.5 \times p_1 & = & 1.5 \times 150 & = & 225 \\
p_3 & = & 1.5 \times p_2 & = & 1.5 \times 225 & = & 337.5 \\
p_4 & = & 1.5 \times p_3 & = & 1.5 \times 337.5 & = & 506.25
\end{array}
$$

From these calculations, we observe that the numbers p_n get steadily larger, and that the increases are also growing. This is a general feature of geometric growth. Remember that we are adding a fixed percentage of the population at each step. As the population gets larger, this fixed percentage does too. So the amount of increase at each step is greater than for the previous step.

This pattern shows up in the graph of the sequence. In Fig. 9.1 the values for p_1 up to p_9 are shown. The points are joined by straight lines, but the graph appears to curve

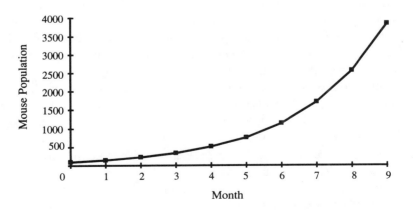

FIGURE 9.1
Geometric Growth with a Growth Factor of 1.5

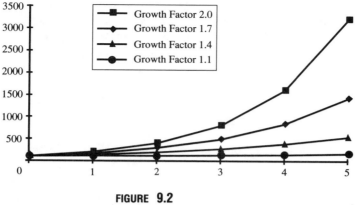

FIGURE 9.2
Different Growth Factors

smoothly upward from left to right. This appearance occurs because the straight lines get steeper and steeper as you look across the curve from left to right. This is the visual equivalent of the idea that the amounts of increase keep getting larger. The upward curve is a characteristic of geometric growth.

The higher the growth factor is, the more rapidly the lines will curve upward. In Fig. 9.2 several different geometric growth models are shown in one graph. Each model has a different growth factor. For the one that curves upward most steeply, the growth factor is 2. That corresponds to the difference equation $p_{n+1} = 2p_n$, and represents 100% growth each month. At the other extreme, the lowest growth factor is 1.1. For that model, the difference equation is $p_{n+1} = 1.1p_n$, representing a population that increases by 10% each month. (Remember that with an increase of 10% or .10, the growth factor is 1.10.) All of the curves show the same characteristic shape, differing only in the steepness of the curves.

In all of the examples so far considered, the growth factor has been greater than 1. There are situations where a growth factor less than 1 makes sense. That is, there are models where the difference equation is $a_{n+1} = ra_n$ with $r < 1$. Let us review one that was introduced in Chapter 2. A water tank with a drain is illustrated in Fig. 9.3. Notice that, when the drain is opened, not all of the water can drain out. For the sake of discussion, let us say that the tank holds 100 gallons of water, and that when the drain is opened, 90 gallons of water will drain out, leaving 10 gallons still in the tank. Now suppose someone accidentally spills some salt into the tank. To remove this contaminant, we drain the tank and refill it with fresh water. Unfortunately we can't drain out all of the water. We can only remove 90 gallons, or 9/10 of the total. Assuming that the salt was completely dissolved in the tank, the water that is removed will only contain 9/10 of the salt, leaving 1/10 in the tank. As a result, the process of draining the tank and refilling it with clean water will leave one-tenth of the salt in the tank. Now imagine doing this procedure several times. Let s_n be the amount of salt in the tank after n repetitions of the draining and refilling process. Then $s_{n+1} = .1s_n$. This difference equation simply says that when you drain and refill, the amount of salt in the tank is

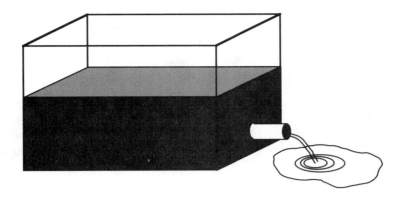

FIGURE **9.3**
A Water Tank

reduced from s_n to one-tenth that amount, or $.1s_n$. If the initial salt spill was 50 pounds, then after draining and refilling the tank once, 5 pounds of salt remain dissolved in the water. That is one-tenth of the original amount. Now with 5 pounds of salt dissolved in the water, we drain and refill again. This leaves one-tenth of the salt, or .5 pounds, still in the tank. As we continue to drain and refill, we reduce the amount of salt to .05 pounds, .005 pounds, .0005 pounds, and so on.

We can change some of the numbers in the model without changing the nature of the difference equation. If we start with 80 pounds of salt instead of 50 pounds, the equation remains $s_{n+1} = .1s_n$, we simply have a different starting amount s_0. What if the tank holds 75 gallons, instead of 100? In this case, each time we drain it the 10 gallons left in the tank is not one-tenth of the total. Rather, the fraction is given by $10/75$. This same fraction of the salt remains in the tank after one draining and refilling. So the difference equation becomes $s_{n+1} = (10/75)s_n$. Or, expressing $10/75$ as a decimal (approximately .1333), $s_{n+1} = .1333s_n$. Once again the difference equation is of the same form with a growth factor that is less than 1.

As the preceding example suggests, the water tank model always leads to a difference equation with a growth factor that is less than one. That means s_n is not growing larger as n increases. On the contrary, it keeps getting smaller and smaller. This is true for any geometric growth model with $r < 1$. Each time you multiply by r, the result is smaller than what you started with. For example, if $r = .5$ and we start with $a_0 = 80$, then the difference equation leads to $a_1 = 40$, $a_2 = 20$, $a_3 = 10$, and so on. Here, the $a's$ are growing smaller. So perhaps it would make more sense to refer to r as a shrinkage factor, or a contraction factor. However, to simplify the discussion, we will continue to call r a growth factor. The point to keep in mind is this. In a model with a difference equation of the form $a_{n+1} = ra_n$, if $r > 1$, the a's grow larger, while for $r < 1$ the a's grow smaller.

We can examine geometric growth with a growth factor less than one both numerically and graphically, just as we did earlier for growth factors greater than one. As a first example, we will use the tank model again. The first several values of s_n are recomputed in a systematic way below:

$$
\begin{aligned}
s_1 &= .1 \times s_0 = .1 \times 50 = 5 \\
s_2 &= .1 \times s_1 = .1 \times 5 = .5 \\
s_3 &= .1 \times s_2 = .1 \times .5 = .05 \\
s_4 &= .1 \times s_3 = .1 \times .05 = .005
\end{aligned}
$$

There is a kind of symmetry to this table of data. Reading down the table, each new number is .1 times the preceding number. But if you start at the bottom and work your way up, each new number is 10 times the size of the one below it. This may be easier to see if you simply list the data:

$$50 \quad 5 \quad .5 \quad .05 \quad .005$$

In this list we observe a growth factor of .1: each number is .1 times the preceding number. If the data in the list are reversed we have a new list with a growth factor of 10:

$$.005 \quad .05 \quad .5 \quad 5 \quad 50$$

In general, reversing the order of data in a geometric growth model results in another geometric growth model, but with the reciprocal of the growth factor. If the original data have a growth factor of 1.25, then reversing the order of the data corresponds to a growth factor of $1/1.25$. Put another way, when you go in one direction, each new number is found by multiplying the preceding number by 1.25. When you go the opposite direction, each new number is found by *dividing* by 1.25.

This reversal property shows up in the graphs, too. In Fig. 9.4 the first several values for the tank model data are shown. Notice that the shape of the graph looks like the mirror image of the graphs considered earlier: instead of sloping upward from left to right, the graph slopes upward from right to left. This is the visual representation of the idea that reversing the order of the data changes the growth factor to its reciprocal. Visually, reversing the order of the data has the effect of flipping the graph across a vertical line. The points on the left are flipped to the right, and vice versa. Then, the

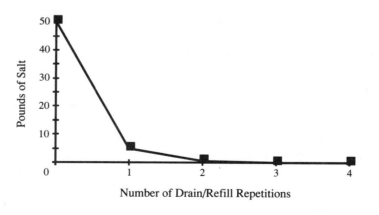

FIGURE 9.4
Amount of Salt in Tank Model

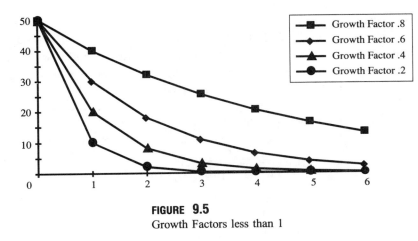

FIGURE 9.5

Growth Factors less than 1

growth factor will be 10 rather than 1/10, and that is the kind of geometric growth shown in Fig. 9.2.

When the growth factor is less than one, there is a characteristic shape to the graph. As you trace it from left to right, the graph flattens out and approaches the x axis. Also, as before, when we use growth factors that are less than 1, different values lead to curves that flatten out more or less rapidly. Several different choices are illustrated in Fig. 9.5.

From the numerical and graphical examination of geometric growth examples, several general characteristics have now been found. For growth factors greater than 1, the data grow larger and larger, with a graph that slopes steeply up to the right. The bigger the growth factor, the more rapidly the curve slopes upward. For growth factors less than one, the data grow less and less with slopes that flatten out as you go the right. The smaller the growth factor, the more rapidly the curve flattens out. The curves with growth factors greater than one are the mirror images of the curves with growth factors less than one. In the next section, we will see that there is a functional equation that can be worked out for any geometric growth model.

Functional Equation

The functional equation for arithmetic growth can be found by observing a simple pattern generated by the difference equation. The same is true of geometric growth. As an illustration of this idea, consider the mouse population difference equation

$$p_{n+1} = 1.5p_n$$

with an initial population of 100. Compute the first several values, but do not actually perform any of the multiplications:

$$
\begin{aligned}
p_1 &= 1.5 \times p_0 &= 1.5 \times 100 \\
p_2 &= 1.5 \times p_1 &= 1.5 \times 1.5 \times 100 \\
p_3 &= 1.5 \times p_2 &= 1.5 \times 1.5 \times 1.5 \times 100 \\
p_4 &= 1.5 \times p_3 &= 1.5 \times 1.5 \times 1.5 \times 1.5 \times 100
\end{aligned}
$$

The pattern is clear, and can be used to find any p_n. For example, p_7 should be equal to 7 factors of 1.5, all multiplied by 100. This pattern makes sense, too. If we start with 100 and multiply by 1.5 every month, then after seven months we will have multiplied by 1.5 seven times, which is just what the pattern says. Of course, it becomes awkward to write out the pattern longhand: $1.5 \times 1.5 \times 1.5 \times 1.5 \times 1.5 \times 1.5 \times 1.5 \times 100$. Imagine writing out the pattern for p_{25}! Using exponents simplifies things considerably. Since an exponent indicates repeated multiplication, we can write the pattern for p_7 as $1.5^7 \times 100$. Similarly, $p_{25} = 1.5^{25} \times 100$. Using this notation in the pattern above leads to

$$\begin{aligned}
p_1 &= 1.5 \times 100 \\
p_2 &= 1.5^2 \times 100 \\
p_3 &= 1.5^3 \times 100 \\
p_4 &= 1.5^4 \times 100
\end{aligned}$$

As usual we can summarize the pattern with a single equation:

$$p_n = 1.5^n \times 100$$

This equation fits all the examples above. But for $n = 0$ it says

$$p_0 = 1.5^0 \times 100$$

This time, the exponent has a slightly different meaning than before. When we write 1.5^2 it means to multiply two 1.5's together. That same idea doesn't work for 1.5^0; you cannot multiply zero 1.5's together—there is nothing to multiply. Instead, 1.5^0 is understood to mean 1. This is a convention that allows the functional equation to make sense when $n = 0$. In fact, this convention extends to any growth factor. The rule is: any positive number raised to the 0 power equals 1, or in symbols, $b^0 = 1$ for all $b > 0$. This is probably a rule you remember from previous studies.

Now we want to generalize from the example to any geoemtric growth model. For the example we have been using, the functional equation is $p_n = 1.5^n \times 100$. Observe that the numbers 1.5 and 100 appear in the original problem as the growth factor and the initial value p_0. This leads to the general equation $a_n = r^n \times a_0$, although it is more customary to write it in the form $a_n = a_0 \times r^n$. This is a functional equation for the sequence specified by the difference equation $a_{n+1} = ra_n$. Using the functional equation, we can immediately compute a_n for any n without computing all the preceding a's. Since this gives the functional equation for any geometric growth model, we restate it below for emphasis.

> If a geometric growth model has the difference equation $a_{n+1} = ra_n$ with r a constant, then the functional equation is $a_n = a_0 r^n$ for $n \geq 1$.

Now that we have a functional equation, we can use it to answer questions about the models. Consider again the water tank model. The difference equation is $s_{n+1} = .1s_n$, and the initial amount of salt is $s_0 = 50$. The functional equation can therefore be written as $s_n = 50(.1)^n$. If we drain and refill the tank 10 times, how much salt will remain? The answer is $s_{10} = 50(.1)^{10} = .000000005$ pounds of salt. It is also easy now to consider variations on the model. Using the modification discussed earlier, with a

75-gallon tank, the difference equation became $s_{n+1} = .1333s_n$. The functional equation now gives $s_n = 50(.1333)^n$. After 10 drain and refill cycles, the amount of salt in the tank will be $s_{10} = 50(.1333)^{10}$, or approximately .0000000886. This is easily computed on a calculator.[2]

As in earlier discussions, we can also ask questions that require inverting the functional equation. For example, suppose that the water in the tank is going to be used for some purpose, and we determine that the amount of salt has to be reduced to a level of less than .001 pounds per gallon. That means we want less than .1 pounds in the one hundred gallon tank. How many times must the tank be emptied and refilled? This corresponds to the question, what is the smallest n for which $s_n < .1$? Instead of giving the value of n and trying to find the value of s_n, which is what the functional equation is designed for, we now want to reverse the process. We want to specify the value of s_n and try to determine n. At this point in the course, all we can do is apply a numerical method. By systematic trial and error, we can find the value of n. Using the original model with a growth factor of .1, compute $50(.1^n)$ for several different n's, looking for the first time that the result is less than .1. For $n = 2$ we have $s_2 = 50(.01) = .5$, and that is too high. With $n = 3$ the result is $s_3 = .05$, which is acceptable. So we have to empty and refill the tank 3 times. Later we will see theoretical methods for dealing with this kind of problem.

Exponential Functions

In the discussion of linear growth, once the functional equation was found, we extended our model to a continuous time variable. Would that make sense here? For the mouse population model, for example, we have the functional equation $p_n = 100(1.5)^n$, where p_n is the population after n months. This makes sense for n a whole number. Can we set $n = 3.5$ to find the population after three and a half months? As before, when we start to use fractional values for n, we will change the notation from p_n to $p(n)$. So, what is $p(3.5)$? According to the functional equation, the result should be $100(1.5)^{3.5}$. This doesn't make much sense if we think of the exponent as a shorthand for repeated multiplication. $(1.5)^3$ actually means $1.5 \times 1.5 \times 1.5$, and $(1.5)^4$ means $1.5 \times 1.5 \times 1.5 \times 1.5$. What does an exponent of 3.5 mean? We can't multiply 1.5 by itself 3.5 times. On the other hand, if your calculator has an exponent key, you can have it compute $(1.5)^{3.5}$, and the result is 4.1335137. In fact, you can use the calculator to make a graph of the equation $p(n) = 100(1.5)^n$ using decimal values for n such as 1.1, 1.2, 1.3, and so on. The result will be a smooth curve that connects the points for whole number n's. For this reason, using the calculator to extend the functional equation to fractional values of n at least provides an appealing graph. But what does it mean?

---

[2] The answer may appear on the calculator in the form 8.86e-8. That is a shorthand way of writing .0000000886. To convert 8.86e-8 to a decimal expression, write down 8.86, and then move the decimal point 8 places to the left. Similarly, to convert 3.34e+5 to a decimal expression, write down 3.34 and move the decimal point 5 places to the *right*.

Just as the linear growth difference equation leads to the more general idea of proportional reasoning, so the geometric growth difference equation leads to a more general kind of reasoning. Initially, geometric growth was defined in terms of time periods of a specific length, say one month, or one week. The more general idea can be stated as follows:

> In any two periods of time of equal length, the growth factor will be the same.

This statement says, for example, in any two one-month periods the growth factor will be the same. But it says more. For example, it implies that in any two one-half-month periods the growth factor must be the same. This leads to a continuous model for growth.

Here is a sample of how this reasoning works. In the mouse population model, we have $p(3) = 100(1.5)^3 = 337.5$. Similarly $p(4) = 100(1.5)^4 = 506.25$. Now let's look at the population after 3.5 weeks. The functional equation is $p(3.5) = 100(1.5)^{3.5}$, and the calculator gives this as 413.35137. If this is correct, we should have the same growth factor in the first half-month (from $n = 3$ to $n = 3.5$) as we do in the second half-month (from $n = 3.5$ to $n = 4$.) Geometric growth says that in any two periods of equal length the growth factors should be equal. For the first half-month, the growth factor is $p(3.5)/p(3) = 413.35137/337.5 = 1.2247448$. For the second half-month the growth factor is $p(4)/p(3.5) = 506.25/413.35137 = 1.2247449$. These are very nearly equal. In fact, the exponent operation on the calculator is specifically programmed so that fractional values make sense for geometric growth models in just this way.

Let's explore this idea in a slightly different way, using a theoretical approach. The growth factor for one month is 1.5. What should the growth factor be for one half a month? Would you guess $1.5/2 = .75$? That is wrong. If we had a growth factor of .75 in the first half-month and again in the second, then the original population of 100 would go from 100 to 75 in the first half-month, and from 75 to $75 \times .75 = 56.25$ in the second half-month. That doesn't fit the data. We know that after two half-months the population should be 150. In order to find the correct growth factor for half a month, let us use a variable r. Starting from 100, after one half-month, the population would be $100r$. In the next half-month we multiply by the same growth factor of r. This gives us $100r^2$ as the population at the end of a month. But we already know that this should be 150. That is, we know that $100r^2 = 150$ should hold. This is a quadratic equation, and we can solve it for r. Dividing both sides of the equation by 100 gives us $r^2 = 1.5$, so we see that $r = \sqrt{1.5}$. Observe that we have seen the following: If r is the growth factor for one half-month, then r^2 must equal the growth factor for a month. In a similar fashion, you can see that if s is the growth factor for a third of a month, then s^3 must equal 1.5, the growth factor for a month. If z is the growth factor for a tenth of a month, then z^{10} must equal 1.5, the growth factor for a month. Using this reasoning, we can figure out what the population should be for any fraction of a month. The power key on the calculator does just that kind of calculation to compute fractional powers. In particular, the .5 power means the same as square root, the $1/3$ power means the same as cube root, and so on. Also, note the way that polynomial equations have popped up again in an unexpected context—geometric growth.

The foregoing discussion shows that an equation like $p(n) = 100(1.5)^n$ can be interpreted for fractional values of n. It is more customary to use the variable t in this setting. Whereas n is a *number* of months, and thought of as a whole number, t is a continuous measure of time, given in units of months. Accordingly, we write the equation

$$p(t) = 100(1.5)^t$$

to express the population as a function of time. With this equation, given any value for the time, we can easily compute the predicted population size using a calculator. Here the time variable is actually a measure of elapsed time from the starting point of the model, at which time $t = 0$.

In general, for a geometric growth model with an initial value of a_0 and a growth factor of r (for one unit of time), the continuous model is given by

$$a(t) = a_0 r^t$$

Functions of this form are called exponential functions, because the variable is in the exponent. In the next chapter you will learn more about the properties of these functions.

As one variation on this subject, there are situations in which the growth factor is not given for one unit of time. For example, in a previous discussion we considered an experiment in which we found the growth factor was 2.8 for a 6-week period. If we use the equation $a_n = a_0(2.8)^n$, the variable n counts six-week periods. With $n = 3$, for instance, we are talking about the population after 3 6-week periods. If we wish to express this as a continuous model, it would be more natural to use the variable t as the time in weeks rather than the time in units of 6 weeks. The correct formulation for this model would be

$$a(t) = a_0(2.8)^{t/6}$$

Notice how the length of the period, 6 weeks, appears in the exponent. This works out correctly, because when $t = 6$, the exponent on 2.8 is 1. This gives us the desired growth factor of 2.8 for a 6-week period. Similarly, if t is any multiple of 6 weeks, say 12, or 18, when we substitute t into the formula we end up with a whole number of 6-week periods. For $t = 12$ the exponent is 2, for $t = 18$ the exponent is 3, and so on. We summarize this with a new boxed comment:

> In a continuous model for geometric growth, suppose the variable a is a_0 at time $t = 0$, and suppose that over d units of time there is a growth factor of r. Then $a(t) = a_0 r^{t/d}$.

As an illustration, in the example we have been discussing a population grows by a factor of $r = 2.8$ in a time period of length $d = 6$ weeks. Substituting those figures into the equation in the box gives

$$a(t) = a_0(2.8)^{t/6}$$

with the variable t representing the number of weeks from the starting point for the model.

This completes the discussion of the functional equations that arise out of geometric growth models. We conclude the chapter with a discussion of a few more applications of geometric growth.

Additional Applications

Compound Interest. One very important application of geometric growth is to the calculation of compound interest. Here, we are in a slightly different modeling context. Previously, we imagined some natural process which produced data, and we wanted to use models that fit the data. However, for compound interest, there is no natural process at work. Rather, a model is adopted by financial institutions, and used as a basis for interest computations. There is no question whether the model is correct or incorrect. We simply have to learn to make calculations that are of interest to us, based on the model that has been adopted.

Most banks make loans and maintain savings plans based on the idea of compound interest. This is defined in terms of a geometric growth model. In a specified period of time, the amount owed on a loan, or the balance of an account, grows by a fixed percentage. If there are no payments into the account or on the loan, the account balance grows geometrically. For example, suppose the bank pays 1% interest on its accounts each month. That means that each month the amount in the account increases by 1%. This corresponds to a growth factor of 1.01. If the account is started at $500, and no further deposits or withdrawals are made, the balance will increase at the rate of 1% per month. Our knowledge of geometric growth now lets us write immediately

$$a_n = 500(1.01)^n$$

for the amount in the account after n months. Usually, the bank will not advertise the amount of interest on a per-month basis. Instead, it will display an annual interest rate. This is really only a nominal rate. You must divide it by the number of interest payments per year to find the percentage by which the account grows with each interest period. So, if the nominal rate is 8% and the interest payments are made monthly, each interest payment is actually one-twelfth of 8 percent, which is computed as $(8/12)$%, or about .006667. Then each month the growth factor is 1.006667. Similarly, if the interest is paid daily, say 365 days per year, then each day you would receive $(8/365)$% in interest. The annual rate is not a true measure of how much interest you would earn over a year. It actually indicates how the account would grow if a linear growth model were used. To find your true earnings over a year, use a geometric model to compute the amount in the account after one year. Using that amount, compute a growth factor for the year, and that will lead in turn to a percentage of growth. Let's do this for the example above. We start with $500. The variable n counts the number of months. So after one year, $n = 12$. The amount in the account is then $a_{12} = 500(1.01)^{12} = 500 \times 1.127$ (approximately). Now this is a growth factor of 1.127, so the percentage of increase is .127. That is 12.7 percent. So although the annual rate is expressed as 12%, what you actually earn over a year is 12.7%. This actual figure is called the *yield*, or the *effective rate*.

Tank Models. The tank model leads to several other applications. Here are two. Suppose a lake contains a certain amount of pollution. The total number of gallons of water in the lake is known. Also, we know how many gallons of clean water flow into the lake from rivers per day, and the same number of gallons of water flow out per day. Over the course of the day, the lake is like the water tank. A fraction of the polluted water flows out and is replaced by clean water. As a result, each day the pollution in the lake is reduced by a fixed fraction. That is, the pollution is flushed from the lake in the same way that the salt was flushed from the tank. This leads to a geometric growth model for the amount of pollution remaining in the lake after n days. This example is further explored in the exercises.

Another variation on the tank model concerns the way the human body removes a drug that is present in the blood. As before, it is reasonable to assume that each hour a certain percentage of the medication is removed. This leads to a geometric growth model for the amount of medicine in the blood. A further variation is to consider the effect of periodic doses of medication. By studying this kind of model it is possible to figure out what dose to administer each time in order to maintain a desired level of medication in the blood stream. This application is also discussed further in the exercises.

Radioactive Decay. Another application that is quite accurate involves atomic radiation. Some elements, such as uranium, occur in nature in several different forms, called isotopes, some of which are radioactive. As a radioactive isotope gives off radiation, it turns into either another isotope, or another element. For this application, we model just the amount of the radioactive isotope, even though it does not exist in a pure form. It has been found that each radioactive isotope follows its own particular geometric growth pattern. In each case, some fixed percentage of the material is transformed to something else in each unit of time. That means the amount of the radioactive element left after each unit of time is a fixed percentage of what was there at the start of the unit of time. This is geometric growth. Because the amount is actually getting smaller over time, this is usually referred to as decay rather than growth.

Empirical study can be performed to find out how fast each element decays. This is commonly expressed by giving the half-life of the element. For example, the half-life of uranium 237 is 6.75 days. This means that an initial amount of the substance will be half gone after 6.75 days. Suppose you have an ore sample made up of a mixture of uranium 237 and other elements. Say it contains 100 grams of uranium 237 when it is first measured. Then after 6.75 days the sample will contain only 50 grams of uranium 237. The total weight of the ore sample will be essentially the same as it was originally. The 50 grams of uranium 237 doesn't disappear, it just changes into something else. But the total amount of uranium 237 within the sample of ore will decrease from 100 grams to 50 grams in 6.75 days. Note that we can immediately write down an equation for the amount left after t days, because we know the growth factor is $r = .5$ in a period of $d = 6.75$ days. Using the boxed statement on page 195, the amount after t days will be $100(.5)^{t/6.75}$.

Geometric growth models can be used to analyze radioactivity. For example, the Chernobyl incident in the former Soviet Union released radioactive elements into the environment. These will continue to pose a threat until the amount of radioactive isotopes is reduced to a safe level. Once the current amount of contamination is known, we can figure out how long it will be before the radiation dissipates to a safe level, using a geometric decay model. Notice that this involves inverting the functional equation: we know how much is a safe level; we want to know when that level will be reached.

Population Growth.   As a model for population growth, geometric growth models are only accurate for limited periods of time. As we have seen, these models involve rapid acceleration of growth, and in any biological application, sooner or later something will act to limit the growth. Nevertheless, there are ways to modify the geometric growth model to make it more true to life. One very important example of this is called a logistic model. We will return to that topic in a future chapter. Interestingly, although logistic models can be very useful, they also lead to chaos in some situations.

Heat Transfer.   As a final example, geometric growth models can be applied to questions concerning the way heat flows from hot to cold objects. Whenever you see someone in a movie estimate the time of a death from the temperature of a body, it is based on a geometric growth model. There are many variations on this subject. For this discussion, we will use a single special context.

Imagine a building in which the temperature is at 70 degrees. The heat is turned off and left off, and the outside temperature is at 0 degrees. We will model the way the temperature in the building cools off. Imagine the building is filled with a large number of air molecules. For simplicity, let us say 1 million. Over time some of the warm air in the building leaks out, and is replaced by cold air from outside the building leaking in. We will assume that this is the only effect that causes the building to cool off. Furthermore, we will assume that each molecule in the building is at 70 degrees, and that each molecule outside the building is at 0 degrees. Finally, we assume that the number of molecules that leak out in any period of time, say an hour, is constant. For the sake of discussion, let us say that 10,000 molecules leak out in an hour. The question is this: if 10,000 warm molecules are replaced by an equal number of cold molecules, what happens to the average temperature inside the building?

Originally there are 1,000,000 particles all at 70 degrees. After an hour, there are 990,000 warm molecules and 10,000 cold molecules. So let's add up all their temperatures and divide by the number of molecules, 1,000,000. For the 10,000 molecules at 0 degrees, the temperatures add up to a total of 0; for 990,000 molecules at 70 degrees, the temperatures add up to $70 \times 990,000$. So the average is

$$\frac{70 \times 990,000}{1,000,000} = 70 \times .99$$

Now if you repeat this calculation assuming that the temperature inside the building is 50 degrees instead of 70, you will see that the average temperature after one hour will be $50 \times .99$. More generally, whatever temperature you start with, after one hour the average temperature of the molecules in the building will be .99 times the starting temperature. That is, the temperature inside the building is subject to geometric growth with a growth factor of .99 for each hour.

In our example, the temperature in the building was cooling down to the temperature of the surroundings, namely 0 degrees. In general, it is the difference between an object's temperature and the surrounding temperature that follows a geometric growth model. So if the temperature outside the building is 20 degrees, then the building starts out 50 degrees warmer than the outside. After an hour it will be $.99 \cdot 50 = 49.5$ degrees warmer than outside. Another hour reduces this to $.99 \cdot 49.5 = 49.005$. To use this model, it is most natural to define the variable d_n to be the difference between the temperatures of the building and the surroundings after n hours. Then, if you know that $d_n = 30$ and it is 20 degrees outside, that means the building temperature is $50°$. One of the exercises uses this kind of model to describe the way a car cools off after it has been driven.

This example illustrates again, in a similar way to the water tank example, that simple models for physical processes can often lead to geometric growth. In this case, we did not start with the assumption that the temperature would be reduced by an equal percentage in any hour. Rather, we started by imagining some simple explanation for the way the temperature goes down. That led to a geometric growth type of difference equation, so we conclude that geometric growth might be a good model. The final step would be to test the model under actual conditions with real buildings. It has been found that a geometric growth law gives a very good model for heat transfer in many situations.

Summary

This chapter has presented the idea of geometric growth models. These can be compared to linear growth in two ways. First, in linear growth, something of interest increases or decreases by a fixed amount in a given period of time. For geometric growth, the variable of interest increases or decreases by a fixed *percentage* in a given period of time.

Alternatively, geometric growth can be explained in terms of growth factors. A growth factor is a number you multiply one data value by to obtain the next data value. Geometric growth occurs when a variable of interest increases or decreases by a fixed growth factor in a given period of time. This leads to the second comparison with arithmetic growth. In an arithmetic growth model, we compute each new data value by *adding* a constant to the preceding value. In geometric growth we compute each new data value by *multiplying* a constant times the preceding data value.

In all cases of geometric growth, there is a difference equation that governs how the variable changes. The difference equation always has a standard form: $a_{n+1} = ra_n$ for some constant r. The numerical and graphical properties of these kinds of difference equations were discussed. For $r > 1$, the difference equation leads to larger and larger values, and has a graph that slopes up to the right. For $r < 1$ the values get smaller and smaller. In this case the graphs slope up to the left and level off along the horizontal axis to the right. There is a symmetry between these two situations. If the data in a geometric growth model with growth factor r are reversed, the result is the same as a growth model with the reciprocal growth factor $1/r$.

There is a standard functional equation for geometric growth as well. It has the form $a_n = a_0 r^n$. Although this is defined in terms of whole number values of n, there is a way to extend the idea of geometric growth to fractional values as well. This leads to a model with a continuous time variable. The functional equation that occurs for this situation is called an exponential function, and will be studied in a later chapter.

Geometric growth models have a large number of applications. In this chapter, some of these applications were presented, including compound interest, dissipation of pollutants in a lake, absorption of medicine from the blood stream, population growth, radioactive decay, and heat transfer.

Exercises

Reading Comprehension

1. Write brief paragraphs explaining the meaning of each of the following:
   a. geometric growth
   b. fixed percentage of growth
   c. growth factor
   d. exponential function
   e. compound interest
   f. half-life

2. Explain the difference between an amount of change and a percentage of change. How does this difference relate to the concepts of arithmetic and geometric growth?

3. In the first and second definitions of geometric growth, the phrase *period of a given length* appears. What does that mean?

4. What is the general form of the difference equation that describes geometric growth? Explain why this equation is correct.

5. What is the general form of a functional equation for geometric growth? Explain why it is correct.

6. Give a brief qualitative description of the graphs that occur in geometric growth. In your answer, comment on the significance of the growth factor in determining the shape of the graph.

7. Suppose a population of mice is found to grow by a factor of 16 in one year. Assuming geometric growth, what should the growth factor be in $1/2$ of a year? Why?

8. Continuing the preceding problem, what should the growth factor be in $1/4$ of a year? Why?

9. Continuing the preceding problem, suppose the functional equation for the mouse population model is $p_n = 1,000 \cdot 16^n$. What does the calculator give you for the population if you set $n = 1/2$? How does that relate to question 7?

10. Continuing the preceding problem, what does the calculator give you for the population if you set $n = 1/4$? How does that relate to question 8?

11. Explain briefly how the calculator programming for fractional exponents is related to geometric growth models.

12. In advertisements, banks often list interest rates for loans and savings accounts in terms of some number of percent per year. Write a short paragraph explaining how these annual rates are used to actually figure out interest payments. Include in your answer an explanation of *effective rate* and *yield*.

Mathematical Skills

1. The following numbers follow a geometric growth law: 3.2, 4.8, 7.2, 10.8. What is the growth factor?

2. What is the percentage of increase for the preceding problem?

3. The following numbers follow a geometric growth law: 1.25, .25, .05, .01. What is the growth factor?

4. In the preceding problem, each number is a fixed percentage less than the number before it. What is the fixed percentage?

5. Do these numbers follow a geometric growth law? 1.4641, 1.331, 1.21, 1.1. How can you tell?

6. Do these numbers follow a geometric growth law? 1.08, 1.13, 1.18, 1.23. How can you tell?

7. Do these numbers follow a geometric growth law? 1/3, 1/4, 1/5, 1/6, 1/7. How can you tell?

8. Do these numbers follow a geometric growth law? 1/4, 1/8, 1/16, 1/32, 1/64. How can you tell?

9. A number sequence follows a geometric growth law with a percentage increase of 72%. What is the growth factor?

10. A number sequence follows a geometric growth law with a percentage increase of 18%. What is the growth factor?

11. A number sequence follows a geometric growth law with a growth factor of 1.08. What is the percentage of increase?

12. A number sequence follows a geometric growth law with a growth factor of 3.2. What is the percentage of increase?

13. A number sequence starts out with $a_0 = 14$ and satisfies the difference equation $a_{n+1} = .3a_n$. What is the functional equation? What is a_{12}?

14. A number sequence starts out with $g_0 = 14$ and satisfies the difference equation $g_{n+1} = 1.4g_n$. What is the functional equation? What is g_{12}?

15. Suppose a number sequence has the functional equation $a_n = 4 \cdot 1.3^n$. What is the difference equation for this number sequence?

16. Suppose a number sequence has the functional equation $a_n = 2.84 \cdot 16^n$. What is the growth factor for this sequence?

17. A number sequence starts out with $a_0 = 1,000$ and has the difference equation $a_{n+1} = 1.08a_n$. Use the difference equation to find a_5. Use a functional equation to find a_5.

18. Continuing the preceding problem, use a numerical approach to figure out which a_n equals 2,000. You may use either the difference equation or the functional equation.

19. A geometric growth model is used to describe the amount of pollution in a lake. The functional equation is $p(t) = 4.23(.87)^t$ where t is the time, in years, from the start of the model. What is the level of pollution 18 months after the start of the model? 3 months?

20. In a continuous model for geometric growth, the variable s starts out at 7.56 at time 0, and grows by a factor of 3.12 over the next 14 hours. What is the equation that gives $s(t)$ where t is the time measured in hours? (see page 195)

21. In a continuous model for geometric growth, the variable h starts out at 1.078 at time 0, and decays by a factor of .88 over the next 3 days. What is the equation that gives $h(t)$ where t is the time measured in days?

Problems in Context

1. An article in the Washington Post, 10/18/94, concerns a surgical procedure called radial keratotomy (RK), which is supposed to reduce or cure nearsightedness. According to the article, the number of patients who have had this surgery each year has "grown exponentially in the past five years." (That is the same as saying that the number of patients has grown geometrically.) The article reports that 30,000 surgeries were performed five years ago, and that 250,000 will be performed this year. Use your knowledge of geometric growth to analyze this situation. Find the number of patients in each of the past five years, and project for the next 5 years. Find a difference equation and a functional equation that goes with your model. Draw a graph showing the number of RK procedures performed each year. The article says that one in four Americans is nearsighted. If there are 280 million people in the US, how long will it be, according to your model, before the potential market for RK surgeries is used up?

2. Geometric growth is often used in population models. In Table 9.1 the population of England and Wales is shown as reported in census data from 1801 to 1911.[3]

    a.   Compute a growth factor for each pair of population values. For example, from 1801 to 1811 the growth factor was $10.16/8.89 = 1.14$. Do the same for the growth from 1811 to 1821, 1821 to 1831, and so on.

    b.   Do the growth factors seem pretty constant? What would you say is a good average value for the growth factor?

    c.   By what percentage did the population grow each decade in the data?

---

[3] Taken from *Exploratory Data Analysis* by John W. Tuckey, Addison–Wesley, Reading, Mass, 1977, page 138

Year	1801	1811	1821	1831
Pop.	8.89	10.16	12.00	13.9

Year	1841	1851	1861	1871
Pop.	15.91	17.93	20.07	22.71

Year	1881	1891	1901	1911
Pop.	25.97	29.00	32.53	36.07

TABLE 9.1
England and Wales Population in Millions

d.   Define a geometric model for the population growth, using your average growth factor, and the starting population of 8.89. How closely does the model fit the data? Look at the errors graphically by plotting the true data and the model data on one graph. Look at the errors numerically by computing the difference between each true data value and the corresponding model data value. What is the maximum error?

3. A young couple receives a $10,000 inheritance from the will of a grandparent. They deposit the money in a bank account that advertises 8.3 percent interest per year.

   a.   If the interest is paid monthly, how much will the couple have after 3 years?

   b.   What about if the interest is paid daily?

   c.   Which would be a better deal for this couple: an account that pays 8.3 percent per year, paid daily; or one that pays 8.5 percent per year, paid only monthly?

   d.   The couple wants to save up enough money for a down payment on a house. They figure they will need $18,000 for the down payment. How long will it take for the original $10,000 to gather enough interest to give them a total of $18,000?

4. One of the major concerns about above-ground nuclear testing was that it produced a radioactive element called strontium 90. The fallout from a nuclear test would be deposited on grass, which would be eaten by cows, and the strontium 90 would get into the milk the cows produced. Scientists have determined that the half-life of strontium 90 is 28.1 years. Suppose that a single above-ground test causes the level of strontium 90 to be 10 times greater than the maximum safe level in one agricultural area. Use a geometric growth model to figure out how long it will take before the level of strontium 90 is again at a safe level. For simplicity, let the safe level be the base unit for measurement. Then after the test, the level of strontium 90 is 10 and the safe level you want to reach is 1 or below. Using the boxed statement on page 195, find an equation for the strontium 90 level t years after the above-ground test. Look at a graph to get a rough idea of how long it will take for the strontium 90 to get down to a level of about 1. Use a numerical method to get a more accurate answer.

5. In this problem you will use a model similar to the one for a water tank in connection with pollution in the great lakes[4]. It has been determined that approximately 38 percent of the water in Lake Erie is replaced each year. That is, in the course of a year, an amount of water flows out of the lake and is replaced by an equal amount that enters the lake; the amount that is replaced in this way is 38 percent of the total water in the lake. Suppose that we could magically stop all pollution from entering the lake. How long would it be before the pollution in the lake was reduced to 10 percent of the current level? To 1 percent? To answer these questions, we start with a difference equation. As in the water tank model, we imagine that 38 percent of the pollution in the lake drains out with the water that leaves the lake each year. That leaves 62 percent of the pollution in the lake. The difference equation is then

$$p_{n+1} = .62p_n$$

where p_n is the total pollution in the lake n years after the start of the model. We also need a starting value, p_0, the total amount of pollution initially in the lake. This is probably thousands of tons, but for ease of reference, we will make up a new unit: eries. We will say that the total amount of pollution in the lake right now is one erie. We want to know when it will be one-tenth as much, or .1 eries. We also want to know how long it will take to get down to 1 percent of the starting amount, or .01 eries. To answer these questions, follow this outline:

   a. Use the information $p_{n+1} = .62p_n$ and $p_0 = 1$ to compute p_1 through p_5, to get a feel for the model.

   b. Use the computer to graph the data and see whether you can find when the value gets to .1 approximately. When it gets to .01 approximately.

   c. Find the functional equation for p_n in this model.

   d. Use the functional equation and a numerical approach to get a better estimate of your answers.

6. When a drug is introduced into the blood stream, the body has mechanisms to eliminate it. For example, some drugs are removed by the kidneys. How does the amount of the drug in the blood diminish over time? A commonly used model to describe this process starts with the assumption that a fixed percentage of the drug is removed every so many hours[5]. For example, in the case of aspirin, about half is removed from the blood every half hour. If you take two aspirins, that is 750 milligrams (abbreviated mg). Let a_n be the amount of aspirin in your blood after n half hour periods. Then the difference equation for a_n is

$$a_{n+1} = .5a_n; \qquad a_0 = 750$$

---

[4] This is a simplified version of a more involved problem in *Discrete Dynamical Modeling*, by James T. Sandfur, Oxford University Press, New York, 1993. See Example 4.2, pp. 152–154.

[5] This is really an extension of the water tank model. We imagine that in each period of time, some fraction of the blood is purified, just as in the water tank model, each time we empty and refill the tank we purify a fixed fraction of the water.

How much aspirin remains in the blood after 4 hours? What is a functional equation that gives the amount in the blood after t hours? (Hint: Use the boxed comment on page 195.)

7. This problem is closely related to the preceding problem. At the Olympics, drug tests are used to check for performance enhancing drugs. These tests have limited sensitivity; there must be some minimum amount of the drug in the blood in order for the test to detect it. Suppose that a test can detect steroids in amounts of 1 mg or more. Suppose also that the body removes about $1/4$ of the steroids in the blood every 4 hours. And finally, suppose that an athlete takes a 16mg dose. How long will it be before the drug is reduced to an undetectable level in the blood? That is, if the athlete takes the blood test immediately after taking the drug, there will be 16mg in the blood, and the test will detect that. But the longer the delay between taking the drug and the blood test, the less of the drug will remain in the blood. Eventually there will be less than 1mg of the drug left in the blood, and the test will not be able to detect that. The question is, how long will that take? As a second question: Suppose we can make the blood test 100 times more effective, so that it can detect steroids in the amount of .01mg or more. Then how long after taking the dose would the test be effective? For this second question, you should form the functional equation that gives the amount of steroids in the blood t hours after taking the drug, and use a graphical and/or numerical approach.

8. This problem is also closely related to the two previous problems. As you are aware, it is very common to take medicine on a regular basis. For example, you may take two aspirin every four hours. The model for the way drugs are removed from the body can be modified to include the effect of taking additional doses. Suppose that $1/4$ of a drug is removed from the blood every four hours. An initial dose of 100mg is taken. Four hours later, one quarter of the drug has been eliminated, leaving 75mg. At that point another dose of 100 mg is added, giving a total of 175mg in the blood. After four more hours, one quarter of that amount is removed, and an additional 100 mg is added. This leads to the following difference equation:

$$d_{n+1} = .75d_n + 100; \qquad d_1 = 100$$

where d_n is the amount of drug in the body after taking n doses. What will be the long-term effect of taking repeated doses? Will the drug level keep going up until it reaches an unsafe level? Use a numerical method to explore the behavior of d_n in this model. What happens if each dose is 200 mg rather than 100 mg? What if each dose is 50 mg? If the doctor would like to keep the amount of drug in the blood at about 150 mg, how much medicine should be given at each dose?

9. When a car is running, the engine temperature remains fairly constant at about 300 degrees. Once the car is parked, the engine will cool off. A detective wants to use this information to predict how long it has been since a suspect's car was last driven. When the detective got to the suspect's house, at 2 p.m., she checked the temperature of the radiator fluid, and found it to be 270 degrees. At 3 p.m., the temperature was measured again, this time it was 90 degrees. Also, the outside

temperature was around 70 degrees all afternoon. Using this information, create a model for temperature as a function of time, and estimate how long the car had been parked when the detective made the measurement at 2 p.m. The following steps will help.

   a.   Let d_n be the difference between the car temperature and the outside temperature n hours after 2 p.m. What is d_0? What is d_1?

   b.   Assume that the d_n's follow a geometric growth law. What is the growth (actually, reduction) factor from d_0 to d_1?

   c.   What is the difference equation for d_n?

   d.   What is the functional equation that gives d_n as a function of n?

   e.   Rewrite the equation from the previous question using t as the variable (instead of n) and in the form used for a continuous model. This will allow you to determine the value of d at any time t, measured in hours after 2 p.m.

   f.   When the car was parked, the temperature was around 300 degrees. What was d at that time?

   g.   For the value of d in the preceding question, find the time t using the equation you found in part e.

   h.   How long had the car been parked when the detective measured the temperature at 2 p.m.? (Hints: use a numerical and/or graphical approach; expect the answer to be a negative number because we are interested in something that happened *before* 2 p.m.)

Solutions to Selected Exercises

Reading Comprehension

7. The growth factor for $1/2$ a year should be 4. To see why, suppose the starting population is 100, and suppose the growth factor for each half-year is 4. Then starting at 100, after half a year the population would be $100 \cdot 4 = 400$, and after a second half a year the population would grow to $(100 \cdot 4) \cdot 4 = 100 \cdot 16 = 1,600$. That shows that the growth factor for a full year would be $4^2 = 16$, and that is what the growth factor for the year was given as in the problem.

8. Using a similar argument, the growth factor for $1/4$ year should be 2. That means the population would double every $1/4$ year. To see that this is right, just check what happens after a full year: it doubles in the first quarter, doubles again in the second, again in the third, and again in the fourth. That is, after a year, the population has grown by a factor of $2 \cdot 2 \cdot 2 \cdot 2 = 2^4 = 16$.

9. With $n = 1/2$ the equation says $p_{1/2} = 1,000 \cdot 16^{1/2}$. The calculator computes $16^{1/2}$ to be 4. So the result will be $p_{1/2} = 4,000$. That is the same answer that would have been found using the answer from problem 7.

10. The calculator will give $p = 2,000$ if $n = 1/4$. This is the same answer that would be found using the growth factor from problem 8.

11. In a geometric growth model, the functional equation will have the form $a_n = a_0 r^n$. If fractional values are used for n, the calculator can be used to compute an answer. That answer goes along with the idea of geometric growth. That is, if data values are computed using fractional exponents on the calculator, those values will adhere to assumption of equal growth factors in equal time periods.

Mathematical Skills

1. Since $4.8/3.2 = 7.2/4.8 = 10.8/7.2 = 1.5$, the growth factor is 1.5.

2. The growth factor is 1.5. To find the percent increase, subtract 1 to get .5, and express in percent as 50%. That is, each number is 50% more than the preceding number.

3. Since $.25/1.25 = .05/.25 = .01/.05 = .2$, the growth factor is .2.

4. With a growth factor of .2, the percentage decrease is $1 - .2 = .8 = 80\%$

5. Compute growth factors: $1.331/1.4641 = .909090\cdots$, $1.21/1.331 = .909090\cdots$, $1.1/1.21 = .909090\cdots$. Since these are all equal, it *is* geometric growth.

6. No. Unequal growth factors

7. No. Unequal growth factors

9. Growth factor $= 1$ plus percentage, so $1 + 72\% = 1 + .72 = 1.72$

11. Growth factor $= 1$ plus percentage, so $1.08 = 1 +$ percentage. That shows that the percentage is .08, or 8%.

13. $a_n = 14(.3^n)$, $a_{12} = 14(.3^{12}) = .0000074402$ which may be shown on the calculator as $7.4402E - 06$.

15. $a_{n+1} = 1.3a_n$

17. Using the difference equation:

$$
\begin{aligned}
a_0 &= 1,000 \\
a_1 &= 1.08 \cdot 1,000 & &= 1,080 \\
a_2 &= 1.08 \cdot 1,080 & &= 1,166.4 \\
a_3 &= 1.08 \cdot 1,166.4 & &= 1,259.712 \\
a_4 &= 1.08 \cdot 1,259.712 & &= 1,360.48896 \\
a_5 &= 1.08 \cdot 1,360.48896 & &= 1,469.32808
\end{aligned}
$$

Using the functional equation: $a_5 = 1,000(1.08^5) = 1,469.32808$.

18. By trial and error, you can find that $a_9 = 1,999$, and $a_{10} = 2,158$. So the closest a_n to 2,000, with n a whole number, is a_9.

19. The t is supposed to be the time in units of years; 18 months is 1 and a half years, so $t = 1.5$. The pollution level is $4.23(.87^{1.5}) = 3.43256877$. For the second part, 3 months is one quarter of a year, so $t = .25$. The pollution level is $4.23(.87^{.25}) = 4.0852645$.

20. $s(t) = 7.56(3.12^{t/14})$

21. $h(t) = 1.078(.88^{t/3})$

Problems in Context

1. We can make a model for the number of RK surgeries performed each year. Define s_n to be the number of surgeries (in thousands) performed in year n, where year 0 is 1994. Then $a_0 = 250$. Since this is going to be a geometric growth model, we will have a difference equation of the form $a_{n+1} = ra_n$ where r is the growth factor for 1 year. Now we can determine from the article that the growth factor for 5 years is $250/30 = 8.33333$. That means a_5 will be $250 \cdot 8.3333 = 2{,}083.3333$. Now we can figure out r in several ways. First, by trial and error. Guess $r = 1.2$ say. With that r, the next five years of the model would be $a_0 = 250$, $a_1 = 250 \cdot 1.2 = 300$, $a_2 = 360$, $a_3 = 432$, $a_4 = 518.4$, and $a_5 = 622.08$. That is too low, because we know that a_5 should be $2{,}083.33$. So now try a larger r. Using trial and error you can work out what r should be, approximately. An alternative approach is to leave r as a variable. We know that $a_5 = 250r^5$, but we also know that $a_5 = 2{,}083.33$. Combining these facts gives $250r^5 = 2{,}083.33$. This equation can be solved for r using graphic and numerical methods, or using algebra: $r^5 = 2{,}083.33/250 = 8.33333$ so r must be the fifth root of 8.33333. Finally, an alternative is to use the continuous model. The growth factor for 5 years is 8.33333 so the number of surgeries in year t will be $250(8.3333^{t/5})$. In particular, after 1 year, we would have $250(8.3333^{1/5})$ and that gives a growth factor for one year of $r = 8.33333^{1/5} = 1.52814214$. In any case, the functional equation for this model can be written either as $a_n = 250(1.52814214^n)$ or $a(t) = 250(8.3333^{t/5})$.

   With 280,000,000 people in the US, only one-fourth of them, or 70,000,000 people are nearsighted. In units of thousands, that would be 70,000. So we want to see when our model predicts that all of these people will have been RK-ed. One way to do this is just to compute a_n for many years into the future, and add up the number of surgeries each year as you go along. For example, $a_0 = 250$, and $a_1 = 382.03$, so that is a total of 632.03 for the first two years. Next $a_2 = 583.80$ so that brings the total to 1,215.83. Continuing in this fashion, you will soon reach $a_{11} = 26{,}530.2454$ and that will bring the total number of surgeries performed to 76,290.04. That is more than the 70,000 maximum so according to the model, the RK market will dry up within 11 years.

2. a. The growth factors are listed below:

$$1.1429 \quad 1.1811 \quad 1.1583 \quad 1.1446 \quad 1.1270 \quad 1.1194$$
$$1.1315 \quad 1.1435 \quad 1.1167 \quad 1.1217 \quad 1.1088$$

   b. They are pretty constant for the first 2 digits—1.1. The average of all the growth factors is 1.136.

   c. The percentage of growth is found by subtracting 1 from the growth factor, and then expressing the decimal as a percent. Here are the figures:

$$14.29 \quad 18.11 \quad 15.83 \quad 14.46 \quad 12.70 \quad 11.94$$
$$13.15 \quad 14.35 \quad 11.67 \quad 12.17 \quad 10.88$$

Looked at this way, the percent growth doesn't look all that constant, ranging from about 11 percent to almost 15 percent.

d. Letting a_n be the modeled population value for the nth data point, so a_0 goes with 1801, a_1 goes with 1811, and so on, the difference equation is $a_{n+1} = 1.136a_n$ with $a_0 = 8.89$. That gives the functional equation as $a_n = 8.89 \cdot 1.136^n$. The graph is shown in Fig. 9.6. In the graph, the individual data points are from the original data, and the curve is based on the model. Numerically, the largest error is 1.1145,

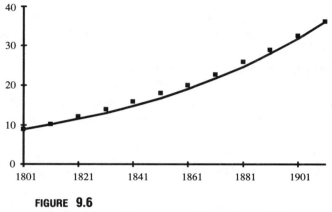

FIGURE 9.6
Wales Population Model: $a_0 = 8.89$, and $r = 1.136$

which means that out of about a population of 18 million, the model underestimated by about 1 million. Looking at the other errors, all but one of the model values are lower than the true value. This suggests modifying the model to raise the curve. In Fig. 9.7 a graph is shown for another model. In this model, $a_0 = 9.4$, the growth factor is 1.132, and the largest error was less than .7.

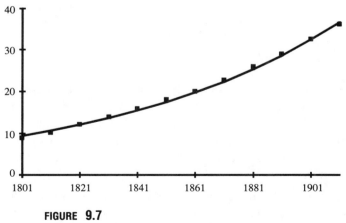

FIGURE 9.7
Wales Population Model: $a_0 = 9.4$, and $r = 1.132$

4. For this problem, there is a geometric growth model for the level of strontium 90. The idea of half-life is this: every 28.1 years the amount of strontium 90 will be reduced by a factor of .5. With a starting value of 10, that gives the following model for the amount of strontium 90 t years after the test:

$$s(t) = 10 \cdot (.5)^{t/28.1}$$

We want to find out when $s(t)$ will equal 1. That means on a graph with years on the x axis and strontium 90 level on the y axis, we want to see where the curve for $y = 10 \cdot (.5)^{x/28.1}$ crosses the horizontal line for $y = 1$. This is shown in Fig. 9.8. In the graph, it looks as if the curve crosses the line $y = 1$ at about $x = 90$. A numerical method can be used to find that the correct answer is actually about 93.35. To check, set $t = 93.35$, and compute $s(93.35) = 10 \cdot (.5)^{93.35/28.1} = .9999$. This shows that it takes over 93 years for the amount of strontium 90 to be reduced to a safe level from a starting amount 10 times that high.

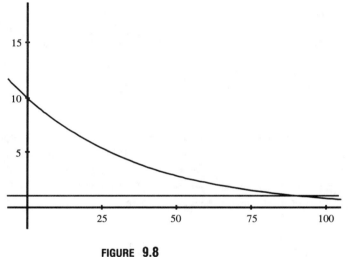

FIGURE 9.8
Graph of Strontium 90 Model

6. Four hours equals 8 half-hour periods, so the amount of aspirin in the blood after 4 hours is the same as a_8. The functional equation for this problem is $a_n = 750(.5)^n$ so $a_8 = 750(.5)^8 = 2.93$ (to two decimal places). The continuous model for this problem has the equation $a(t) = 750(.5)^{t/.5}$ with t in units of hours. Using this equation to check the previous answer, $a(4) = 750(.5)^{4/.5}$, which will give the same answer as before.

7. According to the model, if the original amount of drug is 16 mg, in four hours $1/4$ of the 16 mg, or 4 mg will be removed. That leaves $3/4$ of the 16 mg, or 12 mg. In general, in any four hours, the amount of drug in the blood will be multiplied by $3/4$. So if a_n is the amount of drug after n four-hour periods, then we have $a_{n+1} = (3/4)a_n$ or in decimal form, $a_{n+1} = .75a_n$. Using a numerical method,

start at 16 and keep multiplying by .75 to generate the data for this model: 16, 12, 9, 6.75, 5.06, 3.80, 2.85, 2.14, 1.60, 1.20, .90. These figures are rounded off to 2 decimal places, but the computations were made using the full accuracy of a calculator. After 10 four hour periods, the drug is below 1 mg, and so undetectable. That means it takes about 40 hours for the drug to become undetectable. To be more accurate, use a continuous model: $a(t) = 16(.75)^{t/4}$ and try to find t (in hours) that makes $a(t) = 1$. You know from what we did before that the time will be between 36 and 40 hours. The answer should be 38.55 correct to 2 decimal places.

8. Note: this is not geometric growth, and we don't have a functional equation. About all you can do is investigate numerically what the model does. Starting with $d_1 = 100$, and using the difference equation, each new data value is found by taking the preceding value, multiplying by .75, and then adding 100. Here is what the data look like for this model, rounded off to whole numbers: 100, 175, 231, 273, 305, 329, 347, 360, 370, 377, 383, 387, 390, 393, 395, 396, 397, 398, 398, 399, 399, 399, 399, 400. So the amount of drug slowly climbs up to about 400 and stays there. If the repeated drug dosage is 200 mg, rather than 100, that means we will start at 200, and to get each number we have to multiply by .75 and then add 200 (not 100.) This is what the numbers look like: 200, 350, 463, 547, 610, 658, 693, 720, 740, 755, 766, 775, 781, 786, 789, 792, 794, 795, 797, 797, 798, 799, 799, 799, 799, 800, 800. This time the model slowly climbs to 800, and then stays there. If the repeated drug dosage is 50 mg, rather than 100, that means we will start at 50, and to get each number we have to multiply by .75 and then add 50 (not 100). This is what the numbers look like: 50, 88, 116, 137, 153, 164, 173, 180, 185, 189, 192, 194, 195, 196, 197, 198, 198, 199, 199, 199, 200, 200. This time, the model slowly climbs to 200 and stays there. In each case, the model climbs up to and stays at a fixed level. What is more, that fixed level is 4 times the size of the repeated dose. If the doctor wants the drug to remain at about 150, we should use a dose of $150/4 = 37.5$. Then the steady level will be four times that amount, or 150.

9. a. At 2 p.m. the car temperature was 270, and the outside temperature was 70. The difference in temperature was $d_0 = 200$ degrees. For d_1 we use the car temperature at 3 p.m. and the outside temperature to find $d_1 = 90 - 70 = 20$ degrees.

   b. The growth factor is $20/200 = .1$.

   c. Difference equation: $d_{n+1} = .1d_n$

   d. Functional equation: $d_n = 200(.1)^n$

   e. $d(t) = 200(.1)^t$

   f. If the car temp was 300, then $d = 300 - 70 = 230$ degrees.

   g. Solve $230 = 200(.1)^t$. This can be rewritten in the form $1.15 = .1^t$ (by dividing both sides by 200) but we have not yet covered algebraic methods for solving this kind of equation. You should expect a negative value for t because we know at $t = 0$ that corresponds to 2 p.m., and we are looking for an earlier time, when the engine was hotter. By trial and error, you can find that t should be approximately $-.06$. Checking, $d(-.06) = 200(.1)^{-.06} = 229.63$ which is just about 230. So, the model says it was approximately .06 hours before 2 p.m. when the car was parked. In minutes, that is $.06(60 \text{ minutes}) = 3.6$ minutes. That means when the detective arrived, the car had only been standing for about 3 and a half minutes. So the suspect had just arrived home minutes before the detective.

10

Exponential Functions

In the preceding chapter, geometric growth models were introduced in connection with difference equations. In one example the growth of a population of mice was modeled, using p_n to stand for the population after n months. The difference equation

$$p_{n+1} = 1.5p_n$$

says that the population at the end of each month is one and a half times larger than the population at the end of the preceding month. That is a typical example of geometric growth.

The difference equation for the mouse population example leads to a functional equation for p_n as a function of n:

$$p_n = p_0(1.5)^n$$

With an initial population of 1,000, this equation becomes

$$p_n = 1,000(1.5)^n$$

This is an example of an exponential function. As you know, n is referred to as an exponent; 1.5 is called the *base*. In general, in an exponential function, the variable appears in the exponent attached to a constant base. Although this example defines n as a whole number of months, the functional equation can be used even for fractional values of n. Thus, the model predicts that the population 1.7 months after the start of the study will be $1,000(1.5)^{1.7}$. As this example illustrates, the functional equation allows us to think of time as a continuous variable, taking on all possible fractional values. As in previous chapters, we follow the notational convention of changing n to t and writing $p(t)$ instead of p_t. These notations emphasize the idea that the time variable takes on fractional values, and is not restricted to whole numbers. Accordingly, the functional equation becomes

$$p(t) = 1,000(1.5)^t$$

which we think of as defining p as an exponential function of the continuous variable t.

You have already learned about some of the numerical and graphical characteristics of exponential functions. This chapter continues your study of the subject, paying special attention to four topics:

- Graphs

- Algebraic Properties

- Solving Equations

- The Number e

Along the way we will introduce the idea of a logarithm.

Graphs

In the preceding chapter, graphs were shown for geometric growth models. For example, look back at Fig. 9.1. The functional equation for this graph is $p_n = 100(1.5^n)$, and data points are shown for n equal to each of the whole numbers from 0 to 9, with straight lines connecting the data points. Now we want to reexamine the graphs of this type using fractional and negative values for the variable. In fact, we want to imagine plotting points on the graph for fractional values of n that are so close together that they appear to form a continuous curve. We will use the conventional variables x for the horizontal axis, and y for the vertical axis. In the functional equation, the n will be changed to x and the p_n will be changed to y, resulting in $y = 100(1.5^x)$. This is one particular instance of an exponential function. The general form is $y = Ab^x$ where A and b stand for particular constants. In the example we have been discussing, $A = 100$ and $b = 1.5$.

A graph for the equation $y = 100(1.5^x)$ is shown in Fig. 10.1. If you compare this figure with Fig. 9.1, you will see that the two graphs are very much alike. One difference is that in Fig. 9.1, it is possible to make out corners between the short straight lines joining the plotted points, whereas in Fig. 10.1, the graph is a smooth curve. This is the result of using fractional values for the variable x. Another difference worth noting is the

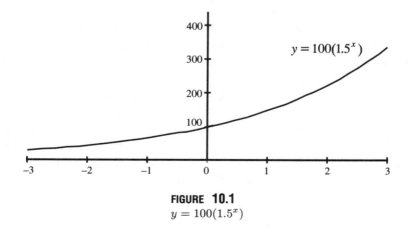

FIGURE 10.1
$y = 100(1.5^x)$

way Fig. 10.1 extends to the left of the y axis. This reflects the use of negative values for x. In spite of these differences, it is clear that the two graphs have the same general shape. In the next few paragraphs we will look at general characteristics shared by the graphs of all exponential functions. We will also see how to predict from the values of A and b what the general nature of a graph will be.

Every exponential function has the same general shape shown in Fig. 10.1: nearly horizontal on one side and curving steeply on the other. If the graph is extended far enough in the direction of the horizontal side, it will appear to lie right on top of the x axis. We saw this kind of situation earlier for rational functions having the x axis as an asymptote. In the same way, the x axis is always an asymptote for one side of an exponential graph.

The flat side of the curve may appear on either the left side or the right side, and the curve may either go steeply up or down. All of these variations are shown in Fig. 10.2. Notice that none of the variations crosses the x axis. This is another important feature of all exponential functions: they have no x–intercept. The entire graph has to stay above the x axis or below the x axis. As the figure indicates, the four different variations depend on the values of A and b in the equation $y = Ab^x$. If A is positive, the entire graph must stay above the x axis. It can be nearly horizontal on the left side and curve steeply up on the right if $b > 1$, or curve steeply on the left and become horizontal on the right if $b < 1$. When $A < 0$, the entire graph must stay below the x axis. Then, for $b > 1$ the curve is nearly horizontal on the left and curves steeply down at the right, while for $b < 1$ the situation is reversed. In all the examples in the preceding chapter, the value of A is positive, and all of those graphs have the same general shape as one of the first two in Fig. 10.2.

The value of A is the y–intercept of the equation $y = Ab^x$. As an example of this, the graph of $y = 100(1.5)^x$ has a y–intercept of 100. In terms of the geometric growth models presented in the previous chapter, this corresponds to the fact that A is the starting

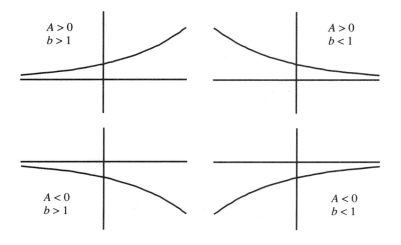

FIGURE **10.2**
General Shapes of Exponential Graphs

value for the model. Algebraically, we know that the y–intercept can be found by setting $x = 0$. In the example equation, this leads to $y = 100(1.5^0)$. But 1.5^0 is 1 (any positive base raised to the zero power is 1), so $y = 100$. This illustrates algebraically that the A in $y = Ab^x$ is the y–intercept. In most applications, A is positive. In this case the y–intercept is above the x axis, and the entire curve must stay above the x axis. If A is negative, the y–intercept is below the x axis, as is the entire curve.

As a final observation about the graphs of exponential functions, the value of b controls how steeply the curve rises or falls. This is what we saw in the discussion of geometric growth, where the base b is the same as the growth factor. Several different growth factors are illustrated in Fig. 9.2 and Fig. 9.5. In those figures, since only whole number values of n were used, the graphs are made up of connected straight lines. In Fig. 10.3 and Fig. 10.4, similar examples are shown using fractional and negative values of the variables. As before, the graphs appear as smooth curves. The point of these curves is to illustrate how the value of b affects the appearance of the graph. In all the curves illustrated, the value of A is 100, so they all have a y–intercept of 100. If some other

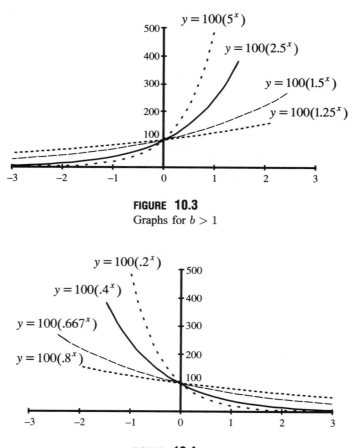

FIGURE 10.3
Graphs for $b > 1$

FIGURE 10.4
Graphs for $b < 1$

positive value of A had been used, the graphs would have looked much the same, differing mainly in the location of the y–intercepts.

This completes the discussion of graphs of exponential functions. The next topic concerns algebraic properties.

Algebraic Properties

In the equation

$$p(t) = 1,000(1.5)^t$$

the symbols on the right side of the equal sign provide a recipe for computing population. For example, if we want to know the population 3 months after the start of the study, the recipe says to raise 1.5 to the third power and then multiply the result by 1,000. There are other possible forms for the recipe. One example is $1,500(2.25)^{.5(t-1)}$. Although this looks quite different from $1,000(1.5)^t$, it will produce the same result for any value of t. You can see some evidence of this by using a calculator to compute both $1,000(1.5)^t$ and $1,500(2.25)^{.5(t-1)}$ for several values of t. Do this now, recording your results in Table 10.1.

Our discussion of algebraic properties of exponential functions will explain how to recognize when one form of an exponential function can be replaced by an equivalent form. As we shall see, there are situations in which one form is easier to use than another. The relationship between different forms will be important both in solving equations as well as in understanding features of a geometric growth model.

The algebra of exponential functions depends on what are usually referred to as rules of exponents. As an example, let us consider what will result from multiplying 2^5 by 2^8. Remember that 2^5 is really $2 \cdot 2 \cdot 2 \cdot 2 \cdot 2$. Similarly, $2^8 = 2 \cdot 2 \cdot 2 \cdot 2 \cdot 2 \cdot 2 \cdot 2 \cdot 2$. Therefore, $2^5 \cdot 2^8 = 2 \cdot 2 \cdot 2 \cdot 2 \cdot 2 \cdot 2 \cdot 2 \cdot 2 \cdot 2 \cdot 2 \cdot 2 \cdot 2 \cdot 2$, or 2^{13}. To express this in words, when we multiply together a group of five twos with another group of eight twos, the result is the same as multiplying together a group of 13 twos. In symbols, this can be expressed in the form of the equation $2^5 \cdot 2^8 = 2^{5+8}$. The same kind of reasoning can be applied for any base (in place of 2) and for any exponents (in place of 5 and 8). The

t	0	1	1.5			
$1,000(1.5)^t$	1,000					
$1,500(2.25)^{.5(t-1)}$	1,000					

TABLE 10.1
Two Versions of an Exponential Function

result, expressed in the equation

$$a^n a^m = a^{n+m}$$

is our first rule of exponents. It can be used with exponential functions as in the following example. In the expression $1{,}500(1.5)^t$ rewrite $1{,}500$ as $1{,}000 \cdot 1.5$. That gives us the alternate version $1{,}000 \cdot 1.5 \cdot 1.5^t$. Now apply the first rule of exponents to show that $1.5 \cdot 1.5^t = 1.5^{t+1}$ (using the fact that $1.5 = 1.5^1$). This leads to $1{,}500(1.5)^t = 1{,}000(1.5)^{t+1}$.

Here is another example of a rule of exponents. Raise 2 to the third power. Now take the result and raise that to the fifth power. What is the final result? In symbols, we wish to consider $(2^3)^5$. Now $2^3 = 2 \cdot 2 \cdot 2$. When we raise that to the fifth power, we multiply together 5 identical copies of the same thing. That gives $(2 \cdot 2 \cdot 2)(2 \cdot 2 \cdot 2)(2 \cdot 2)(2 \cdot 2 \cdot 2)(2 \cdot 2 \cdot 2)$. This is clearly the same as multiplying together 15 twos. That is, if we multiply together 5 groups of 3 twos each, the result is 15 twos, all multiplied. In symbols, $(2^3)^5 = 2^{3 \cdot 5}$. Again, the same reasoning applies for any base (not just 2) and any exponents (not just 3 and 5), so we have a second rule of exponents:

$$(a^m)^n = a^{m \cdot n}$$

The first two rules of exponents were explained using examples in which the exponents were whole numbers. But the same rules apply to any exponent, be it a whole number, fraction, or decimal, positive or negative. You can test this on your calculator. Is $2^{1.6} \cdot 2^{3.2} = 2^{4.8}$? Is $(2^{-1.5})^{2.4} = 2^{-3.6}$? What this indicates is that the calculator has been programmed to compute fractional and negative exponents in such a way that the rules of exponents are followed. You might even say that it is the rules of exponents that tells us how to compute exponents that are not whole numbers.

So far we have covered two rules of exponents. Using the same methods, you should be able to come up with a third rule: What is the result of dividing 2^8 by 2^3? What rule of exponents does this suggest?

$$\frac{a^n}{a^m} = \underline{\hspace{5cm}}$$

The rules of exponents provide ways to modify how exponential functions are written. Let us look at three examples.

Example 1. In the last chapter, we saw the following pattern:

> In a continuous model for geometric growth, suppose the variable a is a_0 at time $t = 0$, and suppose that over d units of time there is a growth factor of r. Then $a(t) = a_0 r^{t/d}$.

Suppose that we observe the mouse population and find that it quadruples in six months. In terms of the statement in the box, $r = 4$ and $d = 6$. If the initial population is $1{,}000$, that gives the equation

$$p(t) = 1{,}000 \cdot 4^{t/6}$$

Now using the fact that $4 = 2^2$, we can rewrite the equation in the form

$$p(t) = 1,000 \cdot (2^2)^{t/6}$$

and using the second rule of exponents, we derive

$$p(t) = 1,000 \cdot 2^{2t/6}$$

In this case, a rule of exponents allows us to change an exponential equation in which the base is 4 into one in which the base is 2. This provides a glimpse of a much more general phenomenon. Any base can be transformed to any other base. This means that, if we wish, we can express every exponential function using the base 10. We will go into this idea in greater depth in a later section of the chapter. Simplifying the exponent in the last equation above leads to

$$p(t) = 1,000 \cdot 2^{t/3}$$

This form of the function can again be related to the statement in the box. This time notice that $r = 2$ and $d = 3$, indicating that the mouse population doubles every 3 months. In the original form of the equation we saw that the population quadrupled every 6 months. Of course, it is clear that these two descriptions are equivalent. We merely point out how the different forms of the equation lead to different interpretations.

Example 2.   We will start again with the equation

$$p(t) = 1,000 \cdot 4^{t/6}$$

This time, write the exponent as $\frac{1}{6} \cdot t$, and use the second rule of exponents again. We then have

$$p(t) = 1,000 \cdot (4^{1/6})^t$$

Now we can approximate $4^{1/6}$ using the calculator as 1.26. This gives our equation the new form

$$p(t) = 1,000 \cdot 1.26^t$$

In this form we can see that the population grows by a factor of 1.26 each month.

Example 3.   As a final example, we will convert the equation into an exponential with the base 10. There is one bit of information that is needed for this process: numerically $10^{.30103}$ is a very close approximation to 2. (You can verify that this is correct using your calculator. Later in the chapter we will discuss methods for discovering this kind of information.) Using the equation we found before

$$p(t) = 1,000 \cdot 2^{t/3}$$

replace the 2 with $(10^{.30103})$. That gives

$$p(t) = 1,000 \cdot (10^{.30103})^{t/3}$$

Using the second rule of exponents again, the equation can be re-expressed as

$$p(t) = 1,000 \cdot 10^{.30103t/3}$$

or, simplifying the exponent,

$$p(t) = 1,000 \cdot 10^{.100343t}$$

In this way we have expressed our function as a base 10 exponential function. Why is that of interest? Many calculators have a 10^x button. The base 10 version of the function is especially convenient to use with that type of calculator.

Summary of the Examples.   The examples showed that the following equations are all equal:

$$p(t) = 1,000 \cdot 4^{t/6}$$

$$p(t) = 1,000 \cdot 2^{t/3}$$

$$p(t) = 1,000 \cdot 1.26^{t}$$

$$p(t) = 1,000 \cdot 10^{.100343t}$$

The first equation allows us to see at a glance that the population quadruples every 6 months. Similarly, we see from the second equation that the population doubles every 3 months, and from the third that the population grows by a factor of 1.26 each month. The last equation uses base 10, and would be convenient to use for numerical investigation on a calculator with a 10^x button.

The point of this section has been to introduce three rules of exponents, and show how they allow exponential functions to be expressed in several different forms. These algebraic skills will be important in the next section, which focuses on solving exponential equations. Before proceeding to that topic, the three rules of exponents are repeated below for easy reference.

For any positive base b and for any exponents r and s, the following rules hold:

$$b^r b^s = b^{r+s}$$

$$(b^r)^s = b^{rs}$$

$$b^r / b^s = b^{r-s}$$

Solving Equations

An exponential equation is one for which a variable occurs in an exponent. The simplest form of such an equation has a single exponential expression on one side of the equal sign, and a number on the other. A typical example is

$$10^x = 2 \tag{1}$$

As usual, we can approach a problem of this type graphically and numerically. In Fig. 10.5 the graph of the equation $y = 10^x$ is shown. To solve the equation $10^x = 2$ graphically, we need to find a point on the curve where the y coordinate equals 2. A horizontal line is shown in the figure crossing the y axis at 2. The point we seek is

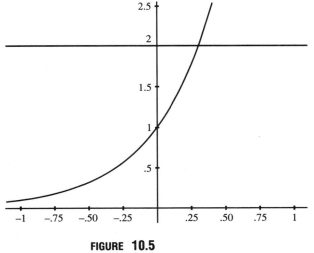

FIGURE 10.5
Graph for Solving $10^x = 2$

where the horizontal line and the curve meet. Carefully draw a vertical line in the figure from the intersection point down to the x axis. That will tell you the x coordinate for the intersection point. You should come up with something very close to .3. That is the value of x we wanted, where $10^x = 2$.

Now let's switch to a numerical method. Using a calculator, you will find that $10^{.3}$ is a little less than 2. On the other hand, $10^{.31}$ is a bit more than 2. So the solution to the equation $10^x = 2$ must be somewhere between .3 and .31. This means that it must start out as .30. Using the calculator, compute $10^{.301}$, $10^{.302}$, $10^{.303}$, and so on. Then fill in these blanks: the solution to the equation $10^x = 2$ is between .30____ and .30____. By using systematic trial and error in this fashion, you can eventually determine the solution to the equation to as many decimal places as the calculator can display. This is the best that can be hoped for. There is no exact solution that can be written down in a finite number of decimal places.

Exponential equations are inevitably encountered in geometric growth models. This is simply another instance of the inversion of a functional equation. For example, in the mouse population model, the properties of geometric growth lead to an equation that gives the population as a function of time:

$$p(t) = 1,000(1.5)^t$$

With this equation, we can predict the population at any time, simply by substituting that time for t and carrying out the computations indicated by the right side of the equation. This step is referred to as *evaluating* the function. It is very natural to ask questions in which the population size is given, and the time is unknown. *When will the population reach* 5,000? That kind of question requires us to invert the function. We wish to start with $p(t)$ (in this example, 5,000) and figure out the value of t. That is, we want to solve the equation

$$5,000 = 1,000(1.5)^t$$

for t. This is not quite the same form as Eq. (1). However, if we divide both sides of the equation by 1,000, the result is

$$5 = 1.5^t$$

which is of the same form as Eq. (1), although with the two sides reversed. With the equation in this form we can again apply graphical and numerical methods. That would be a time-consuming process. These equations occur so frequently in applications that a streamlined method has been developed.

The streamlined approach uses what are called *logarithmic* functions. In theory, there are different logarithmic functions for each base. For now, we will focus on the base 10 logarithm, also referred to as a *common* logarithm. On most calculators there is a built-in function for this operation, either as a key labeled *log* or as an item in a menu. The log key automatically calculates accurate approximate solutions to equations of the form $10^x = r$. For example, to solve the equation we studied earlier, $10^x = 2$, you must instruct the calculator to apply the log operation to 2. On some calculators this is done by entering 2 and then pressing the log button. On others you must first press the log button, then enter 2 and press =. Figure out how your calculator operates. The result you are looking for is approximately .30103.

Similar operations can be used to solve an equation with any other number in place of 2. Consider this equation: $10^x = 7$. The solution can be found by applying the log operation to 7. The result should be .845098 to the first six decimal places. You can check that this is a good approximate solution to $10^x = 7$ by raising 10 to the .845098 power on the calculator. Similarly, to solve $10^x = 15$, you must apply the log operation to 15. The result should be 1.176091.

The log operation is the inverse function for the base 10 exponential function. In the equation $10^x = y$, if you know x and you compute y, that is evaluating the exponential function 10^x. Conversely, if you know y and wish to compute x, we use the log operation. That is inverting the exponential function. We write the solution as $x = \log y$. So, for the equation $10^x = 7$, we express the solution as $x = \log 7$ and use the calculator to find the numerical approximation .845098. The log operation can be applied to any positive number. It cannot be applied to a negative number, or to 0, because there are no solutions to the equation $10^x = r$ if r is negative or 0. That is easy to verify by reviewing the graphical method used to solve $10^x = 2$ earlier. For this reason, we say that there is no log of a negative number or of 0.

How can logarithms actually be computed? One approach is to use trial and error as in the numerical approach used at the beginning of this section to solve $10^x = 2$. Alternatively, you could compile a very complete table of values for the function 10^x. The table might start out like this:

x	10^x
0.001	1.0023052
0.002	1.0046158
0.003	1.0069317
⋮	⋮

Now suppose you want to find out when 10^x equals 2. You look down the table until you find something as close to 2 as possible in the right-hand column.

x	10^x
⋮	⋮
0.300	1.9952623
0.301	1.9998619
0.302	2.0044720
⋮	⋮

For the data in the table, .301 is the exponent that comes closest to giving the desired result of 2. You can use this idea to help you remember what a logarithm is. Think of log 2 as an abbreviation for

LOcate the exponent **G**iving 2

Then, to calculate a numerical value for log 2, you locate the exponent giving 2, and that exponent is your answer. Looked at in this way, a logarithm is an exponent; log 2 is the exponent you must put on 10 to produce 2 as a result[1].

So far, all of this discussion has dealt with a very special case of exponential equations, namely, those with a base of 10. However, the special case can be used to solve any exponential equation through a two step process. First, we write each of the constants as a power of 10. This is done using logarithms. That will lead to a new equation which we can solve easily.

To illustrate, here is how to solve the equation $2^x = 7$. First, express 2 and 7 as powers of 10. You may recall from our earlier discussion that 2 is very closely approximated by $10^{.30103}$. But even if you had forgotten, you can simply set $2 = 10^r$ and solve for r. The answer is $r = \log 2 = .30103$. In either case, we arrive at $2 = 10^{.30103}$. Similarly, $7 = 10^s$ for $s = \log 7 = .84510$, so $7 = 10^{.84510}$. Now use these results to replace the 2 and the 7 in the original equation:

$$2^x = 7$$

$$(10^{.30103})^x = 10^{.84510}$$

$$10^{.30103x} = 10^{.84510}$$

---

[1] This idea of searching for a solution to an exponential equation in a table is related to the history of the development of logarithms. The Swiss mathematician Burgi, credited as one of the inventors or discoverers of logarithms, compiled tables in essentially this way. His tables were published in 1620. However, this approach requires an incredible amount of calculation, a significant problem prior to the development of computers and calculators! More efficient (and more subtle) methods were developed by the Scot Napier. His tables were first published in 1614, and his work was the foundation for tables of logarithms that were essential to scientific computation from his day until the middle of our century. For more information on this subject, see *What is a Naperian Logarithm?* by Raymond Ayoub, **American Mathematical Monthly**, volume 100, number 4, April 1993, page 351.

(Notice that we again used the second rule of exponents in the last step.) Now it should be clear that this equation will be satisfied if the exponents are equal, that is, if

$$.30103x = .84510$$

This leads to the solution $x = .84510/.30103 = 2.8074$. Does this give the answer to the original equation $2^x = 7$? Check using your calculator.

The calculator shows that $x = .84510/.30103 = 2.8074$ is close to a solution to the equation $2^x = 7$, but it is not exactly correct. Remember that the two decimal numbers in the fraction are approximations to logarithms. The top number is an approximation to $\log 7$, and the bottom number is an approximation to $\log 2$. If you use more decimal places in this approximation, your answer will be more accurate. To fully use the accuracy of the calculator, express the answer in this form:

$$x = \frac{\log 7}{\log 2}$$

There is a simple pattern to be observed here. The solution to the equation $2^x = 7$ is $x = \log 7/\log 2$. Can you guess the solution to the equation $3^x = 5$? Following the same pattern, it should be $x = \log 5/\log 3$. Check this on the calculator. In fact, this pattern always holds, and it provides a simple way to solve exponential equations using a calculator with a log button. The pattern is expressed in a general form below:

> The solution to the equation $b^x = c$ is given by $x = \log c/\log b$.

As an illustration, to solve $1.5^x = 4$, simply compute $x = \log 4/\log 1.5 = .60205999/.17609126 = 3.4190226$. To check that this is a good approximation to the correct answer, compute $1.5^{3.4190226}$ and see how close it comes to 4.

The preceding discussion shows that a calculator with a log button can be used to solve exponential equations involving any base. As discussed earlier, this also allows us to express any exponential function with 10 as the base. As an example of this idea, consider again 1.5^x. We can express 1.5 as a power of 10. Specifically, we LOcate the exponent Giving 1.5, and so recall that $\log 1.5$ is the correct power. Then, since $1.5 = 10^{\log 1.5}$, we can rewrite 1.5^x as $(10^{\log 1.5})^x = 10^{(\log 1.5)x} = 10^{.17609126x}$. In this way, we can do any computations required for exponential functions using two keys, 10^x and $\log$. In this regard, there is nothing special about the base 10. We could choose any other number to be our favorite base instead. For example, there are some applications where the most natural base to use is 2. Using very similar reasoning to what was presented above, we could introduce the idea of a base 2 logarithm, defined as follows: $\log_2 x$ *is the exponent you must put on 2 to produce a result of x.* Using these base 2 logarithms, we could solve equations just as above. The solution to $2^x = 7$ is simply $\log_2 7$. The solution to $5^x = 7$ is $\log_2 7/\log_2 5$. The point of these remarks is simply to show that what worked for base 10 would work equally well for any other base.

Actually, for most applications, the preferred base is neither 2 nor 10. It is an irrational number that is approximately 2.718281828. Because it is irrational, it can not be exactly expressed using a finite number of decimal places. So a special symbol has been given to this number, in just the same way that π is used as a symbol for the irrational number

$3.14259\cdots$. The symbol is e. On your calculator you should find a key that is marked e^x. That can be used to compute powers of the special base e. For example, if you enter 1 and push the e^x key (or push e^x and then 1 and = on some calculators), the calculator will compute $e^1 = e$ and display a decimal approximation like 2.7182818. Similarly, there is a key marked *ln* that computes the base e logarithm of any number. The *ln* label has an *l* for *log* and an *n* for *natural,* and the base e logarithm is often referred to as the *natural log.* There will be more discussion below about why this particular base is preferred for so many applications. Before going on to that topic, let us look at a couple of examples showing the use of base e exponential and logarithm functions.

As a first example, we will convert the function $p(t) = 1{,}000(1.5)^t$ to an expression using base e. As usual, we begin by expressing 1.5 as a power of e. We need to find an exponent r so that $1.5 = e^r$. By definition, the solution is $\ln 1.5$. The calculator gives this approximately as .405465. Then, replacing 1.5 with $e^{.405465}$, we have $p(t) = 1{,}000(e^{.405465})^t = 1{,}000e^{.405465t}$. This is an approximation. We can also leave it in the form $p(t) = 1{,}000e^{(\ln 1.5)t}$. In this example, the original equation comes from a geometric growth model, and is naturally expressed in terms of powers of 1.5. Then, when we use the functional equation to describe a continuous model, we convert into a base e exponential. An alternate approach to this kind of modeling leads directly to a base e exponential. Once we adopt a geometric growth model, we know that the end result will be some functional equation of the form $p_0 b^t$. What is more, we know that this can be re-expressed in the form $p_0 e^{ct}$ for an appropriate constant c. So, after collecting some data showing the population at various times, we choose the values of p_0 and c that give the best agreement between the data and the equation. This is similar to fitting the best straight line to a set of data for a linear model. In the end we have a model equation of the form $p(t) = p_0 e^{ct}$, fit directly to the data.

Here is a second example of base e calculations. In the preceding chapter compound interest was discussed as an example of a geometric growth model. Suppose that you put \$10,000 in a bank account that pays 9 percent per year, compounded monthly. Then each month you will be paid interest in the amount of one twelfth of 9 percent, or $.09/12 = .0075$. That means your account balance will grow by a factor of 1.0075 each month. Letting a_n be the account balance after n months, the foregoing explanation shows that $a_{n+1} = 1.0075a_n$. This leads to the functional equation $a_n = 10{,}000(1.0075)^n$. Using this equation, we can predict how long it will take to double your money. That is, we would like to find n so that $a_n = 20{,}000$. We must solve

$$20{,}000 = 10{,}000(1.0075)^n$$

for n. Dividing both sides of the equation by 10,000, we have the equation

$$2 = (1.0075)^n$$

That can be solved immediately, using the methods we discussed earlier, but this time using the natural logarithm instead of the base 10 logarithm. The solution is $n = \ln 2/\ln 1.0075 = 92.766$ (approximately). This is the same solution we would have found using base 10 logarithms. To solve an exponential equation, you can use any base logarithms you please. To emphasize this fact, we repeat the previous boxed comment using the natural logarithm.

> The solution to the equation $b^x = c$ is given by $x = \ln c / \ln b$.

To complete the discussion of the problem, we turn again to the solution $n = 92.7€6$. Remember that this is in units of months. The account will be less than twice the original amount after 92 months, but more than twice the original amount after 93 months. This is a little less than 8 years. In this case, it may not be valid to use a continuous model, because the bank might not pay interest for fractional parts of a month. However, the continuous function analysis gives us a quick way to get the answer of 92.766. It is then necessary to round the answer off to a whole number of months.

As a final example, we return again to the topic of radioactive decay. It is conventional to describe radioactivity in terms of half-life. For example, the half-life of strontium 90 is 28.1 years. This means that in a sample that contains strontium 90, the amount of strontium 90 will be reduced by half in any 28.1 year period. Using our knowledge of geometric growth, we can immediately write an equation of the form

$$a(t) = a_0(.5)^{t/28.1}$$

for the amount of strontium 90 after t years, where a_0 is the original amount of strontium 90. We will change this into an equation that uses e for the base. Following previous examples, observe that $.5 = e^{\ln .5}$. Therefore, $a(t) = a_0(e^{\ln .5})^{t/28.1} = a_0 e^{t \ln .5/28.1}$. This is approximately equal to $a(t) = a_0 e^{-.024667t}$.

In many areas of application, it has become standard practice to use e as the base for exponential functions. Most calculators include special built-in operations for e^x and $\ln$. In our approach to the subject, we began with geometric growth models, which led in a natural way to exponential functions with various bases. The preceding examples show how to express these exponential functions using the base e. But what is so special about e, and why is it used so much? A complete understanding of this question requires some knowledge of calculus. However, in the next section, we will try to give some idea of the significance of e.

The number e

One aspect of the significance of e is connected with the computation of exponential functions. Consider the base 2 exponential 2^x. For whole number values of x, this is simply a matter of repeated multiplication. So, 2^4 just means multiply 2 by itself a total of 4 times. The result is $2 \cdot 2 \cdot 2 \cdot 2 = 16$. But what about something like $2^{1.35}$? How is that computed? As discussed in the preceding chapter, the programming used by calculators has been designed to give results that are consistent with the idea of geometric growth. An equivalent explanation is that the programming is based on the rules of exponents given earlier. But that is not much help in coming up with a specific answer. Here is one approach that is based on methods of calculus. It uses a polynomial to approximate an exponential function. For small values of the exponent x, we use

$$2^x \approx 1 + \frac{.693x}{1} + \frac{(.693x)^2}{2 \cdot 1} + \frac{(.693x)^3}{3 \cdot 2 \cdot 1}$$

For larger values of x we have to add more terms in the same pattern. Next would come $\frac{(.693x)^4}{4 \cdot 3 \cdot 2 \cdot 1}$, then $\frac{(.693x)^5}{5 \cdot 4 \cdot 3 \cdot 2 \cdot 1}$, and so on. But no matter what x we use, the polynomial will give a good approximation if we use enough terms. To illustrate, let's approximate $2^{.2}$ using the equation above, which goes out to the $(.693x)^3$ term. With $x = .2$, we have $.693x = .1386$, so we must compute $1 + .1386 + .1386^2/2 + .1386^3/6 = 1.1486$ to four decimal places. Compare that with the calculator's answer for $2^{.2}$, which is 1.1487 to four places. As another example, to compute $2^{1.3}$ we first calculate $.693 \times 1.3 = .9009$. This time we will use one more term of the polynomial approximation. That is, compute

$$1 + .9009 + .9009^2/2 + .9009^3/6 + .9009^4/24 = 2.4560$$

showing the answer to four decimal places. Is that a good approximation to $2^{1.3}$? It is not as good an approximation as before, but it is still pretty good. There are ways to predict how many terms of the polynomial are needed to obtain a desired level of accuracy, but we won't go into that here. The main point of this discussion has been to illustrate how 2^x can be approximated with a polynomial.

Interestingly, the identical pattern can be used to estimate an exponential function for *any* base. Here is how it works for 3^x : The equation we use is

$$3^x \approx 1 + \frac{1.0986x}{1} + \frac{(1.0986x)^2}{2 \cdot 1} + \frac{(1.0986x)^3}{3 \cdot 2 \cdot 1}$$

This polynomial follows the same pattern as before, but instead of powers of $.693x$ we use powers of $1.0986x$. To approximate $3^{.25}$, for example, we first compute $1.0986 \times .25 = .27465$, then calculate

$$1 + .27465 + .27465^2/2 + .27465^3/6 = 1.3158$$

That is a pretty good approximation for the correct value of 1.3161. Try some other small values of x, and you will see that similar kinds of approximations can be obtained.

To approximate any base exponential, we always use the same pattern — only the constant multiplied by x changes. The constant is $.693$ for 2^x, 1.0986 for 3^x, 2.3026 for 10^x, 1.61 for 5^x, $.9163$ for 2.5^x, and so on. Of course, the easiest situation would be if the constant is 1. In that case the approximating polynomial is just

$$1 + \frac{x}{1} + \frac{x^2}{2 \cdot 1} + \frac{x^3}{3 \cdot 2 \cdot 1} + \cdots$$

What base exponential does that approximate? To find out, we can just take $x = 1$. Then the polynomial will approximate the first power of this special base. Using your calculator, compute

$$1 + \frac{1}{1} + \frac{1^2}{2 \cdot 1} + \frac{1^3}{3 \cdot 2 \cdot 1} + \frac{1^4}{4 \cdot 3 \cdot 2 \cdot 1} + \frac{1^5}{5 \cdot 4 \cdot 3 \cdot 2 \cdot 1}$$

After simplifying the fractions, the result is

$$1 + 1 + \frac{1}{2} + \frac{1}{6} + \frac{1}{24} + \frac{1}{120} = 2.717$$

This approximation is correct to two decimal places. The exact value of this special base is e. One of the things that is special about e is that the approximating polynomial for e^x just uses powers of x, not powers of some constant multiplied by x. In fact, e can be defined as the result of the infinite sum

$$e = 1 + \frac{1}{1} + \frac{1^2}{2 \cdot 1} + \frac{1^3}{3 \cdot 2 \cdot 1} + \frac{1^4}{4 \cdot 3 \cdot 2 \cdot 1} + \frac{1^5}{5 \cdot 4 \cdot 3 \cdot 2 \cdot 1} + \cdots$$

There are many other special properties of e. We will mention two more here. One has to do with the graphs of curves of the form $y = b^x$ for various choices of the base b. As we saw in the previous chapter, all of these graphs cross the y axis at the same place, namely at the point $(0, 1)$. However, the curves cross with different slopes. That is, if you looked at a magnified graph that just showed points very near the y axis, the curves would look like straight lines. Some are very steep, like 10^x. Others are very flat, like 1.01^x. In between, for some base, the curve crosses the y axis at a 45-degree angle. What is this base? It is e.

Another special property has to do with comparisons of the following sort: which is greater, $.6^{1.4}$ or $1.4^{.6}$? Questions like this can be analyzed by comparing the curves $y = .6^x$ and $y = x^{.6}$. The first curve is an exponential function with the base .6. The second is the power function with the power .6. If you graph these two curves, you will see that for some x's the power function is greater, and for some x's the exponential function is greater. This makes it difficult to predict by inspection whether $.6^{1.4}$ is greater than $1.4^{.6}$. A similar situation occurs if we change .6 into another number, say 2.5. Because 2.5^x is greater than $x^{2.5}$ for some values of x and less for other values of x, it is difficult to say right off whether or not $2.5^3 < 3^{2.5}$. There is just one exception. For just one number b, the exponential function b^x is never less than the power function x^b for positive x's. That one number is e. That is, for all positive x,

$$e^x \geq x^e$$

For any other base b, there are some positive x values for which $b^x > x^b$, and there are other x values for which $b^x < x^b$. This is another of the many properties that distinguish e as a special choice for the base of an exponential function.

Summary

In this chapter we have expanded our knowledge of exponential functions. Three rules of exponents were presented for use in algebraic operations. These rules give us the ability to write an exponential expression in many different forms. In particular, with these rules it is possible to use any number we wish as the base for a given exponential expression.

We also discussed methods to solve exponential equations. The concept of a logarithm was introduced to simplify this process. The base 10 logarithm of a number x (written $\log x$) is defined as the exponent that must be attached to 10 to produce a result of x.

Most calculators have a button for $\log x$. Using logarithms, the solution of any equation of the form $b^x = c$ can be expressed as $x = \log c / \log b$.

There is a special number called e that is often used as a base for exponential functions. Because e is irrational, it cannot be written down exactly in a finite number of decimal places; the value of e is 2.71828 to five decimal places. A complete understanding of the significance of e requires a knowledge of calculus. However, there are a number of properties of e that can be understood without calculus. Several of these have been presented. Most calculators feature buttons for computing e^x and the logarithm with base e. This latter is usually represented as $\ln x$ where the letters *ln* come from *logarithm* and *natural*. The natural logarithm can be used to solve exponential equations in the same way that the base 10 logarithm is used. The solution of any equation of the form $b^x = c$ can be expressed as $x = \ln c / \ln b$.

Exercises

Reading Comprehension

1. Explain what is meant by an exponential function. By an exponential equation. How do they differ?

2. Explain what the rules of exponents are. Give examples.

3. Give an explanation of why the rule $a^n a^m = a^{n+m}$ holds.

4. Give an explanation of why the rule $(a^n)^m = a^{nm}$ holds.

5. Explain how one exponential function can be written in several forms that look quite different.

6. Explain how the equations $p(t) = 1{,}000 \cdot 8^{t/6}$, $p(t) = 1{,}000 \cdot 2^{t/2}$, and $p(t) = 1{,}000 \cdot (1.414)^t$ can each be interpreted at first glance. (For example, if t is in units of hours, the first equation indicates that p increases by a factor of 8 every 6 hours.) Also explain why it is clear from these interpretations that these three formulas all express the same geometric growth model.

7. What is meant by *evaluating* a function? What is meant by *inverting* the function?

8. Explain the concept of the logarithm function. As part of your answer, explain the conceptual meaning of $\log 6$.

9. What is the meaning of $\log_2 7$? How can you compute this approximately using a numerical method?

10. Write an explanation of why there is no logarithm of a negative number, nor of 0, using the graph in Fig. 10.5.

11. What is e? What are some of the special properties of e?

Mathematical Skills

1. Let $p(t) = 1{,}000 \cdot 16^{t/4}$ for the parts of this problem.

    a. Replace 16 by 4^2 and express $p(t)$ as an exponential function with base 4.

    b. Replace 16 by 2^4 and express $p(t)$ as an exponential function with base 2.

    c. Approximate 16 by $10^{1.204}$ and express $p(t)$ as an exponential function with base 10.

    d. Approximate 16 by $e^{2.773}$ and express $p(t)$ as an exponential function with base e.

    e. Writing $16^{t/4}$ as $(16^{1/4})^t$, express $p(t)$ as an exponential function for which the exponent is simply t.

2. Let $p(t) = 1,000(9)^{t/2}$ for the parts of this problem.

    a. Express $p(t)$ as an exponential function with base 3.

    b. Express $p(t)$ as an exponential function with base 10.

    c. Express $p(t)$ as an exponential function with base e.

    d. Express $p(t)$ as an exponential function with exponent equal to t.

3. Use a numerical method to approximately solve the equation $10^t = 13$.

4. Use a numerical method to approximately solve the equation $10^t = 78$.

5. Redo problem 3, this time using the log function, and compare your new answer with what you got the first time.

6. Redo problem 4, this time using the log function, and compare your new answer with what you got the first time.

7. Use a numerical method to approximately solve the equation $7^t = 5$.

8. Use a numerical method to approximately solve the equation $1.06^t = 2$.

9. Solve the equation in problem 7 using the log function, and compare your answer to what you got before.

10. Solve the equation in problem 8 using the log function, and compare your answer to what you got before.

11. Solve the equation $3^{1.2t} = 7$ by first writing $1.2t = (\log 7)/(\log 3)$, and then solving for t. Check your answer by substituting it in the original equation.

12. Solve $1.35^{t/2.3} = 25$.

13. Use a polynomial approximation to estimate $2^{.126}$ as on page 227. Compare with the answer the calculator gives for $2^{.126}$.

14. Use a polynomial approximation to estimate $3^{-.178}$ as on page 227. Compare with the answer the calculator gives for $3^{-.178}$.

Group Activities.   Generally when you compute the log of something, the answer is a decimal expression. The whole number that is in front of the decimal point is called the *integer part,* and the decimal expression is called the *fractional part.* For example, $\log 25 = 1.39794$, with 1 being the integer part and .39794 being the fractional part.

These exercises will guide you in exploring the meaning of these parts.

1. Compute the integer part of the log of each of the following numbers: 12.56, 56.32, 84.1, 123.45, 996.43, 602.45, 1,234.5, 6,345.78, 1,739.3, 66,666.6, 35,247.0, 10,000.1.

2. Look for a pattern that you could use to predict the integer part of the log without computing the log exactly.

3. Make up some more numbers and test your pattern.

4. Does your pattern hold for these numbers? .01234, .0754, .05674

5. Does your pattern hold for these numbers? .005546, .00992, .0056

6. Use your pattern to predict the integer part of the log for each of the following, then check with the calculator: .000713, .00000567, .00000001.

7. Go back to the original list of numbers in problem 1. Express each of these numbers in scientific notation. For example, in scientific notation 12.56 is given as $1.256 \cdot 10^1$. For each scientific notation expression, calculate the log of each part, and compare to the logs of the original numbers. Make a guess about the meaning of the fractional part of the logarithm of a number.

8. Write a short report on what your group discovered.

Solutions to Selected Exercises

Mathematical Skills

1.   a. $p(t) = 1,000(4^2)^{t/4} = 1,000(4^{2t/4}) = 1,000(4^{t/2})$

    b. $p(t) = 1,000(2^4)^{t/4} = 1,000(2^{4t/4}) = 1,000(2^t)$

    c. $p(t) = 1,000(10^{1.204})^{t/4} = 1,000(10^{1.204t/4}) = 1,000(10^{.301t})$

    d. $p(t) = 1,000(e^{2.773})^{t/4} = 1,000(e^{2.773t/4}) = 1,000(e^{.693t})$

    e. $p(t) = 1,000(16^{1/4})^t = 1,000(2^t)$

2.   a. Since $9 = 3^2$, $p(t) = 1,000(3^2)^{t/2} = 1,000(3)^t$

    b. Since $9 = 10^{.95424}$ approximately, $p(t) = 1,000(10^{.95424})^{t/2} = 1,000(10)^{.47712t}$

    c. Since $9 = e^{2.1972}$ approximately, $p(t) = 1,000(e^{2.1972})^{t/2} = 1,000e^{1.0986t}$

    d. $p(t) = 1,000(9^{1/2})^t = 1,000(3)^t$

3. $10^{1.1} = 12.589$ is too small, and $10^{1.2} = 15.849$ is too big. So the correct answer should be 1.1 something. Using similar methods, you can find that $10^{1.11} = 12.88$ is too small and $10^{1.12} = 13.18$ is too big. So the correct answer is 1.11 something.

5. The answer is $t = \log 13 = 1.1139434$

7. Using trial and error, $7^{.82} = 4.93$ which is too small, and $7^{.83} = 5.03$ is too small. So the answer is $t = .82$ something.

9. The answer is $t = (\log 5)/(\log 7) = .82709$

12. $t/2.3 = (\log 25)/(\log 1.35)$, so $t = 2.3(\log 25)/(\log 1.35)$

13. $2^{.126}$ can be approximated as

$$2^{.126} \approx 1 + \frac{.693 \cdot .126}{1} + \frac{(.693 \cdot .126)^2}{2 \cdot 1} + \frac{(.693 \cdot .126)^3}{3 \cdot 2 \cdot 1}$$

Or, since $.693 \cdot .126 = .087138$, as

$$2^{.126} \approx 1 + \frac{.087138}{1} + \frac{.087138^2}{2 \cdot 1} + \frac{.087138^3}{3 \cdot 2 \cdot 1}$$

This calculation comes out to $2^{.126} \approx 1.09124\cdots$. Using the calculator directly, $2^{.126} = 1.09126\cdots$. so the approximation is pretty accurate.

11

More On Logarithms

In the preceding chapter, logarithmic functions were introduced to simplify the problem of solving exponential equations. However, there is much more to be said about these functions. In addition to simplifying the solution of exponential equations, logarithms are applied in three important ways:

- Logarithmic Models
- Logarithmic Scales
- Transforming Data

Not much will be said here about the first of these. In earlier chapters you have already seen several groups or families of functions: linear functions, polynomial functions, and exponential functions. In each case, you have seen applications where properties of the data suggest using a certain kind of function as an approximation. So, if the data appear to fall approximately on a straight line, we may decide to use a linear model, and attempt to choose the best possible line for the data. Similarly, if the data follow a curve that wiggles up and down, a polynomial function might be selected as an appropriate model. Again in the case of exponential functions, if the data remind us of the characteristic shape of an exponential graph, we can choose to use an exponential function in our model. So too for logarithms, there are occasions when the appearance of the data leads us to use a logarithmic model. Although this step uses our most recently studied type of function, a logarithmic function, the approach is the same as for the families of functions studied earlier in the course. For this reason, no specific example of this type will be presented. The student should be aware that logarithms are intrinsically useful in this context, just as linear functions, polynomial functions, and exponentials function are.

The second type of application of logarithms involves what is called a logarithmic scale. Here the basic idea is to consider the logarithms of the data values, rather than using the original data directly. In one of the following sections we will discuss this topic, including the motivation for and advantages of using logarithmic scales. There will also be a section on one particular logarithmic scale, the pH scale.

The third area of application, transforming data, seeks to represent in a linear form a relationship that is originally nonlinear. If the transformed data do fall on or near a straight line, we can use the equation of that line to find an equation relating the original variables. This topic will be presented following the section on the pH scale.

Before proceeding to these applications, it will be helpful to examine a few of the algebraic properties of logarithms. As we have already seen many times, it is often helpful to change an equation from one form to another. The rules of algebra tell us how to make these changes. There are algebraic rules for polynomials, for rational functions, for exponents, and it should come as no surprise that there are algebraic rules for logarithms. That is the next subject that will be presented.

Rules for Logarithms

To begin, let us review the meaning of logarithms. The common logarithm, represented by $\log$, is the logarithm that goes with problems involving powers of 10. The expression $\log x$ stands for the exponent that must be put on 10 to produce x. For example, the $\log 3$ is the exponent that must be applied to 10 to produce 3 as a result. This idea can be expressed in the equation

$$10^{\log 3} = 3$$

This makes a kind of convoluted sense: Find the exponent that you need to raise 10 to to get 3—now go ahead and raise 10 to that exponent. The result is 3. In this example, there is nothing special about 3; the same idea works with any positive number. So the first rule of logarithms says, for any positive number a,

$$10^{\log a} = a$$

There is a variation on this idea. What happens when you take the logarithm of something that is already expressed as a power of 10? For example, what is the $\log 10^3$? This asks, what exponent do you have to put on 10 to get 10^3. Clearly, the exponent that must be used is 3. So in equation form, that says

$$\log 10^3 = 3$$

Again, there is nothing special about 3 in this equation. The second rule of logarithms is

$$\log 10^a = a$$

for any a.

The other rules of logarithms are closely related to the rules of exponents. Remember that a logarithm *is* an exponent. So, just as there are rules that have to do with adding, subtracting, and multiplying exponents, so there are rules that involve adding, subtracting, and multiplying logarithms. Here they are:

$$\log a + \log b = \log(ab)$$

$$\log a - \log b = \log(a/b)$$

$$r \cdot \log a = \log(a^r)$$

Here are some examples to illustrate the rules. Suppose you are working with the equation $y = 100(1.02^x)$. Applying the logarithm to each side we obtain a new equation:

$$\log y = \log[100(1.02^x)]$$

Now we will use the first of the three rules above. Put 100 in place of a and 1.02^x in place of the b. That tells us

$$\log 100 + \log(1.02^x) = \log[100(1.02^x)]$$

Combining that with the preceding equation, we obtain

$$\log y = \log 100 + \log(1.02^x)$$

Next, use the third of the three rules of logarithms, the one with r. In place of r, put x; and in place of a, put 1.02. The rule then says

$$x \cdot \log 1.02 = \log(1.02^x)$$

Now that can be substituted into the earlier equation for $\log y$ with the result

$$\log y = \log 100 + x \cdot \log(1.02)$$

Finally, a calculator shows that $\log 100 = 2$ and $\log 1.02 = .0086$ (approximately). So the equation for $\log y$ can be rewritten in the form

$$\log y = 2 + x \cdot .0086$$

or with a slight rearrangement

$$\log y = .0086x + 2$$

Compare that with the original equation

$$y = 100(1.02^x)$$

and you can see that the two forms look quite different. One has an exponential function on the right side, the other has a linear function on the right side. As we will see, there are situations where the linear version is preferable. The rules of logarithms allow us to go back and forth from one version to the other.

The rules presented above for base 10 or common logarithms have counterparts for any base logarithm. In particular, there is a variation on each rule for base e or natural logarithms. For future reference, all of the rules for both common and natural logarithms are repeated in the box on the next page.

Here is a final example using the rules of logarithms. Compare the logarithms of 123 and 1.23. These numbers have the same digits, differing only in the placement of the decimal point. That means that you can get one by multiplying the other by a power of 10: $123 = 10^2 \cdot 1.23$. Now apply logs to both sides of the equation, and use the rule for

For any positive a and b and for any r, the following rules hold:

$$\log 10^r = r$$

$$10^{\log a} = a$$

$$\log a + \log b = \log(ab)$$

$$\log a - \log b = \log(a/b)$$

$$r \cdot \log a = \log(a^r)$$

$$\ln e^r = r$$

$$e^{\ln a} = a$$

$$\ln a + \ln b = \ln(ab)$$

$$\ln a - \ln b = \ln(a/b)$$

$$r \cdot \ln a = \ln(a^r)$$

adding logs:

$$123 = 10^2 \cdot 1.23$$

$$\log 123 = \log(10^2 \cdot 1.23)$$

$$= \log 10^2 + \log 1.23$$

$$= 2 + \log 1.23$$

If you use the calculator to find both the log of 123 and the log of 1.23, you will see the result above is correct. What is more, the log of 1.23 is not very large, so the answer is pretty close to 2. This reveals a general rule for common logarithms: to approximate the common logarithm of any number, just count the number of digits in front of the decimal point. The logarithm will be less than the number of digits, but more than the next lower whole number. For example, since 34,738 has 5 digits in front of the decimal point, the log of 34,738 will be less than 5 and more than 4. This is because $\log 34,738 = \log(10^4 \cdot 3.4738) = \log 10^4 + \log 3.4738 = 4 + \log 3.4738$. This relationship between the log of a number and the number of digits in front of the decimal point is closely connected to the next topic of discussion, logarithmic scales.

Logarithmic Scales

There are many situations in which data cover so wide a range of values that it is difficult to comprehend the relationships in the data. To illustrate this idea, here is a rough comparison of several different population groups. The smallest group we consider is the family, which might have 4 members. Next, in a single class at school, there might be about 30 members. Continuing in this fashion, we can imagine the size of a school

Group	Family	Class	School	Town	Urban Area	USA
Size	4	30	4,000	20,000	5,000,000	250,000,000

TABLE 11.1
Various Population Group Sizes

(4,000), of a small town (20,000), of a major urban area (5,000,000), and of the entire country (250,000,000). For reference, these values are shown in Table 11.1.

The numerical values in the table cover a wide range. To fully appreciate what this means, imagine trying to create a number line for population size, with a mark for each of the items in the table. We begin the line at 0, and put a mark down somewhere for the size of a family. For concreteness, let us say that we will put that mark one-eighth of an inch to the right of 0. Now the mark for *class* should be about 8 times as far from 0, because 30 is about 8 times as large as 4. So the mark for *class* is made one inch to the right of zero. The size of the school is over 100 times the size of one class. So the mark for *school* should be at least 100 inches to the right of 0. That is more than 8 feet. The town figure is 5 times the size of the school figure, so the mark for that entry should be about 40 feet away from 0. Continuing in this fashion, the marks for *urban area* and *USA* have to be placed two miles and 100 miles away from 0. So the entire number line is 100 miles long, with some of the marks within the first few feet, and only two marks in the last 98 miles. Looked at in this way, you can see that these numbers are practically impossible to compare directly in a meaningful way.

In fact, when you compare two numbers from the table, it is much more significant to look at the number of digits than the actual numbers involved. For example, the school size is in the thousands (4 digits) whereas the urban area size is in the millions (7 digits). The difference between the number of digits in these numbers is 3, indicating that one is about 1,000 times the size of the other[1]. That leads to a different way of making a number line. Make several marks evenly spaced out (perhaps an inch apart), and label these marks 1, 10, 100, 1,000, 10,000, and so on. On this type of line, the mark for a family of size 4 would fall between the 1 and the 10; a class of size 30 is between 10 and 100; the school of 4,000 is between 1,000 and 10,000, and so on. This is illustrated in Fig. 11.1. Because the different powers of 10 are spaced equally in this graph, the position of each data point mainly reflects how many digits it has.

What does this have to do with logarithms? The number line in Fig. 11.1 is called a logarithmic scale because it is equivalent to graphing the logarithms of the population values on a normal number line. As we saw at the end of the preceding section, the log of a number is closely related to the number of digits. The log of a 6-digit number is 5 point something; the log of an 8-digit number is 7 point something, and so on. So on a plot

---

[1] When numbers are compared in this way, there is often a reference to the idea of an *order of magnitude*. Quantities with different numbers of digits (to the left of the decimal point) are described as having different orders of magnitude. For example, we might describe the school size (4,000) as being 3 orders of magnitude less than the urban area size (5,000,000), because one has 4 digits and other has 7.

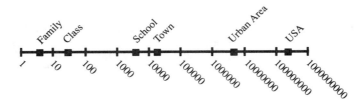

FIGURE 11.1
Alternate Number Line for Population

showing the logarithms of the population data, the whole numbers will correspond to data values of 1, 10, 100, 1,000, and so on, and the data points will fall between these whole numbers, just as in Fig. 11.1. In Table 11.2 the logarithms of the population sizes are shown for each data value used earlier. Graphing these logarithms on a normal number line gives Fig. 11.2. This is essentially the same as Fig. 11.1, except for the labels below the number line. That is, plotting the logarithms of the raw data on a normal number line has the same effect as graphing the raw data on the new kind of number line used earlier, with equally spaced points indicating 1, 10, 100, 1,000, and so on.

This new kind of number line, with the different powers of 10 evenly spaced, is called a logarithmic scale. Sometimes data are graphed on such a scale, and it is possible to purchase graph paper that is already marked out in this way on one or both axes. However, it is also common to use logarithms of the data values and plot these on a normal number line. There are a few examples of this kind of plot that are part of everyday language: The Richter scale for earthquake intensity; the decibel scale for sound intensity; and the pH scale for measuring the strength of acidity. Each of these is defined using logarithms, and each is also referred to as a logarithmic scale. So for example, when you hear that an earthquake measures 7 on the Richter scale, that means that the logarithm of an energy measurement, when plotted on a normal number line, would come out at 7.

The preceding discussion shows that *logarithmic scale* is used in two slightly different ways. First, a number line in which evenly spaced marks are labeled with the powers of

Group	Family	Class	School	Town	Urban Area	USA
Log of Size	0.60	1.48	3.60	4.30	6.70	8.40

TABLE 11.2
Logarithms of Population Group Sizes

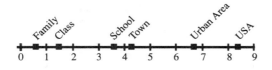

FIGURE 11.2
Number Line for Logarithm of Population

10 is a logarithmic scale. But a logarithmic scale can also be a normal kind of number line that is used to plot values derived from the logarithms of data. To make these ideas more concrete, let us apply them to the population data introduced in Table 11.1. We have already seen a plot of the data on a logarithmic scale—that is, on a logarithmic number line—in Fig. 11.1, as well as a plot of the logarithms of population size on a regular number line (Fig. 11.2). Now we will make up a fancy name for this latter approach to population measurement, the Gump scale.[2] The Gump scale value for a population figure is just the logarithm of the population. So, in a town of 70,000, the population measures 4.85 on the Gump scale because the log of 70,000 is approximately 4.85. Any population can be expressed using the Gump scale. To parody the way that *Richter scale* is heard in popular usage, this is what a reference to the Gump scale might sound like: *Tokyo is one of the largest cities in the world, measuring 7.43 on the Gump scale.* That simply means that the logarithm of the population of Tokyo is 7.43. With this idea in mind, observe that the data in Table 11.2 are the readings on the Gump scale for the populations shown in Table 11.1, and Fig. 11.2 is a graph of these Gump scale readings. The Gump scale is a logarithmic scale for population because it is defined in terms of the logarithm of population. And if you make a normal number line plot of data defined using the Gump scale (Fig. 11.2), you get the same picture as plotting the population figures directly on a logarithmic scale number line (Fig. 11.1). Only the labels under the number line are different.

In comparing numbers in a logarithmic scale, it is important to keep in mind that each increase of 1 unit corresponds to multiplying by 10. For example, in Table 11.2, the value for *Urban Area* is 6.70 and the value for *Town* is 4.3, a difference of about 2. But remember that these are logarithms, and each corresponds roughly to the number of digits in one of the original data values. Accordingly, the difference of 2 indicates that the original data value for the *Urban Area* population had about two more digits than the value for the *Town* population. That is, the urban area has about 100 times as many people as the town. In a similar way, a difference of 3 on the logarithmic scale corresponds to a factor of 1,000 in the original data; a difference of 4 to a factor of 10,000, and so on. Not all logarithmic scales are defined as simply as the Gump scale. For example, the pH scale is defined not as the logarithm of the strength of acidity, but as the negative of that logarithm. Similarly, the decibel scale is not defined simply as the logarithm of sound intensity, but rather by a formula of the form

$$\text{decibels} = 10 \log I \tag{1}$$

where I is the sound intensity.[3] For example, if a sound has intensity $I = 10^3$, that corresponds to a decibel value of $10 \log(10^3) = 10 \log 1,000 = 30$.

---

[2] Named in honor of Forrest Gump, a fictional character who could relate to all kinds of people.

[3] The exact form of this equation depends on the units used to measure the sound intensity. For example, when the intensity is expressed in units of watts per square centimeter, the expression for decibels becomes $10 \log(10^{16} I)$. This is the form used in the exercises. The simpler form of Eq. (1) is correct for sounds measured in units of microwatts per square kilometer. That is a very uncommon choice for most applications, but we will use it for this discussion so that the equations are as simple as possible.

Although the pH and decibel scales are not simply computed as the logarithms of acidity or sound intensity, they still share the basic properties of a logarithmic scale, reflecting primarily the order of magnitude of a data value, and simplifying the comparison of values with widely different magnitudes. However, the extra operations that appear in the definitions of pH and decibels do have an effect on how comparisons should be interpreted. In the earlier discussion, it was shown that a difference of 1 on the Gump scale corresponds to a 10-fold increase in a population, with a difference of 2 indicating a 100-fold increase, a difference of 3 indicating a 1,000-fold increase, and so on. The same is true for Richter scale readings. But things are a little different for pH. Because pH is defined as the *negative* of the logarithm of acid strength, larger pH values indicate *weaker* acids and smaller pH values indicate stronger acids. For a specific example, if one acid has a pH of 2 and another has a pH of 4, that is a difference of 2, so one of them is 100 times stronger. But it is the smaller pH that indicates the more powerful acid. That is, the acid with the pH of 2 is 100 times more powerful than the acid with a pH of 4. These ideas will be explored more fully in the next section.

Here is a similar kind of comparison using the decibel scale. Suppose one sound measures 46 on the decibel scale, and another measures 16. That is a difference of 30. But remember that in the decibel scale, each logarithm is multiplied by 10, so that the difference between two decibel scale values is actually 10 times the difference of the corresponding logarithms. For the two sounds we are considering, since the decibel readings differ by 30, the corresponding logarithms only differ by 3. And that means one of the sounds is 3 orders of magnitude, or 1,000 times more than the other.

When this idea is described verbally, it is somewhat confusing. It may be easier to understand using equations and numbers. Let's look again at the two decibel readings: 16 and 46. We can use this information to compute the actual intensity of each noise. Using I for sound intensity and Eq. (1), for the first noise we have

$$16 = 10 \log I$$

so dividing both sides of this equation by 10,

$$1.6 = \log I$$

In order to isolate I, apply each side of this equation as an exponent on 10:

$$10^{1.6} = 10^{\log I}$$

Now the second rule of logarithms (see page 236) can be applied on the right side of this equation, resulting in

$$10^{1.6} = I$$

So the sound intensity for the first noise is $10^{1.6}$. Following similar steps, the sound intensity for the second noise can be calculated as $10^{4.6}$. This is larger than the preceding result. How many times larger? To find out, divide the two intensities:

$$\frac{10^{4.6}}{10^{1.6}} = 10^{4.6-1.6} = 10^3 = 1,000$$

This shows again that the louder noise has an intensity 1,000 times as great as the quieter noise. As a general rule, if two decibel readings differ by an amount d, then the sound intensities differ by $d/10$ orders of magnitude. For this example, the decibel readings differed by 30, and the sound intensities differed by $30/10 = 3$ orders of magnitude.

This section has been concerned with explaining what logarithmic scales are, and why they are used. A made-up example called the Gump scale was developed to illustrate this concept. Then a brief introduction was given to three common logarithmic scales, the Richter scale, the decibel scale, and the pH scale. Now we will look at one of these, the pH scale, in greater detail. Additional attention will be paid to the Richter and decibel scales in the exercises.

The pH Scale

An acid is a solution that contains a certain kind of dissolved hydrogen, referred to as ionic hydrogen. The amount of this hydrogen determines how strong the acid is. In Table 11.3, the amount of ionic hydrogen in several different common substances is shown.

These figures are in units of moles of hydrogen ion per liter. We are not concerned with the exact meaning of this unit here. The main idea is that the measurements indicate the number of hydrogen ions in a certain quantity of fluid[4].

As in the population example, these figures vary so widely that the *number of* decimal places is a more meaningful comparison than the actual digits *in* those decimal places. For this reason, acidity is generally reported in terms of pH, which is defined as the negative of the (base 10) logarithm of the amount of ionic hydrogen in each liter of a substance. The information from the previous table has been repeated in Table 11.4, with an additional column for pH. The first two entries have been placed in this column. You should complete the table by filling in the final column. For each number in the

Substance	H = Amount of Ionic Hydrogen
lime juice	.0126
lemon juice	.00501
apples	.000794
beer	.00000316
milk	.000000355
eggs	.0000000158

TABLE 11.3
Hydrogen Ion Concentrations in Common Substances

---

[4] A unit of one *mole* represents a quantity of 6.02×10^{23}, or about six hundred billion trillion. The entry in the table for lime juice is about a hundredth of a mole per liter. That means each liter of lime juice contains about six billion trillion hydrogen ions.

Substance	H	$-\log H$
lime juice	.0126	1.9
lemon juice	.00501	2.3
apples	.000794	
beer	.00000316	
milk	.000000355	
eggs	.0000000158	

TABLE 11.4
Hydrogen Ion Concentrations and pH

second column, compute the negative of the (base 10) logarithm and write that in the third column. Round your results off to one decimal place. So for example, for *apples*, compute $-\log .000794 = 3.10018$, and round that off to 3.1.

You can see in the extended table that the pH values of the different substances are compared much more easily than the original data values. As in the population example, it is essential to recall what such a comparison means. If the pH of one substance is 2.3, and another substance is 4.3, that is only a difference of 2. But it means that one is one hundred times more acidic than the other. That is, each liter of one substance has 100 times the amount of ionic hydrogen as a liter of the other substance.

This completes the discussion of logarithmic scales. The main ideas of this discussion are

- Logarithmic scales are used when actual data cover a vast numerical range.

- Logarithmic scales are defined using the logarithms of the data, rather than the data themselves.

- Fixed differences on a logarithmic scale correspond to multiplications by fixed factors in the raw data. For example, each difference of 1 on the pH scale represents a factor of 10 in acidity.

- Commonly used logarithmic scales include pH, the decibel scale for sound intensity, and the Richter scale for earthquake strength.

When logarithmic scales are used, logarithms of data values are of interest because they make the comparison of widely different values more convenient. Sometimes taking logarithms of data values is useful for an entirely different reason, involving the discovery of linear relationships. That is the topic for the next section.

Transforming Data

One of the most important uses of logarithms in applications is to *linearize* data that would not otherwise fall on a straight line. Here is how it works. The starting point is a set of data points, apparently not closely approximated by a straight line. The data are transformed by taking the logarithms of one or both coordinates of each point, and the transformed data are used to create a new graph. If the new graph seems to fit closely to a straight line, the best fitting line is found. From the equation of the best fitting line, an equation for the original data can be derived.

This technique is extremely common in models in the sciences. In one example presented at a recent conference, the data came from a study of the foraging patterns of beavers. For each tree that the beavers had chewed through, the diameter of the tree was recorded, as well as the distance from the tree to the water. The idea was to find a relationship between diameter and distance, based on the set of data points. For each data point (x, y), x was the diameter of a tree and y the distance of the tree from the water. When the points were graphed, they clearly did not fit on a straight line. In this example, the data were transformed by taking the logarithms of both coordinates of each point. That produced a new set of data points of the form $(\log(x), \log(y))$. The transformed points were then used to find the best fitting straight line, using the approach described in Chapter 8. The equation of that line was then used to find an equation for the original variables, x and y, representing diameter and distance.

Why does this approach work? What is really going on? Linearizing a set of data in this way is really just the same as deciding to adopt one of two kinds of models, exponential or power functions. Exponential functions you have already seen. Power functions express one variable in terms of a fixed power of another variable. For example, the equations $y = x^3$, $y = 3x^{1.2}$, and $y = .56x^{-.78}$ all express y as a power function of x. Now there are two fundamental facts about these functions:

- If one variable is an exponential function of another, then the logarithm of the first variable is a *linear* function of the second.

- If one variable is a power function of another, then the logarithm of the first variable is a *linear* function of the logarithm of the second.

As a consequence of the first fact, if a set of data points is closely approximated by the graph of an exponential function, then taking the logarithm of the second coordinate of each point produces a set of points that fall nearly on a straight line, and the equation of that line can be used to find the equation of the original exponential function. In a similar way, if the original data points are closely approximated by the graph of a power function, then transforming the data by taking the logarithms of each coordinate will result in a set of data points that fall nearly on a straight line, and as before, the equation of that line can be used to find the equation of the original power function.

Generally, one cannot tell by looking whether a data set seems to follow a power or exponential curve. But you can certainly recognize a straight line! The logarithm function allows us to transform the functions that are not so easily recognized into a form that we *can* recognize. Often, data are not closely fit by either an exponential or power function.

planet	d (in millions of miles)	t (in days)
Mercury	36.0	97.97
Venus	67.1	224.70
Earth	92.9	365.26
Mars	141.7	686.98
Jupiter	483.4	4,332.59
Saturn	886.1	10,759.20
Uranus	1,782.7	30,685.93
Neptune	2,793.1	60,187.64
Pluto	3,666.1	90,885.00

TABLE 11.5
Data on the Planets

Then transforming the data will not lead to something that resembles a straight line. On the other hand, if we transform the data and see a close match with a straight line, that is evidence of an underlying relationship, either exponential (if we transformed just one coordinate) or power (if we transformed both coordinates). So the point of logarithmic transformations is to make certain kinds of nonlinear relationships recognizable. What is more, we can actually obtain equations for these nonlinear relationships by applying our techniques for finding the best fitting straight line to the transformed data.

As an example of this process, you will work through its application using some data about the planets of our solar system. For each planet, you are given data for two variables. The variable d stands for the distance from the planet to the sun, the variable t is the time it takes the planet to complete an orbit. For each planet, these two values are given in Table 11.5. These data are plotted in Fig. 11.3. If you check the graph with a straight edge, you will see that the data do not lie on (or very close to) a straight line. The steps for using logs to obtain a linear form for the graph are presented below.

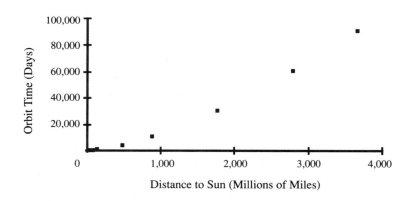

FIGURE 11.3
Planetary Distance and Orbit Time

planet	$x = \log d$	$y = \log t$
Mercury	1.5563	1.9911
Venus		
Earth		
Mars		
Jupiter		
Saturn		
Uranus		
Neptune		
Pluto		

TABLE 11.6
Transformed Data

Step 1. We will take the logarithm of each coordinate of each data point. If the resulting data fall on a straight line, that will indicate that the data are closely approximated by a power relationship.

As part of this process, two new variables will be introduced, to represent the logarithms of the original variables. For each data point we have values for d and t. Let $x = \log(d)$ and $y = \log(t)$. Complete Table 11.6 showing the x and y values that come from the original data. For example, in the first row of the table the entries are $\log(36.0)$ and $\log(97.97)$.

Step 2. The x and y data from the new table have been plotted in Fig. 11.4. Check a few of the points to verify that they are plotted correctly. On this graph, draw a straight line as close as possible to the data points. Notice that the data points really seem to fit a straight line quite well.

Step 3. Figure out the equation of the straight line. One approach to this step is to estimate two points on the line. Then work out the slope and use the point-slope formula. Do this and write your equation in the box below.

For the line I drew, I found that the points $(1.2, 1.5)$ and $(2.2, 3)$ seemed to be just about right on the line. Those points lead to a slope of 1.5 and the equation $y - 1.5 = 1.5(x - 1.2)$. Algebra can be used to rearrange this equation to the form

$$y = 1.5x - .3 \qquad (2)$$

For an even more accurate line, you can use a calculator or computer to find the *best* line

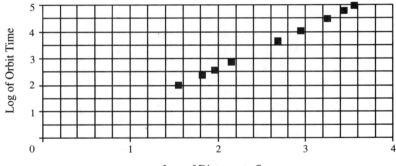

FIGURE 11.4
Transformed Planet Data

in the sense of Chapter 8. That leads to the equation

$$y = 1.49x - 0.36 \tag{3}$$

Step 4. Now we want to use the equation relating x and y to find an equation for the original variables, d and t. We will do that using the Eq. (2). In fact, a very simple equation can be obtained using algebra and the rules of logarithms. However, there are several steps involved. It is important to understand these steps, because they can be applied any time we use logarithms to transform data as in this example. To make the process as clear as possible, it is shown below with an explanation of each step.

$y = 1.5x - .3$ Starting Equation

$1.5x - y = .3$ This is a simple rearrangement of the original equation using the algebra of linear equations.

$1.5 \log d - \log t = .3$ Here we have just used the definitions of x and y as the logarithms of d and t.

$\log d^{1.5} - \log t = .3$ Here we have used the rule $r \cdot \log a = \log a^r$, with 1.5 representing r and d taking the place of a.

$\log\left(\dfrac{d^{1.5}}{t}\right) = .3$ Here we have used the rule $\log a - \log b = \log a/b$, with $d^{1.5}$ taking the place of a and t standing in for b.

$10^{\log\left(\frac{d^{1.5}}{t}\right)} = 10^{.3}$ Each side of the equation has been used as an exponent on 10. The results are still equal, since the exponents used on 10 are equal.

$\dfrac{d^{1.5}}{t} = 10^{.3}$ This time the rule $10^{\log a} = a$ was used, with a replaced by $d^{1.5}/t$.

$\dfrac{d^{1.5}}{t} = 1.995$ A calculator was used to determine $10^{.3}$.

$d^{1.5} = 1.995t$ Simple algebra: multiply both sides of the equation by t.

This completes the derivation of an equation involving d and t. It also completes the process of transforming data to find a linear relation. To summarize what was done, we began with a set of data points whose graph appeared not to fit closely to a line. We created a new data set by taking the logarithm of each coordinate in the original data. A plot of the transformed data seemed very close to a line, and we were able to find the equation of a line that closely approximated the transformed data. Finally, the equation was manipulated so that the original variables could be reintroduced. As a result of this entire process, we obtained the equation $d^{1.5} = 1.995t$.

There are a few comments that should be made about this process. First, the significance of the final equation is that it gives insight about the nature of planetary motion. At the outset, it might appear that the distance of a planet from the sun and the length of time it takes the planet to complete an orbit are totally unrelated. After completing this analysis, we know that there is a very specific relationship. Moreover, we actually have an equation with which, given the distance of a planet from the sun, we can predict how long it takes the planet to complete an orbit. This in turn can be used in further study of the solar system. For example, a theoretical model for describing planetary motion can be tested for agreement with the equation we found. If the theoretical model fails to predict motions in accord with the equation, we will suspect that the model is incorrect. If the model leads to the same equation we found from analyzing the planetary data, that makes the model more believable. In fact, that is very close to what actually happened in history. An equation like the one we found was first discovered by Johannes Kepler in 1619, although he used different methods than the ones we used above. Later, in 1687, Sir Isaac Newton proposed a universal law of gravitation that was supposed to describe the motions of all the planets, their moons, and the other heavenly bodies. Newton was able to show that Kepler's equation was a consequence of the universal law of gravitation. This was one piece of evidence that led scientists to believe that Newton's law was correct.

Second, the process illustrates the power and importance of algebra. The original form of the equation, the one that involves the transformed variables, is easily derived from our knowledge of straight lines and linear equations. The final form is the one that most directly gives us insight about the behavior of the planets. It is the rules of algebra that provide the ability to successively change one form into another until we eventually reach the equation in its final form.

Third, this method can be applied anytime we find a linear relationship between the logarithms of two variables. The same steps of algebra can be applied: start with a linear equation relating the transformed variables; reintroduce the original variables in the form of logarithms; use the rules of logarithms to combine all of the variables into a form with a single log term; and then, to eliminate the log, use each side of the equation as an exponent on 10. The final result will always be an equation in which one variable is given as a constant multiple of a power of the other variable. That is, the relationship between the variables is defined by a power function.

Finally, keep in mind that this entire process only works out if the original data points are closely approximated by some kind of power equation. It is quite possible to transform data and find that the new points still don't fit a line very well. However, experience has shown that many phenomena do follow power equations fairly closely, and the transformation approach has been used very effectively in a wide range of applications. A variant of the approach can be used when the data appear to follow an exponential curve, with the characteristic shape of the graphs in Chapter 9. In this case, only the y coordinate of each data point is transformed by taking the logarithm. If the resulting data fit closely to a line, the equation of that line can be found, and can be used to find an exponential equation for the original data.

Summary

In this chapter, two main ideas have been presented. The first concerns the use of logarithmic scales. These are useful when data cover a very wide range of values, so that how many digits a data value has can be more significant than what those digits are. In this case, it is often useful to introduce a new variable that is defined in terms of the logarithm of the original variable. The new variable is said to be defined on a logarithmic scale. Examples of logarithmic scales are pH (which measures acidity), the decibel scale (which measures sound intensity), and the Richter scale (which measures the strength of earthquakes).

The second main idea of the chapter is a procedure for transforming data to find linear patterns. The basic idea of this technique is that many phenomena are closely approximated by exponential or power equations, and that the data for these phenomena can be transformed into nearly linear patterns using logarithms. For a power relationship, both coordinates of each data point have to be transformed; for an exponential relationship, just the second coordinate of each data point has to be transformed. Once the transformed data are graphed, if there is a linear pattern, an equation for the line can be found as in past work. This equation provides a relationship between the coordinates in the transformed data. It can be used to find an equation directly relating the coordinates in the original data. In this way, one obtains appropriate power functions and exponential functions for modeling data.

Exercises

Reading Comprehension

1. What is a logarithmic scale? Why are these scales used?

2. What is a power function?

3. How is the (base 10) logarithm of a number related to the number of decimal digits in the number?

4. What is meant by order of magnitude?

5. Explain what is meant by pH and how it is defined. Why is pH an example of a logarithmic scale?

6. Outline the steps for using logarithms to transform data. Indicate what the goals of this procedure are, too.

7. Suppose you take the logarithms of both coordinates of each data point, and the graph of the new data appears to line up in a straight line. When you use the equation of that line to get an equation for the original variables, what type of equation should you expect to find? Why?

8. Suppose you suspect that a data set follows the graph of an exponential function. How can you use logarithms to test this suspicion?

Mathematical Skills

1. Without using a calculator, determine the nearest integers to each logarithm. Then check your answer with a calculator

    a. $\log 1{,}543.6$ (Solution: $1{,}543$ is between 10^3 and 10^4, so the log is between 3 and 4.)

    b. $\log 4{,}925{,}697.23$

    c. $\log .0000456$

    d. $\log(3.86 \times 10^5)$

    e. $\log(3.86 \times 10^{-5})$

2. In each part of this exercise, an equation is given. Use properties of logarithms to express the equation in a new form that does not involve any logarithms.

    a. $\log y = 1.7 \log x + 2.3$

    b. $\log y = 3 - 2 \log x$

    c. $y = 4.1 \log x - 3.2$

    d. $y = .87 - 2 \log x$

3. In each part of this exercise, an equation is given. Apply log to both sides of the equation, and then use properties of logarithms to express the equation in a simpler form.

    a. $y = 7.9(2^x)$

    b. $y^2 = 100(1.08)^x$

    c. $y = 6.8x^3$

    d. $y^3 = 19x^2$

4. In the example with planets, the equation $t = d^{1.5}/1.995$ is derived. How well does this fit the data? For each data point in Table 11.5, use the equation and the d in the table to estimate t. Then compare to the actual value of t in the table. How well does the equation fit the data?

5. For the planet data, the best line (in the sense of Chapter 8) is given in the chapter as Eq. (3). Following the steps shown in the reading for rewriting Eq. (2), use Eq. (3) to derive an equation for d and t. [Answer: $t = d^{1.49}/2.291$]

6. How well does the equation from the preceding question fit the data in Table 11.5? Compare with your results in question 4

Problems in Context

1. Human spinal fluid has a pH of about 7.4. How many times more ionic hydrogen is in a liter of apple juice than in a liter of spinal fluid?

2. Sound is made up of vibrations in the air, and the strength of the vibrations can be measured in units of watts per square centimeter. This unit describes how the power of the vibration is spread out over an area (for example, over the area of the ear drum). The human ear can hear sounds in an incredible range of intensities, from a whisper (about 10^{-13} watts per square centimeter) to the roar of a jet engine (on the order of .1 watts per square centimeter). A logarithmic scale for sound intensity is the decibel scale. This is defined as $d = 10 \log(10^{16} \cdot p)$ where p is the power of the vibration in units of watts per square centimeter. What is the decibel rating for the roar of a jet engine? If the energy in the sound produced by traffic is about 10^{-8} watts per square centimeter, how loud would that be in decibels?

3. If one sound has a decibel value that is 3 units higher than another, how much louder is it? For example, suppose the first sound is 61.2 decibels and the other is only 58.2 decibels. How do these sounds compare in terms of the actual intensity (or power)?

4. When sound intensity is measured in units of watts per square centimeter, the formula for computing decibels is $d = 10 \log(10^{16} \cdot p)$. Using rules of logs, show that that equation can also be expressed in the form $d = 10 \log p + 160$. Using this equation, show that if two sounds are compared, and if the intensity of the first sound is p_1 and the intensity of the second is p_2, then when you compute the decibel reading for each sound and subtract, the result is $10(\log p_1 - \log p_2)$.

   [This problem illustrates the effect of the choice of units on the decibel calculation. When the units for sound are microwatts per square kilometer, the equation is simply $d = 10 \log p$. For the more commonly used units of watts per square centimeter, the equation is $d = 10 \log p + 160$. Thus, changing units from microwatts per square kilometer to watts per square centimeter requires adding a constant correction of 160 to the decibel formula. In a similar way, making any other choice of units has the effect of adding a constant to the formula for computing decibels. Note that this has no effect when you compute the *difference* between two decibel readings. No matter what units you use, the difference in the decibel readings will always be ten times the difference in the logs of the sound intensities: $10(\log p_1 - \log p_2)$.]

5. The Richter scale is used to measure earthquake intensity. The energy released by an Earthquake that is just noticeable is about equal to 10,000 atomic bombs[5]. Call that energy level I_0. The Richter scale value R for a quake of energy level I is given by $R = \log(I/I_0)$. For example, if a quake is three times as powerful as the just noticeable one, then it has an intensity level of $I = 3I_0$. The Richter scale value

---

[5] Sheldon Gordon, et. al. *Functioning in the Real World,* Class Test Edition, Addison Wesley, 1995.

for that quake is $\log(3I_0/I_0) = \log 3 = .48$. Suppose a quake measures 5 on the Richter scale. How many times more powerful than the just noticeable quake is that? About how many atomic bombs would produce the same amount of energy? Suppose another quake measures 8 on the Richter Scale. How many times more powerful is it than the one that measured 5?

6. As a springboard and tower diver, I am quite interested in impact velocity. This is simply a measure of how fast you are going when you hit the water. Falling from a height of 16 feet, you are going 21.8 miles per hour when you hit the water. The table below shows the impact velocity for several different heights. Using these data, follow the same steps as in the planet data example to find an equation relating height and impact velocity. Is it twice as dangerous to jump from twice as high? That is, if you jump from 80 feet up do you hit the water going twice as fast as when you jump from 40 feet up?

height (feet)	10	16	32	50	100
impact velocity (miles per hour)	17.2	21.8	30.9	38.6	54.5

7. In Chapter 9, in the exercises, there is a problem about population growth in England and Wales. In the original problem, you tried to fit a geometric growth model to the data by looking at the average growth factor. In this exercise, you will use the idea of transforming data to look for the best fitting geometric growth model.

   a. Use the data in Table 9.1. Transform the data to a new data set by taking the *natural* logarithm (*ln*) of each of the population amounts. Do not take the logarithm of the years. For example, in the new data set, there would be a data point $(1801, \ln 8.89) = (1801, 2.185)$. Create a table showing all the transformed data points, (y, L) where y is the year and L is the natural logarithm of the population.

   b. Using the computer or graph paper, make a graph of the transformed data. Do the data points seem to fit on a line?

   c. Find the best line you can for the transformed data. It is recommended that you use a computer module for this. Express the equation for your line in the form $L = ? \cdot y + ?$ but with numerical values in place of the question marks.

   d. For this step, change your equation back into one describing the original data. To do this, note that $L = \ln p$ where p stands for population. Now $\ln p$ is an exponent, namely, it is the exponent you have to put on e to get p. If you put that exponent on e, you have to get p. Therefore $e^L = p$. Now to get the equation relating p and y, start with your equation for L and y. Use each side of this equation as an exponent on e. On the left side of the equation, you will have e^L, and that is actually p. On the other side you will something of the form $e^{? \cdot y + ?}$ but with numbers in place of the question marks. Simplify that side of the equation using the same kinds of steps that were used in the planet example.

   e. How does the equation you found in the preceding question compare to the equation you found in the exercises for Chapter 9?

Solutions to Selected Exercises

Mathematical Skills

1. b. $\log 4{,}925{,}697.23$ is between 6 and 7.

   c. $\log .0000456$ is between -5 and -4.

   d. $\log(3.86 \times 10^5) = \log 3.86 + \log 10^5 = \log 3.86 + 5$, so it is between 5 and 6.

   e. $\log(3.86 \times 10^{-5}) = \log 3.86 + \log 10^{-5} = \log 3.86 - 5$, so it is between -5 and -4.

2. a. $\log y = 1.7 \log x + 2.3 = \log x^{1.7} + 2.3$. So $\log y - \log x^{1.7} = 2.3$. Then $\log(y/x^{1.7}) = 2.3$. Now use each side of this equation as an exponent on 10. That gives $10^{\log(y/x^{1.7})} = 10^{2.3}$. Using $10^{\log a} = a$ leads to $y/x^{1.7} = 10^{2.3}$, which on the calculator is about 199.5. From $y/x^{1.7} = 199.5$, we obtain the final form: $y = 199.5(x^{1.7})$.

   c. $y = 4.1 \log x - 3.2$, so $y + 3.2 = 4.1 \log x$ and $(1/4.1)y + (1/4.1)3.2 = \log x$. Using the calculator, that leads to $.2439y + .7805 = \log x$. That means $10^{.2439y + .7805} = x$. This can be further simplified using the rule for adding exponents: $(10^{.7805})(10^{.2439y}) = x$. Finally, computing $10^{.7805}$ on the calculator, we get $6.03(10^{.243y}) = x$.

3. a. $\log y = \log[7.9(2^x)] = \log 7.9 + \log 2^x = \log 7.9 + x \log 2$. Using the calculator, that gives $\log y = .897627 + x(.30103)$ or putting the right side into $mx + b$ form, $\log y = .30103x + .897627$.

   b. $\log y = .0167x + 1$

   c. $\log y = \log(6.8x^3) = \log 6.8 + \log x^3 = \log 6.8 + 3 \log x = .8325 + 3 \log x$. So, $\log y = 3 \log x + .8325$.

4. In the table, for Mercury $d = 36$. Then the equation gives $t = d^{1.5}/1.995 = 36^{1.5}/1.995 = 108.27$. The t for Mercury in the table is 97.97. This shows that the answer from the equation is off by $108.27 - 97.97 = 10.3$ days. This means that the error is about 10 days out of the true answer of 98 days, or about 10 percent. Repeating the calculations for the other planets, all the other errors are right around 22 percent. In fact, the model is consistently about one-fifth too large. So for Neptune, the true value of t is roughly 60,000 days, and the model overestimates this by somewhat more than one-fifth, or 12,000 days.

5. Starting from $y = 1.49x - 0.36$, replace x with $\log d$ and y with $\log t$; that gives $\log t = 1.49 \log d - 0.36 = \log d^{1.49} - .36$. Apply each side of this equation as an exponent on 10: $10^{\log t} = 10^{\log d^{1.49} - .36} = 10^{\log d^{1.49}}/10^{.36}$, which then becomes $t = d^{1.49}/2.291$.

6. As in question 4, we begin with Mercury, using $d = 36$. This time the equation is $t = d^{1.49}/2.291 = 36^{1.49}/2.291 = 90.96$. The error is about 7 days, or 7 percent of the true figure. For the next two planets the error is about 2 and a quarter percent, and for the rest of the planets the errors are all under 2 percent. Looking again at the example of Neptune, the model underestimates the true t (roughly 60,000) by

less than 700 days. That is actually just a little more than a one-percent error. That is much better than the other model we looked at, for which the error was 12,000 days. These results show that, although the two lines have very similar equations, and their graphs both seem to fit the data in Fig. 11.4 very well, the line that was found by the best-fit method provides much greater accuracy than the one found by eye.

Problems in Context

1. The pH for spinal fluid is 7.4; for apple juice it is 3.1. The difference in these is 4.3, so one is about 10,000 times stronger than the other. The lower pH, for apple juice, indicates the stronger acid. So apple juice has about 10,000 times more ionic hydrogen than spinal fluid.

2. For a jet engine, the sound intensity is given as $p = .1$ so using the equation $d = 10\log(10^{16} \cdot p) = 10\log(10^{16} \cdot .1)$. Now $10^{16} \cdot .1 = 10^{15}$, and the log of that is just 15, so we get $d = 10 \cdot 15 = 150$. For traffic we repeat the same steps starting with $p = 10^{-8}$. The final result is $d = 80$, so the jet is at about 150 decibels and traffic is at about 80 decibels.

3. A difference of 3 on the decibel scale means a difference of .3 in the logs of sound intensity. That means one noise is $10^{.3}$ (or about 2) times louder than the other. We can check that directly for the example numbers of 58.2 and 61.2. Starting with $d = 58.2 = 10\log(10^{16} \cdot p)$, we can solve for p. Dividing by 10, $5.82 = \log(10^{16} \cdot p)$. Apply each side of this equation as an exponent on 10. That gives $10^{5.82} = 10^{\log(10^{16} \cdot p)}$ or $10^{5.82} = 10^{16}p$. This leads to $p = 10^{5.82}/10^{16}$. Following similar steps for a decibel value of 61.2 will lead to the similar result $p = 10^{6.12}/10^{16}$. Now compare these two figures for power by dividing one by the other. We find $(10^{6.12}/10^{16}) \div (10^{5.82}/10^{16}) = 10^{6.12}/10^{5.82} = 10^{6.12-5.82} = 10^{.3} = 1.995$

4. The decibel rating for the first sound is $10\log(10^{16} \cdot p_1)$. Similarly, the decibel rating for the second sound is $10\log(10^{16} \cdot p_2)$. Subtracting one of these from the other gives $10\log(10^{16} \cdot p_1) - 10\log(10^{16} \cdot p_2) = 10(\log 10^{16} + \log p_1) - 10(\log 10^{16} + \log p_2) = 10(\log 10^{16} + \log p_1 - \log 10^{16} - \log p_2) = 10(\log p_1 - \log p_2).$

5. If the Richter scale value is 5, then we know $5 = \log(I/I_0)$, so $I/I_0 = 10^5$. That means the quake is 100,000 times more powerful than the just noticeable one, and so is about equal to the energy of 1,000,000,000 atomic bombs. The quake that measures 8 on the Richter scale is 3 units higher, and so $10^3 = 1,000$ times more powerful, than the one that measured 5.

6. Use the variables h for height and v for velocity. Taking the logs of these variables, define $x = \log h$ and $y = \log v$. Then the transformed data will be given by the following table:

log of height (x)	1.0000	1.2041	1.5051	1.6990	2.0000
log of impact velocity (y)	1.2355	1.3385	1.4900	1.5866	1.7364

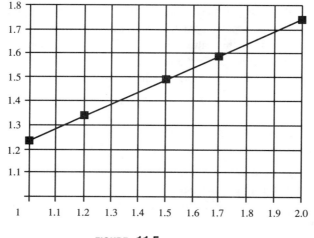

FIGURE 11.5
Transformed Diver Data

The graph of these data points is shown in Fig. 11.5. The best-fit line for the transformed data is given approximately by the equation $y = .5x + .735$. Replacing x with $\log h$ and y with $\log v$, this equation becomes $\log v = .5 \log h + .735$. Multiply both sides by 2: $2 \log v = \log h + 1.47$. Now use rules of logs to get $\log v^2 = \log h + 1.47$, then use algebra, $\log v^2 - \log h = 1.47$, and finally $\log(v^2/h) = 1.47$. To remove the log, apply each side as an exponent on 10. That gives $v^2/h = 10^{1.47} = 29.5$, so $v^2 = 29.5h$, and finally $v = \sqrt{29.5h}$. The second question is about linearity. The relationship between v and h is not linear, so proportional reasoning doesn't apply. In particular, doubling the height does not double the impact velocity. You can see that in the table, comparing the impact velocity for heights of 50 and 100 feet. Using the formula, for a height of 20 feet the impact velocity will be $v = \sqrt{29.5 \cdot 20} = 24.3$ miles per hour. If you jump from four times as high, from 80 feet, you hit the water going $v = \sqrt{29.5 \cdot 80} = 48.6$ miles per hour. This is not four times as fast; it is only twice as fast. In general, the effect of doubling the height is to multiply the impact velocity by $\sqrt{2}$ which is less than one and a half. So if you jump from twice as high, you hit the water going about one and a half times as fast.

12

Geometric Sums and Mixed Models

In the beginning of the course, the models we looked at were very simple. Both arithmetic growth and geometric growth involve difference equations that describe extremely simple patterns of change. In arithmetic growth, there is a numerical constant that can be added to each data value to obtain the next data value. In geometric growth, each new data value is found by multiplying the preceding value by a numerical constant. Because these patterns of change are so simple, it is easy to find functional equations for these models. We have also seen that these very simple models can be applied in the study of many real problems.

Now we are going to begin to look at some more complicated models. These will be developed by making minor modifications to the models we have already studied. This process has already been hinted at in earlier chapters, and is illustrative of the way models are developed in general. At the outset, a problem will be greatly simplified and studied with a very simple model. This leads to information both about the problem being studied, as well as about the kind of model that has been introduced. Next, the model will be refined to make it agree better with the true problem context. The refined model is generally more complicated than the original model, but, as we study it, we can apply what we have learned about the original model.

In this chapter, we will reconsider three problems from earlier work. In each case, a model that was used earlier will be extended or modified to produce a more involved model. The first example extends the geometric growth model for elimination of a drug from the body, by considering the effects of taking additional doses on a regular basis. This model is referred to as a mixed model because it mixes both arithmetic and geometric growth. It was mentioned in an exercise in Chapter 9. Here we will look at it more systematically. The second example involves a geometric growth model for the number of new cases of a disease each day. In this example we will be interested in projecting the total number of cases by adding up the new cases from each day. Finally, for the third example, we will combine aspects of both of the preceding examples to develop a new model for oil reserves. In this last example, a mixed model will be used for daily oil consumption. That will then be extended to a model for oil reserves. We will find, as we did in an earlier chapter, that the oil reserves model depends on adding up the terms from the oil consumption model. At this step, we will use the ability to define a functional equation for a mixed model, as well as the ability to add the terms of a geometric growth model, to obtain a functional equation for the oil reserves model. Then we will revisit the question of predicting when the oil will run out.

In addition to these examples that involve previously considered models, an example connected with financial calculations will be presented. In that example, the effects of making monthly loan payments will be analyzed.

Interestingly, although these examples are apparently quite different, they all involve one kind of pattern, namely, a pattern that comes from adding up the terms in a geometric growth sequence. Before we discuss the examples, it will be helpful to investigate this kind of pattern.

Geometric Sums

A geometric growth model always generates a similar pattern of data. Let's look at some examples. If the difference equation is

$$a_{n+1} = 2.3a_n$$

then, starting with a_0, the first few numbers in the pattern are

$$a_0 \quad a_0(2.3) \quad a_0(2.3)^2 \quad a_0(2.3)^3 \quad a_0(2.3)^4$$

Each number in the pattern is found by multiplying the preceding number by 2.3. Of course, if the geometric growth involves a growth factor different from 2.3, the same kind of pattern will emerge. Consider the difference equation

$$b_{n+1} = \frac{1}{2}b_n$$

For this pattern, each new number is found by multiplying the preceding number by $1/2$. That leads to the sequence

$$b_0 \quad b_0\left(\tfrac{1}{2}\right) \quad b_0\left(\tfrac{1}{2}\right)^2 \quad b_0\left(\tfrac{1}{2}\right)^3 \quad b_0\left(\tfrac{1}{2}\right)^4$$

We can describe this pattern generally by using a parameter for the growth factor. The difference equation

$$a_{n+1} = ra_n$$

leads to the number sequence

$$a_0 \quad a_0(r) \quad a_0(r)^2 \quad a_0(r)^3 \quad a_0(r)^4$$

This type of sequence is called a geometric progression. That is just another name for the sequence of numbers that occurs in a geometric growth model.

Now we are interested in what happens when these numbers are added together. More specifically, what is the total if we add the first 10 terms of a geometric progression? Or the first 20? Or the first 100? Adding together the terms of a geometric progression in this way produces what is called a *geometric sum*.

As so often in the past, there is a pattern that emerges when this question is systematically investigated. For the present discussion, we will be content to show what that pattern is, and how it can be used. To illustrate the pattern, here are some specific examples:

$$1 + .3 + (.3)^2 = \frac{1 - (.3)^3}{1 - .3}$$

$$1 + .3 + (.3)^2 + (.3)^3 = \frac{1 - (.3)^4}{1 - .3}$$

$$1 + .3 + (.3)^2 + (.3)^3 + (.3)^4 = \frac{1 - (.3)^5}{1 - .3}$$

In the same pattern,

$$1 + .3 + .(3)^2 + \cdots + (.3)^9 = \frac{1 - (.3)^{10}}{1 - .3}$$

Let's verify that the first few lines of the pattern are correct. Starting with the first example, we can compute $1 + .3 + (.3)^2 = 1 + .3 + .09 = 1.39$. On the other side of the equation we find $(1 - (.3)^3)/(1 - .3) = (1 - .027)/.7$. Compute that on your calculator, and you will see that it does in fact equal 1.39. Similarly, it is easy to compute directly $1 + .3 + (.3)^2 + (.3)^3 = 1.39 + .027 = 1.417$. The pattern says this is supposed to be the same as $(1 - (.3)^4)/(1 - .3) = (1 - .0081)/.7$. Your calculator will verify that this is equal to 1.417, as required. To add up all the powers of $(.3)$ from 1 up to $(.3)^9$ directly would be pretty tedious. The pattern allows us to compute this as $(1 - (.3)^{10})/.7 = 1.4286$ (to four decimal places). The point of the pattern is that it allows us to compute in one formula the result of adding up any number of terms of a geometric progression. This provides a shortcut method for computing sums from a geometric growth model.

The general form of the pattern is this:

$$1 + r + r^2 + \cdots + r^n = \frac{1 - r^{n+1}}{1 - r} \tag{1}$$

Using the sum notation of Chapter 5, this becomes

$$\sum_{k=0}^{n} r^k = \frac{1 - r^{n+1}}{1 - r}$$

Notice that on the left we are adding up the powers of a fixed number, r, from $1 = r^0$ up to the nth power. On the right, the top of the fraction involves the next higher power of r. Another equivalent form of this pattern is

$$1 + r + r^2 + \cdots + r^n = \frac{r^{n+1} - 1}{r - 1} \tag{2}$$

The first pattern is more convenient to use if r is less than one, while the second is preferred when r is greater than one, although they both give the same answer. For example, take $n = 6$ and $r = 5$. The pattern says that we can compute $1 + 5 + 5^2 + 5^3 + 5^4 + 5^5 + 5^6$ as either $(1 - 5^7)/(1 - 5)$ or $(5^7 - 1)/(5 - 1)$. The first of these gives $(1 - 78,125)/(1 - 5) = (-78,124)/(-4)$. The second gives $(78,125 - 1)/(5 - 1) = 78,124/4$. Both end up with the same result, but the second version avoids negative numbers in the top and bottom of the fraction.

Notice in the example with powers of 5 that the top of the fraction gets very large. In fact, compared to 5^7, the extra 1 that is subtracted is really insignificant. So for r bigger than 1 and even moderately large values of n, the formula is closely estimated by $r^{n+1}/(r - 1)$. On the other hand, when r is less than 1, the powers of r get very small. In the example with $r = .3$, we computed $(.3)^{10}$, and the result was only .000006. In this case, $1 - r^{n+1}$ will be very nearly equal to 1, so a good estimate for the pattern is $1/(1 - r)$. The higher n is, the better this estimate becomes. So for example, we can estimate

$$1 + \frac{1}{4} + \left(\frac{1}{4}\right)^2 + \cdots + \left(\frac{1}{4}\right)^{20} \approx \frac{1}{1 - \frac{1}{4}} = \frac{1}{\frac{3}{4}} = \frac{4}{3}$$

How was the geometric sum pattern discovered? Probably in just the same way as the other patterns presented in the course. Many examples were worked out, and a relationship was observed between the parameter r and the form of the answer. Can we be sure that the pattern is generally correct? The answer is yes. To be sure that the same pattern holds no matter what numbers are used in place of n and r, we can use algebra to see that the two sides of the equation are equal. This is one of the main advantages of algebra. By allowing us to work with parameters in a way that does not depend on a particular numerical value, algebra reveals patterns that must hold for all numbers. Rather than use algebra in its fullest generality here, we will look at the special case $n = 5$. By examining the algebraic steps for that case, you will get a feeling for why the pattern holds for any choice of n.

To proceed, we want to use algebra to see that the following equation is true:

$$1 + r + r^2 + r^3 + r^4 + r^5 = \frac{1 - r^6}{1 - r}$$

To simplify the algebra, let us use the letter L for what appears on the left side of the equation. Therefore

$$L = 1 + r + r^2 + r^3 + r^4 + r^5$$

Now multiply both sides of this equation by r :

$$rL = r + r^2 + r^3 + r^4 + r^5 + r^6$$

Therefore, subtracting these equations gives

$$L - rL = 1 - r^6$$

This equation can be rewritten in the form

$$(1 - r)L = 1 - r^6$$

and dividing both sides of the equation by $1 - r$ leads to

$$L = \frac{1 - r^6}{1 - r}$$

This is exactly what we wished to show. Keep in mind that the whole point has been to see how the properties of arithmetic make the pattern hold up, no matter what numbers are used in place of r and n. Of course, we don't think about this argument every time we use the pattern. On the contrary, we simply use the pattern without giving much thought to why it is valid. But some such argument as the one given here is necessary to feel confident that using the pattern will not lead us astray.

Now we will turn to the three examples that were described earlier. As you shall see, each of these examples will lead us to some kind of geometric sum. Now that we have a formula for computing such a sum, we will be able to obtain functional equations for the examples. That in turn will provide the means for answering questions that arise in each example.

A Medicine Dosage Model

Suppose that $1/4$ of a drug is eliminated from the body every four hours. If an initial amount of the drug is taken, say 100 mg, then after four hours, after removing $1/4$ of the drug, $3/4$ will remain. That will leave 75 mg in the blood. After another 4 hours, $1/4$ of the remaining drug will be removed, leaving $3/4$ of 75 mg. Continuing in this fashion, it is clear that the amount of drug after each additional four-hour period will be $3/4$ of the amount at the start of that period.

This situation is described by a geometric growth model. We will use the notation d_n to represent the amount of drug in the body after n four hour periods, starting with an initial amount $d_0 = 100$. The geometric growth model has the difference equation

$$d_{n+1} = \tfrac{3}{4}d_n; \qquad d_0 = 100$$

So far this is a typical example of a geometric growth model.

Now we wish to add to the model the effects of taking additional doses every 4 hours. For example, suppose that every four hours 100 mg are added to the amount already in the body. Initially, there are 100 mg in the body. After four hours that amount is reduced by a factor of $3/4$, and then an additional 100 mg are added. So, immediately after taking the second dose, the amount of drug in the blood is

$$\tfrac{3}{4}100 + 100 = 175$$

Now repeat the process. After four hours, $3/4$ of the 175 mg remain, and then another 100 mg are added. Immediately after taking the third dose the amount is

$$\tfrac{3}{4}175 + 100 = 231.25$$

This process can also be represented using a difference equation. Immediately after taking a dose of the medicine, whatever amount of the drug is present in the body will be multiplied by $3/4$ and have 100 added by the time the next dose has been taken. That is,

$$d_1 = \tfrac{3}{4}d_0 + 100$$
$$d_2 = \tfrac{3}{4}d_1 + 100$$

$$d_3 = \tfrac{3}{4}d_2 + 100$$

$$d_4 = \tfrac{3}{4}d_3 + 100$$

and so on. This pattern can be expressed in general as the difference equation

$$d_{n+1} = \tfrac{3}{4}d_n + 100$$

What happens to the total amount of drug in the body according to this model? As a first approach to answering that, we can apply numerical and graphical methods. First, let us simply compute d_n for the first several doses, and see what the numbers look like. At the same time, let us graph the data and see what patterns we observe. See Table 12.1 and Fig. 12.1.

Both the data and the graph reveal an interesting pattern. The amount of drug in the body increases rapidly at first, then levels off and seems to remain nearly steady at about 400 mg. A doctor might have several immediate questions. Is the long-term behavior really steady? If the patient is on the drug for many months, will the level remain at about 400, or will it slowly creep higher and higher? Assuming that the drug level does remain

n	0	1	2	3	4
d_n	100.00	175.00	231.25	273.44	305.08

n	5	6	7	8	9
d_n	328.81	346.61	359.95	369.97	377.47

n	10	11	12	13	14
d_n	383.11	387.33	390.50	392.87	394.65

n	15	16	17	18	19
d_n	395.99	396.99	397.74	398.31	398.73

TABLE 12.1
Data from Mixed Model

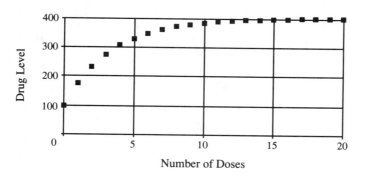

FIGURE 12.1
Drug Levels from Mixed Model

steady, how can the level be predicted? For example, if a steady of dose of 250 mg is desired, what dosage should be taken every four hours? Is there some way to reach the steady dosage sooner? More generally, what is the effect of changing either the initial dosage or the repeated dosage?

Although we will not go into all of these questions, it will be interesting to look at just a few variations on this model. First, since the dosage seems to level off at 400, it seems reasonable to ask what would happen if we started out with an initial dose of 400. That is, give the patient 400 mg to start, and then give repeated doses of 100 mg every four hours. What will happen then? Note that the difference equation for this variation is

$$d_{n+1} = \tfrac{3}{4}d_n + 100; \qquad d_0 = 400$$

Work out the first few data values according to this difference equation. What do you observe?

Next, suppose we give a much larger initial dose, say 1,000 mg, and then the repeated doses of 100 mg every four hours, as before? That is described by this difference equation:

$$d_{n+1} = \tfrac{3}{4}d_n + 100; \qquad d_0 = 1,000$$

You can work out the first several data values for this one. The graph is shown in Fig. 12.2.

What happens if we change the amount of the repeated doses? If we give an initial dose of 50 mg and repeated doses of 50 mg, the difference equation is

$$d_{n+1} = \tfrac{3}{4}d_n + 50; \qquad d_0 = 50$$

The graph for this situation is shown in Fig. 12.3.

These examples show some of the intriguing questions that can come up in a model. In this case, there are several parameters that we can change: the initial dose, the repeated dose, the length of time between doses, and the amount of medicine eliminated by the body between doses. A thorough understanding of the behavior of this kind of model would take into account the effects of these parameters on the long-range drug levels in the body.

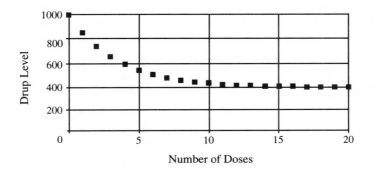

FIGURE 12.2
Drug Level with Large Initial Dose

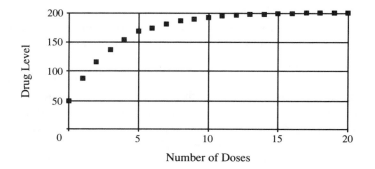

FIGURE 12.3
Initial and Repeated Doses of 50 mg

The difference equation we have been looking at is what might be called a mixed model.[1] It combines elements of both arithmetic growth and geometric growth. Arithmetic growth involves adding the same amount each time you compute the next data value. The general form of the difference equation for arithmetic growth is $a_{n+1} = a_n + d$. Geometric growth involves multiplying by the same amount each time. The difference equation for geometric growth has the form $a_{n+1} = ra_n$. A mixed model combines both steps, multiplication by a constant, and addition of a constant. The difference equation then takes the form $a_{n+1} = ra_n + d$. Any difference equation that is in this form, with numerical constants in place of the parameters r and d, defines a mixed model.

> A growth model that combines both arithmetic and geometric growth is called a **mixed model.** In any mixed model, the difference equation is of the form $a_{n+1} = ra_n + d$ where r and d are parameters.

There is another way to compare arithmetic and geometric growth. Both use special kinds of linear functions to define their difference equations. Each of these cases can be thought of as fitting the pattern

$$a_{n+1} = f(a_n)$$

where $f(a_n)$ indicates that some operation is to be performed on a_n. For arithmetic growth, the form of the f is defined by $f(a_n) = a_n + d$. For geometric growth it is $f(a_n) = ra_n$. Both of these are linear functions. The mixed model has the most general form of a linear function: $f(a_n) = ra_n + d$.

Since we can find equations to express a_n as a function of n in both of the special cases, arithmetic growth and geometric growth, it is natural to see whether we can do the same for the mixed model. Then we could make predictions easily, as well as solving inverse problems. For example, we might want to know whether the amount of drug in the body will ever reach 500 mg, and if so, when. Moreover, a functional equation would allow us to see the effects of changing parameters in the model.

---

[1] Although the expression *mixed model* will be used throughout the chapter, the reader is cautioned that this is not a standard terminology in the literature.

One approach to seeking a functional equation is to work through several terms from the difference equation and look for a pattern. For the difference equation

$$d_{n+1} = \tfrac{3}{4}d_n + 100$$

the first term is given by

$$d_1 = \tfrac{3}{4}d_0 + 100$$

For the next term, we compute

$$
\begin{aligned}
d_2 &= \tfrac{3}{4}d_1 + 100 \\
&= \tfrac{3}{4}\left(\tfrac{3}{4}d_0 + 100\right) + 100 \\
&= (\tfrac{3}{4})^2 d_0 + 100 \cdot \tfrac{3}{4} + 100 \\
&= (\tfrac{3}{4})^2 d_0 + 100\left(\tfrac{3}{4} + 1\right)
\end{aligned}
$$

Next, we have

$$
\begin{aligned}
d_3 &= \tfrac{3}{4}d_2 + 100 \\
&= \tfrac{3}{4}\left[(\tfrac{3}{4})^2 d_0 + 100 \cdot \tfrac{3}{4} + 100\right] + 100 \\
&= (\tfrac{3}{4})^3 d_0 + 100 \cdot (\tfrac{3}{4})^2 + 100 \cdot \tfrac{3}{4} + 100 \\
&= (\tfrac{3}{4})^3 d_0 + 100\left[(\tfrac{3}{4})^2 + \tfrac{3}{4} + 1\right]
\end{aligned}
$$

One more:

$$
\begin{aligned}
d_4 &= \tfrac{3}{4}d_3 + 100 \\
&= \tfrac{3}{4}\left[(\tfrac{3}{4})^3 d_0 + 100 \cdot (\tfrac{3}{4})^2 + 100 \cdot \tfrac{3}{4} + 100\right] + 100 \\
&= (\tfrac{3}{4})^4 d_0 + 100 \cdot (\tfrac{3}{4})^3 + 100 \cdot (\tfrac{3}{4})^2 + 100 \cdot \tfrac{3}{4} + 100 \\
&= (\tfrac{3}{4})^4 d_0 + 100\left[(\tfrac{3}{4})^3 + (\tfrac{3}{4})^2 + \tfrac{3}{4} + 1\right]
\end{aligned}
$$

There is a pattern to these results. To make it clearer, the results we have already found are repeated below:

$$
\begin{aligned}
d_1 &= \tfrac{3}{4}d_0 + 100 \\
d_2 &= (\tfrac{3}{4})^2 d_0 + 100\left[\tfrac{3}{4} + 1\right] \\
d_3 &= (\tfrac{3}{4})^3 d_0 + 100\left[(\tfrac{3}{4})^2 + \tfrac{3}{4} + 1\right] \\
d_4 &= (\tfrac{3}{4})^4 d_0 + 100\left[(\tfrac{3}{4})^3 + (\tfrac{3}{4})^2 + \tfrac{3}{4} + 1\right]
\end{aligned}
$$

Can you see what the next term should be? It will start with $(3/4)^5 d_0$ and then add 100 times a sum of powers of $(3/4)$. This sum goes from $(3/4)^4$ down to $(3/4)^0$, which is 1. Similarly, according to this pattern, we expect that d_{10} would appear as

$$d_{10} = (\tfrac{3}{4})^{10} d_0 + 100\left[(\tfrac{3}{4})^9 + (\tfrac{3}{4})^8 + \cdots + (\tfrac{3}{4})^2 + \tfrac{3}{4} + 1\right]$$

Notice that this pattern allows us to write an immediate expression for d_{10} without first computing all the preceding d's. If we use the variable n in place of 10, we obtain a functional equation of sorts:

$$d_n = (\tfrac{3}{4})^n d_0 + 100\left[(\tfrac{3}{4})^{n-1} + (\tfrac{3}{4})^{n-2} + \cdots + \tfrac{3}{4} + 1\right] \tag{3}$$

Why is this called a functional equation *of sorts*? It does tell us how to compute d_n as soon as we specify a value of n, and in that sense it expresses d_n as a function of n. But the part of the equation that requires us to add up all the powers of $3/4$ is not really in the spirit of a functional equation. It takes too much work, and the bigger n is, the more work it takes. This is where we can apply the results of the earlier discussion of geometric sums. According to the geometric sum pattern Eq. (1),

$$(\tfrac{3}{4})^{n-1} + (\tfrac{3}{4})^{n-2} + \cdots + (\tfrac{3}{4})^2 + \tfrac{3}{4} + 1$$

$$= \frac{1 - (\tfrac{3}{4})^n}{1 - \tfrac{3}{4}}$$

$$= \frac{1 - (\tfrac{3}{4})^n}{\tfrac{1}{4}}$$

$$= 4\left[1 - (\tfrac{3}{4})^n\right]$$

This can be substituted in the equation for d_n, producing the result

$$d_n = (\tfrac{3}{4})^n d_0 + 400\left[1 - (\tfrac{3}{4})^n\right]$$

The equation can be simplified a little by combining the two terms that involve $(3/4)^n$. That produces

$$d_n = 400 + \left(\tfrac{3}{4}\right)^n (d_0 - 400) \tag{4}$$

as the functional equation for d_n.

The steps followed above can be applied in any mixed model. If the difference equation is $a_{n+1} = r a_n + d$, then in place of Eq. (3) we would have

$$a_n = a_0 r^n + d(r^{n-1} + r^{n-2} + \cdots + r + 1)$$

The shortcut for adding up powers of r now leads to the equations

$$a_n = a_0 r^n + d\left(\frac{1 - r^n}{1 - r}\right)$$

and

$$a_n = a_0 r^n + d\left(\frac{r^n - 1}{r - 1}\right)$$

Both of these are correct. Either one may be used as the functional equation for a mixed model. For emphasis, this result is restated below.

> If a mixed model has an initial value of a_0 and obeys the difference
> equation $a_{n+1} = ra_n + d$, then the functional equation can be expressed
> in two ways:
>
> $$a_n = a_0 r^n + d\left(\frac{1 - r^n}{1 - r}\right)$$
>
> and
>
> $$a_n = a_0 r^n + d\left(\frac{r^n - 1}{r - 1}\right)$$

In the drug study example, the parameters are $r = 3/4$ and $d = 100$. As shown earlier, additional steps of algebra can be applied to restate the functional equation as Eq. (4). In this form, we can see at a glance that d_n will in fact level off at 400. The bigger n gets, the smaller $(3/4)^n$ will be, and so, as n increases, d_n gets closer and closer to 400. We can say more. If the initial dose, d_0 is more than 400, the factor $d_0 - 400$ is positive, so each d_n is above 400. If d_0 is less than 400, then each d_n will be less than 400. And if the initial dose is exactly equal to 400, every d_n is equal to 400, too. In this way, we obtain a clear understanding of the qualitative behavior of this model. In the long run, the level of medication levels off at 400, no matter what the initial dose is. If the starting dose is higher than 400, then the level will steadily decrease to 400. On the other hand, if the initial dose is less than 400, it will gradually build up to 400. A starting dose of 400 will produce the long-term level immediately. This information would be useful to a doctor in planning the introduction of the medication. In some cases, there might be a reason to build up the level gradually. In other cases, it might be advisable to start out with a large initial dose, but to taper that off to the desired maintenance level.

The functional equation provides another kind of insight. To emphasize it, let us choose a numerical value for d_0, say 1,000. Then Eq. (4) becomes

$$d_n = 400 + 600(\tfrac{3}{4})^n \tag{5}$$

Observe that the expression on the right is made up of two parts, a constant and an exponential function. This is characteristic of mixed models. Whereas a linear growth model has a linear functional equation, and a geometric growth model has an exponential functional equation, the functional equation for a mixed model will always feature a constant added to an exponential function. Graphically, this amounts to one of the familiar exponential function curves we saw in Chapter 10, shifted up or down on the graph. So, for the specific example of Eq. (5), we have the familiar graph of the exponential function $600(3/4)^n$ shifted upward by 400 units. This is shown in Fig. 12.2. That figure has the typical shape for an exponential function with a base less than 1, but it levels off at a value of 400, rather than along the horizontal axis. The added constant of 400 in Eq. (5) has the effect of shifting the exponential curve upward 400 units. In general, a mixed model will always have a graph that appears as a vertically shifted exponential curve.

Recognizing the functional equation as a shifted exponential is one way of seeing that the amount of drug will level off. There is another way to see this using the difference equation. Imagine that the drug level completely levels off, so that there is no change at all from one data point to the next. Then we would have a situation where the difference equation

$$a_{n+1} = (3/4)a_n + 100$$

would produce exactly the same value for a_{n+1} that we already had for a_n. At what point can that occur? Let us use a variable x for this unknown level. When a_n reaches the value of x, then a_{n+1} will also equal x. So, with a_{n+1} and a_n both equal to the unknown level x, the difference equation above becomes

$$x = (3/4)x + 100$$

This is a linear equation that can be easily solved: subtract $(3/4)x$ from both sides to get $(1/4)x = 100$, and that means $x = 400$. That is a much easier way to find the leveling-off point!

In this type of analysis, x is called a fixed point of the difference equation. The idea is this. A difference equation describes a way of operating on a data value to figure out the next data value. For the drug dosage problem the operation on a_n is to multiply by $3/4$ and then add 100, and any other difference equation can be viewed in similar way. At a fixed point, the operation has no effect. That point is left *fixed*, or unchanged by the operation. Except for some strange cases, when a difference equation model levels off somewhere, it does so at a fixed point. So finding a fixed point is one good way of trying to find leveling-off points. And the method for finding a fixed point is always the same. Replace both a_{n+1} and a_n in the difference equation with x, and then solve for x.

The idea of a fixed point is an important one, and enters into the analysis of difference equation models in a surprisingly significant way. Interestingly, in a mixed model (and in many other kinds of models) it is not possible to reach a fixed point exactly, unless the model starts there. Instead, the fixed point serves as a kind of ideal state to which the sequence values are attracted. So when we say that the model levels off, that is only approximately true. It is a very good approximation. For all practical purposes, the model can be assumed to reach a state of equilibrium. But the sequence values cannot ever reach the fixed point with perfect, mathematical, equality. In spite of that fact, fixed points often hold the key to understanding what a model is going to do. We will return to this idea in the next two chapters.

We are just about finished with the discussion of the medicine dosage model. It has provided examples of a number of features that are typical for mixed models. There is just one more feature of these models that needs to be mentioned here, and the drug model will again provide a good example. The point is this: Expressing results about a model in a general form with parameters helps us to see the effect of changing the assumptions of the model. In the drug model, the parameters are: d for the repeated dosage; and r, which denotes the fraction of the drug that remains in the body after each time period. For the specific case that $d = 100$ and $r = 3/4$ we found that the drug dosage levels off at 400. Now we can try changing the assumptions and see what the effect is. If you set the repeated dose at 50, rather than at 100, you can again determine that the model levels off. This can be done either by repeating the analysis of the functional equation, or by looking for a fixed point. Following this second approach, since the difference equation

is
$$a_{n+1} = (3/4)a_n + 50$$

at the fixed point x we would have
$$x = (3/4)x + 50$$

Solving this equation shows that the fixed point is at 200.

Now repeat this process, but use parameters instead of the constants. The difference equation is
$$a_{n+1} = ra_n + d$$

so at the fixed point x we would have
$$x = rx + d$$

Using algebra, we find $x - rx = d$, or $(1 - r)x = d$. Solving for x produces
$$x = d/(1 - r)$$

Now this form of the equation makes it a snap to see what happens when we change the assumptions in the model. If we use $r = 3/4$, in which case $(1 - r) = 1/4$, then the equation for the fixed point becomes
$$x = 4d$$

This shows at once what happens when we change the repeated dosage. If the patient gets 100 units at each dose, the drug will level off at 400. If the patient gets 50 units repeatedly, the drug will level off at 200. Knowing this, if the doctor decides that a long-term dosage of 120 mg is required, the repeated dose should equal 30 mg.

For another variation, suppose that the body only eliminates $1/6$ of the drug between doses. That means that $5/6$ of the drug remains, and so $r = 5/6$. Then the fixed point equation will become
$$x = \frac{d}{1 - r} = \frac{d}{1/6} = 6d$$

In this case the long term drug level will be equal to 6 times the repeated dose. With this information, we can predict what the effects of repeated medication will be as soon as we know how fast the body eliminates the drug. The model can then be used to decide what dosage to set in order to reach a desired long term level of the drug in the body.

This example seems to be much richer than those we have studied earlier. The theoretical approach in this case leads to many interesting relationships, and makes this model very useful in setting drug levels. What is more, the methods we have developed here can be applied in any mixed growth model. We will look at another example next, one that has nothing to do with drug dosages.

A Mixed Model for Repeated Loan Payments

Have you ever wondered how payments are worked out for a purchase on credit? Say you buy a $2,000 stereo and spread the payments over 2 years. How does the store or finance company figure out how much you should pay each month? They consider that you have been loaned $2,000, and every month they require that you pay interest on that loan. Part of each payment you make is to pay off the interest. The rest of your payment is used to repay part of the loan.

Let's work out the math for a few payments. To simplify things, we will assume that the interest charged is one percent per month, and that you bought the stereo on January first. On February first, there is a payment due. You owe one percent of the $2,000 loan, and that is $20 of interest. But your payment will be more than just the interest, otherwise the loan will never get paid off. So say that the monthly payment is $50. For the first payment the interest is $20. The remaining $30 pays off part of the loan. That leaves a balance of $2,000 − $30 = $1,970. Next, on March first, another payment is due. This time you only owe interest on $1,970. That amounts to one percent of $1,970, or $19.70. Your $50 payment includes the $19.70 of interest, plus $30.30 paying off the loan. After the second payment, the amount you still owe is $1,970 − $30.30 = $1,939.70. Now you can see that it will be pretty tedious to work this out all the way to the point where the loan is paid off. Fortunately, we can use difference equations to model the process. And in this case, the model is exactly correct.

To formulate the difference equation, we begin with a decision to focus on how much is still owed after each payment. In symbols, let b_n be the balance after n payments. Then $b_0 = 2,000$, because after no payments you owe the full amount. In the discussion above, we computed $b_1 = 1,970$ and $b_2 = 1,939.30$. Now we need to see a pattern in those calculations showing how the balance owed changes with each payment. Reviewing the steps followed before, observe that we computed one percent of the old balance, subtracted the result from 50, and then subtracted *that* result from the old balance. In words, we can write this as follows

$$\text{next balance} = \text{current balance} - (50 - .01 \cdot [\text{current balance}])$$

That translates to this difference equation:

$$b_{n+1} = b_n - (50 - .01 \cdot b_n)$$

Using a little algebra, we can simplify this slightly

$$b_{n+1} = b_n - (50 - .01 \cdot b_n)$$
$$= b_n - 50 + .01 \cdot b_n$$
$$= 1.01 \cdot b_n - 50$$

This is another mixed model—it has the same form as the difference equation for the drug dosage example. It is not surprising that we can analyze it in the same way.

In fact, we can recognize in this equation the two effects—interest and payments. If you owe b_n, then with interest, that grows to $1.01 b_n$ at the end of a month. That is the familiar form of geometric growth, and the growth factor 1.01 is 1 plus the interest rate. But then you make a 50-dollar payment, so the balance is reduced by 50. The difference equation shows that each month the previous balance gets multiplied by 1.01 and then is reduced by subtracting 50. That is a mixed model.

Using the difference equation, we find the first few b_n's follow a pattern:

$$b_1 = 1.01b_0 - 50$$

$$b_2 = 1.01^2 b_0 - 1.01 \cdot 50 - 50$$

$$b_3 = 1.01^3 b_0 - 1.01^2 \cdot 50 - 1.01 \cdot 50 - 50$$

$$b_4 = 1.01^4 b_0 - 1.01^3 \cdot 50 - 1.01^2 \cdot 50 - 1.01 \cdot 50 - 50$$

As the pattern suggests, if we make 24 payments, the balance that remains will be

$$b_{24} = 1.01^{24} b_0 - 1.01^{23} \cdot 50 - 1.01^{22} \cdot 50 - \cdots - 1.01 \cdot 50 - 50$$

As one more simplification, collect all the terms that include 50, and group them together:

$$b_{24} = 1.01^{24} b_0 - 50(1.01^{23} + 1.01^{22} + \cdots + 1.01 + 1)$$

Once again, we observe the presence of a geometric sum. This time we have

$$1.01^{23} + 1.01^{22} + \cdots + 1.01 + 1 = \frac{1.01^{24} - 1}{1.01 - 1} = 26.9735$$

Therefore, the balance after 24 payments will be

$$b_{24} = 1.01^{24} b_0 - 50 \cdot \frac{1.01^{24} - 1}{1.01 - 1}$$

This is just what we would have found using the functional equation for a mixed model with $r = 1.01$ and $d = -50$. To get a numerical answer, simply compute

$$b_{24} = 1.01^{24} b_0 - 50 \cdot \frac{1.01^{24} - 1}{1.01 - 1}$$

$$= 1.01^{24} \cdot 2,000 - 50 \cdot 26.9735$$

$$= 2,539.47 - 1,348.68 = 1,190.79$$

This result shows that a monthly payment of \$50 is too low. After two years of payments, there will still be some money owed. Using a numerical approach, you could now work through the entire process with a payment of \$75 per month and see whether that is enough to pay off the loan in two years. In that process how would the calculations be modified? Go back through the steps. Do you see that the only place the payment amount shows up is in a single figure of 50? If that is changed to 75, the end result will be

$$b_{24} = 1.01^{24} b_0 - 75 \cdot \frac{1.01^{24} - 1}{1.01 - 1}$$

$$= 1.01^{24} b_0 - 75 \cdot 26.9735$$

$$= 1.01^{24} \cdot 2,000 - 75 \cdot 26.9735$$

$$= 2,539.47 - 2,023.01 = 516.46$$

This shows that a payment of $75 is still too low, so we should try again with an even higher payment amount. By successively refining the estimate of the payment, we would eventually end up with an amount that would just pay off the balance in exactly 2 years.

A theoretical approach will lead to the same conclusion with less work. Introduce the variable P to stand for the monthly payment. This variable shows up in just one place in the calculation, just as 50 did the first time, and 75 the next. In fact, it is just the parameter d from the general mixed model difference equation. Using the functional equation with P in place of d, we find

$$b_{24} = 1.01^{24}b_0 - P \cdot \frac{1.01^{24} - 1}{1.01 - 1}$$
$$= 1.01^{24}b_0 - P \cdot 26.9735$$
$$= 1.01^{24} \cdot 2,000 - P \cdot 26.9735$$
$$= 2,539.47 - P \cdot 26.9735$$

If we choose the payment correctly, the balance after 24 months will be exactly 0. So we set $b_{24} = 0$ and solve the resulting equation,

$$0 = 2,539.47 - P \cdot 26.9735$$

for P. The result is $P = 2,539.47/26.9735 = 94.15$

This example shows how the use of difference equations, and our understanding of mixed models, can be used to determine what the monthly payment should be. Having done this analysis once, we need never do it again. In the equation

$$b_{24} = 1.01^{24}b_0 - P \cdot \frac{1.01^{24} - 1}{1.01 - 1}$$

it is possible to identify recognizable aspects of the original problem. We use 24 because we wish to make 24 payments; 1.01 comes from the interest rate of one percent per month; b_0 is the original amount borrowed; P is the monthly payment. Using the same pattern you can make calculations for any monthly payment problem. If the interest is 1.125 percent per month instead of 1 percent, it should be clear what to change in the equation. Likewise, if you wish to pay the loan off over 5 years rather than 2 years, you should be able to see what to change. This is the same idea that was presented in the discussion of the drug dosage model, concerning the advantages of using parameters to express conclusons about models. In fact, by introducing parameters for interest per month, and number of payments, you can find an equation that can always be used to find the monthly payment amount. That will be left to you to do in the exercises.

It should also be mentioned here that the repeated payments model, unlike the medicine model, does not level off. The difference between the two models is that r is bigger than 1 for the loan payment model, but is less than 1 for the medicine model. For the loan payments model, if you keep making payments, the loan balance will keep going down, eventually becoming negative. But we do not it expect it to level off at a fixed balance that remains the same month after month. Given this understanding, it is interesting to see what happens if you try to find a fixed point. As always, we take the difference equation

$$b_{n+1} = 1.01b_n - 50$$

and set both b_n and b_{n+1} equal to an unknown value x. The result is

$$x = 1.01x - 50$$

Now solve for x. Rearranging the equation to $50 = 1.01x - x$, we find $50 = .01x$, and so $x = 50/.01 = 5,000$. This agrees with the earlier formula $x = d/(1 - r) = (-50)/(-.01)$. Since the loan started at \$2,000 and the balance is going down each month, you never reach this fixed point. On the other hand, if the original loan was above \$5,000, then the payment would not be enough to cover the interest each month, and so the balance would keep growing. The fixed point only appears if the loan starts out at exactly \$5,000. In this case, the monthly payment is just enough to pay the interest charges, and the balance just stays constant at \$5,000. This shows that a fixed point doesn't always have to be a leveling-off point, although a leveling-off point must always occur at a fixed point. For mixed models, the general rule is this: if $r < 1$, the model will level off at $d/(1 - r)$; and if $r > 1$, the model won't level off.

In both of the preceding examples, a mixed model was found by working directly with the difference equation used to describe a problem. Our understanding of how repeated drug doses work, and of how interest and loan payments work, turned out to give the mixed-model difference equation. The shortcut for adding up powers of a number was essential for finding the functional equation. This shortcut can also be applied to find sums in geometric growth models. That will be discussed next.

Geometric Sum Models

It is often of interest to add up the terms of a number sequence. For example, suppose we study an epidemic, using a discrete variable c_n to model the number of *new* cases of the disease each day. At the start of the model, the total number of infected people will be a known figure, say 1,000. To be even more specific, we will say there were 1,000 infected people at the *start* of day 0, and there were c_0 new cases *during* day 0. That leads to a total of $1,000 + c_0$ infected people at the start of day 1. In a similar way there will be $1,000 + c_0 + c_1$ infected people at the start of day 2, $1,000 + c_0 + c_1 + c_2$ infected people at the start of day 3, and so on. The pattern here should be apparent: at the start of day 5 there will be $1,000 + c_0 + c_1 + c_2 + c_3 + c_4$ infected people. In words, this says that the total number of infected people is defined by the 1,000 we started with, plus the number of new cases observed each day.

We have seen this situation before. We can describe it using a difference equation if we introduce a new variable. Let T_n be the total number of infected people at the start of day n. Verbally, we can describe the growth of T_n as follows: the total number of sick people at the start of a new day is the total number at the start of the preceding day, plus the number of new cases during that day. In symbols, this becomes the difference equation

$$T_{n+1} = T_n + c_n$$

Since $T_0 = 1,000$, we can work out several examples using this difference equation observe the same pattern we saw earlier:

$$T_1 = 1,000 + c_0$$

$$T_2 = 1,000 + c_0 + c_1$$

$$T_3 = 1,000 + c_0 + c_1 + c_2$$

$$\vdots$$

$$T_n = 1,000 + c_0 + c_1 + c_2 + \cdots + c_{n-1}$$

Although this framework can be used with any kind of model for c_n, we are especially interested here in the situation that arises when c_n follows a geometric growth model. Then c_n will be of the form $c_0 r^n$ and the sum $c_0 + \cdots + c_{n-1}$ can be worked out using either of the geometric sums Eq. (1) or Eq. (2).

The discussion in this section started out in the context of a model for the spread of a disease. Similar questions about adding up the terms in a model arise in many other situations, including models for the school population in a county, the growth of a regular investment program, or the way oil reserves are used up subject to a growing rate of consumption. Our previous encounters with this type of situation have involved arithmetic growth models. Starting on page 97, a model for the construction of houses is discussed. There, an arithmetic growth model is used for the number of new houses built each year, and these annual construction figures are added up to compute the total number of houses built over several years. If the number of new houses each year is estimated with a geometric growth model, instead, then we end up with a geometric sum. Similarly, the oil reserves model discussed on page 46 assumes that annual oil consumption will grow arithmetically in the future. Using a geometric growth model for consumption instead would lead to a reserves model that includes a geometric sum.

Let us work through a specific example. We will consider a mythical county's school population, assuming geometric growth in annual enrollments.

In Fantasy County, the 1990 population of elementary school students numbered 18,700. The county planners project future enrollments to follow a geometric growth model, increasing 2 percent per year over the next decade. To budget for consumable supplies, it is necessary to add up the total number of students over the decade. Let p_n be the number of students enrolled in year n. Assuming geometric growth, we have

$$p_n = 18,700 \cdot (1.02)^n$$

where the projected growth rate of 2 percent results in a growth *factor* of 1.02. Now we want to add up the populations for years 0 through 9:

$$\begin{aligned} \text{Total} &= p_0 + p_1 + p_2 + \cdots + p_9 \\ &= 18{,}700 + 18{,}700(1.02) + 18{,}700(1.02)^2 \\ &\quad + 18{,}700(1.02)^3 + \cdots + 18{,}700(1.02)^9 \\ &= 18{,}700(1 + 1.02^1 + 1.02^2 + \cdots + 1.02^9) \end{aligned}$$

Here we can use our geometric sum pattern

$$1 + 1.02^1 + 1.02^2 + \cdots + 1.02^9 = \frac{1.02^{10} - 1}{1.02 - 1}$$

which simplifies to give a total of

$$\frac{1.02^{10} - 1}{1.02 - 1} = \frac{.21899}{.02} = 10.95$$

to two decimal places. This gives the total number of students in the schools over the decade as

$$\text{Total} = 18{,}700 \cdot 10.95 = 204{,}765$$

Using this analysis, it is easy to see how the 10 year total is affected by different assumptions. For example, suppose we assume that the population will grow by 10 percent each year, instead of 2 percent. Then the only change in the previous calculations is a replacement of the figure 1.02 by 1.10. The final total for the ten years would then be

$$\text{Total} = 18{,}700 \cdot \frac{1.1^{10} - 1}{1.1 - 1} = 298{,}000$$

We can as easily make a projection for any estimate of the growth rate. Once again, it is profitable to express our results in terms of a parameter. Let r be the percentage of growth each year, expressed as a decimal. So with a 2-percent growth rate we would have $r = .02$. Using the parameter, the total over ten years will be

$$\text{Total} = 18{,}700 \cdot \frac{(1 + r)^{10} - 1}{r}$$

This formula can be used to quickly compare different assumptions about the growth rate. This has been done for several different growth rates and depicted graphically in Fig. 12.4. Note that a value of $r = .1$ on the horizontal axis indicates a 10-percent growth rate. The corresponding value on the vertical axis shows what enrollment the model predicts over 10 years, based on that growth rate. With this kind of graph, the

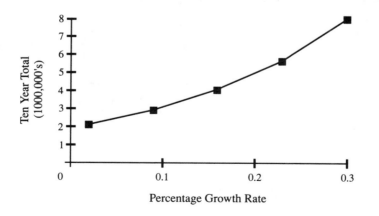

FIGURE 12.4
Projected 10 Year Enrollment vs. Growth Rate

analyst studies how conclusions from a model depend on the assumptions that went into the model.

Two comments are in order here. First, the ability to define such a compact equation rests on our formula for adding a geometric progression. Indeed, using the formula, we find a relatively simple expression for the ten-year total school enrollment as a function of the growth rate r; as soon as a value is specified for r we can readily compute the ten-year total. Second, the use of the geometric sum formula gives an exact projection of what will occur under a geometric growth model. This is quite different from working directly with the population figures and trying to fit a model. Imagine that we somehow obtained the projections of enrollments without knowing how they were computed. At a 2-percent growth rate the projected enrollment for the decade is about 205,000; at a 10-percent growth rate the projection is 298,000. Based on these figures, what would you expect an 18 percent growth rate to produce? Using proportional reasoning, you could argue that increasing the growth rate by 8 percent caused an increase of 93,000 in enrollments, and so another increase of 8 percent should also produce an enrollment increase of 93,000. In that case we would project that an 18-percent growth rate would produce an enrollment total of 391,000. Is that an accurate forecast? As we have seen, that kind of proportional reasoning is equivalent to using a straight-line model. Often that is an adequate approach if we have no other information available. But in the present case we can compute the enrollments projections exactly. As Fig. 12.4 shows, the true relationship between growth rate and enrollment projection does not follow a straight line. Our theoretical approach has given us a much more detailed understanding of the effects of different growth rates. We still must make an initial assumption about the underlying growth model. True population growth will not exactly follow a geometric growth model. But once that assumption is made, we can derive the more complicated relationship between growth rate and enrollment projection exactly. That is the strength of the theoretical approach.

A Mixed Model for Oil Consumption and Reserves

As a final example, we return to a familiar problem: oil consumption. In Table 12.2 and Fig. 12.5 the data and graph we have used before are repeated for easy reference.

Year	1991	1992	1993	1994	1995
Daily Oil Use	66.6	66.9	66.9	67.6	68.4

TABLE **12.2**
Daily World Oil Consumption in Millions of Barrels

Early in the course we looked at an arithmetic growth model for this data set. However, the data really don't look much like a straight line. A more reasonable model might involve geometric growth. After all, oil consumption will be heavily influenced by

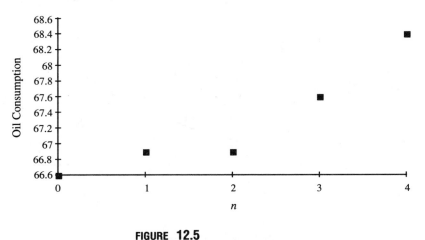

FIGURE 12.5
Daily World Oil Consumption

population growth, and that, we know, is often much more accurately described by geometric growth than by arithmetic growth. Unfortunately, it is not possible to find a geometric growth model that fits the data very well. This should not really surprise us because it is reasonable to expect population growth to be just one of many factors that contribute to the consumption of oil. Other obvious variables include economic development and market forces. Given these insights, a mixed model provides an attractive candidate for trying to approximate the oil consumption data. It involves both arithmetic and geometric influences, and can accurately portray a wider range of growth patterns than either arithmetic or geometric growth alone. Starting with the basic form of the difference equation

$$c_{n+1} = rc_n + d$$

we can attempt to find choices for r and d that provide a good fit to the data. By trial and error, I came up with the following model:

$$c_{n+1} = 1.6 \cdot c_n - 39.765; \qquad c_0 = 66.6$$

where c_n, as usual, is the average daily consumption n years after 1991. The first several values produced by this model are shown in Fig. 12.6 along with the original data points. As you can see from the graph, the model does a good job of approximating the data.

At this point, we can again work with patterns to find the functional equation. However, it is easier to use the general form for the functional equation of a mixed model, as shown in the box on page 265. The parameters we are using are $r = 1.6$ and $d = -39.765$, with $c_0 = 66.6$. The functional equation is therefore

$$c_n = 66.6 \cdot 1.6^n - 39.765 \left(\frac{1.6^n - 1}{1.6 - 1} \right)$$

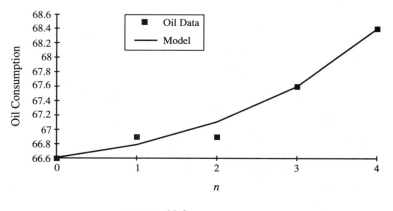

FIGURE 12.6
Model and Data Comparison

Applying a little algebra,

$$\frac{1.6^n - 1}{1.6 - 1} = \frac{1.6^n - 1}{.6}$$

$$= \frac{1}{.6}(1.6^n - 1)$$

$$= 1.6667(1.6^n - 1)$$

$$= 1.6667 \cdot 1.6^n - 1.6667$$

Combined with the earlier equation for c_n this brings us to

$$c_n = 66.6 \cdot 1.6^n - 39.765(1.6667 \cdot 1.6^n - 1.6667)$$

$$= (66.6 - 66.275)1.6^n + 66.275$$

$$= .325(1.6)^n + 66.275$$

In this way, using our best fitting mixed-model difference equation, we obtain a functional equation for daily oil consumption.

Now we can again formulate an oil reserves model based on the consumption model, just as in Chapter 3. As before, we observe that the reserves next year will be what is left after subtracting this year's consumption from this year's reserves. That can be expressed as a difference equation $r_{n+1} = r_n - 365c_n$ (where the factor of 365 reflects the fact that c_n tells how much oil is used up each day of the year). Or, looked at another way, the reserves after n years can be found by adding up the consumption for all those years and then subtracting the total from the original reserves. That is expressed by this equation:

$$r_n = r_0 - 365(c_0 + c_1 + c_2 + \cdots + c_{n-1})$$

Both of these equations convey the same thing. The point to observe, though, is that we have a functional equation for c_n, as determined in our mixed model. This can be used to replace each c in the preceding equation. The replacements will follow this pattern:

$$c_0 = .325 + 66.275$$

$$c_1 = .325(1.6) + 66.275$$

$$c_2 = .325(1.6)^2 + 66.275$$

$$\vdots$$

$$c_{n-1} = .325(1.6)^{n-1} + 66.275$$

Adding these all up, and rearranging the terms we find

$$c_0 + c_1 + \cdots + c_{n-1} = .325(1 + 1.6 + 1.6^2 + \cdots + 1.6^{n-1})$$
$$+ (66.275 + 66.275 + \cdots + 66.275)$$

The part in the first set of parentheses we know how to sum. The second set of parentheses just adds up to $(n)66.275$. These results then lead to

$$r_n = r_0 - 365 \left[.325 \frac{1.6^n - 1}{.6} + 66.275n \right]$$

Algebra can be used to simplify this a little. The final result is

$$r_n = r_0 - [197.708(1.6)^n + 24,190.375n - 197.708]$$

This is a functional equation for the oil reserves n years after 1991. It is interesting that in this analysis we actually had to use the ability to compute a geometric sum twice. First, we used a mixed model for the annual oil consumption, and a geometric sum came up when we found the functional equation. Then, we used oil consumption to model oil reserves, and a geometric sum came up again when we found the functional equation for that. This process also illustrates again the way one model builds on another. We have been steadily filling up a tool box of model types, starting with arithmetic growth, then quadratic growth models, then geometric growth, sums of geometric growth, and mixed models. Here finally we have developed a model featuring a sum of a mixed model. Isn't it amazing how far we have been able to progress using just a few mathematical procedures!

Let us ask the same question we did with the arithmetic growth model for oil consumption: when will the oil run out? That is, for what n will $r_n = 0$? Since $r_0 = 999{,}100$, we would like to solve this equation:

$$999{,}100 - [197.708(1.6)^n + 24,190.375n - 197.708] = 0$$

There is no simple formula or procedure to solve this exactly, but a numerical approach is certainly feasible. Guess an answer: the oil will run out in 30 years. Set $n = 30$ and compute r_{30} using the functional equation for r_n. What do you find? A negative result indicates that the oil will not last that long. So try 20 years. Continuing in this fashion, it won't take you long to discover that the oil runs out between $n = 17$ and $n = 18$. Therefore, according to this model, the oil will run out between 2008 and 2009.

Do you recall what happened when we approached the same question using an arithmetic growth model for oil consumption? In that case, our functional equation for oil reserves was

$$r_n = 999{,}100 - 24{,}254.25n - 54.75n^2$$

To find when the reserves reach 0, we need to solve

$$0 = 999{,}100 - 24{,}254.25n - 54.75n^2$$

This is a quadratic equation, and we can find the solution using the quadratic formula presented in Chapter 7. The formula is

$$n = \frac{-b \pm \sqrt{b^2 - 4ac}}{2a}$$

where $a = -54.75$, $b = -24{,}254.25$, and $c = 999{,}100$. This actually leads to two solutions:

$$n = -480.94$$

and

$$n = 37.94$$

The one that is negative is not valid for our model, which was developed using the assumption that n is positive. The other solution gives the predicted time until the oil runs out as 37.94 years. In summary, with an arithmetic growth model for annual oil consumption, we predict the oil will last nearly 38 years. But with a mixed model for oil consumption, we predict the oil will only last about 18 years.

This illustrates one of the reasons that scientific analyses often produce controversial results. Here, we have used two very simple models to describe oil reserves. In one, we conclude that the oil will last almost 40 years. In the other, the oil is predicted to run out after only 18 years. Of course, for a real analysis, we would use much more data and develop additional theories about why oil consumption grows. But ultimately, any model includes assumptions. Making different assumptions can lead to widely different conclusions. Different scientists can find different models completely believable and reasonable. When the results are controversial, one really should go back to the assumptions of each model and try to decide which seem most plausible. In general, even scientifically proven results can only be as valid as the assumptions on which they are founded. It should also be remembered that often conclusions drawn from a model are used to argue for some kind of change. In that kind of context, we might say, *if oil consumption continues to grow according to our model, we will run out of oil in 18 years.* Here, the point of the model is to point out possible future scenarios. If we don't want the oil to run out so quickly, we will want to see the oil consumption deviate from the model presented here. This deviation might occur as a result of overt action by policymakers, or might occur if nothing is done whatever. In keeping an eye on consumption, we can use the model as a baseline for comparison purposes. If the consumption data continue to follow the model, the conclusions drawn from the model become more plausible.

Summary

This chapter has shown some examples that can be built up from earlier work with geometric growth. One problem that emerges in each example is to calculate a geometric sum—that is, a sum of terms from a geometric growth model. A general pattern holds for all such sums, as given in the following equation:

$$1 + r + r^2 + \cdots + r^n = \frac{r^{n+1} - 1}{r - 1}$$

Using this pattern, we were able to develop a functional equation for a difference equation taking the form

$$a_{n+1} = ra_n + d$$

where r and d represent numerical constants. This kind of difference equation was referred to as a mixed model because it combines elements of both arithmetic and geometric growth, although the mixed model terminology is not a standard one in the literature.

Mixed models were used in three examples. The first example derived a mixed model from a simple description of the way medication is introduced to and removed from the blood stream. In that example, it was an understanding of a physical process that suggested the mixed model. The second example used a mixed model to calculate the cumulative effect of making monthly payments to pay off a loan. In that case, the rules governing the growth of interest charges are completely artificial, reflecting customs and conventions (as opposed to the behavior of some physical phenomena). The model is an exact representation of these rules, and so gives results that are exactly correct, within the limits of the accuracy of computation. The third and final example simply adopted a mixed model as a plausible choice for describing some data. The parameters in the mixed model were chosen to obtain as close an agreement as possible between actual data and the model. In this example we re-analyzed the issue of oil reserve depletion using a mixed model for annual oil consumption. The results were compared to those derived earlier under the assumption of an arithmetic growth of annual oil consumption.

In addition to all of these mixed model examples, there was one example defined in terms of a sum of geometric growth results. In that example, annual school enrollments were modeled using a geometric growth assumption. The annual enrollment figures were then totaled to obtain projections of total student enrollments over a ten-year period. Projections of this type might be useful in developing budgets for consumable supplies, and other expenses that are proportional to total enrollments.

Exercises

Reading Comprehension

1. Explain the pattern for adding the terms of a geometric progression. Give an example of the use of this pattern.

2. Explain what is meant by a mixed model. Why is it called *mixed*?

3. Write a one-page essay about fixed points. Tell what a fixed point is, how it can be found, and what it reveals about a mixed model.

4. Which model for oil consumption do you think is more plausible, the arithmetic growth or mixed model? Why?

5. What are the assumptions in the school population example? How do the conclusions depend on the assumptions?

6. What are the assumptions in the drug example? Are these plausible?

Mathematical Skills

1. Calculate the total of each sum:

   a. $1 + 1/2 + 1/4 + 1/8 + \cdots + (1/2)^{10})$

   b. $1 + 1.2 + 1.2^2 + \cdots + 1.2^{20}$

   c. $\displaystyle\sum_{k=0}^{15} 2^k$

   d. $\displaystyle\sum_{k=0}^{10} 9 \cdot .1^k$

2. A mixed-model difference equation is defined by $a_{n+1} = 1.3a_n + 2.4;\quad a_0 = 5.$ Work out a pattern for a_n, and use it to develop a functional equation for a_n

3. Define a new model using the mixed model from the previous problem. The new model is defined by $b_{n+1} = b_n + a_n$, where a_n is defined in the previous problem and $b_0 = 0$. Work out a pattern for b_n and use it to find a functional equation for b_n

4. **[Note: this problem may be extra difficult for some readers.]** Review the discussion of the monthly payment example. Use A for the original amount of the loan (instead of b_0), i for the monthly interest rate expressed as a decimal, P for the amount of each payment, and m for the number of payments. By generalizing from the example in the text, develop an equation relating A, i, P, and m. Use algebra to arrange this equation so that P is expressed as a function of A, i, and m.

Problems in Context

1. Use the formula you found in the last math skills problem to determine the monthly payment to pay off a credit card balance of $1,200 in 18 months, if the credit card charges 1.5 percent interest per month.

2. Suppose you buy a car on credit. The car costs $18,000. You pay $3,000 cash, and pay off the remaining $15,000 over five years. If the auto maker gives you a loan that charges 3 percent per *year*, how many percent would they charge per month? Using that figure, determine what the monthly payment should be to pay off the $15,000 in 5 years.

3. **[Note: this problem may be extra difficult for some readers.]** A couple buys a condo for $180,000. They use their savings to make a down payment of $30,000,

and take out a 30-year mortgage for the rest. The lender specifies the interest as 9 percent per year (which really means $9/12 = .75$ percent per month). What will the monthly mortgage payment be?

4. Continuing the previous problem, another lender will give the couple a mortgage at only 8 percent per year. However, the couple will have to pay a fee of $1,800 to get that loan. How much lower will the payments be at the lower interest rate? How long will it take the savings in the mortgage payments to equal the extra expense of $1,800 required to get the lower interest mortgage?

5. A doctor is testing a new drug. She knows from the research on the drug that the body eliminates 12 percent of the drug every six hours. She wants to have the patient take the drug every 6 hours, and wants the level of the drug in the blood stream to stay around 250 mg. How much should the patient take for each dose?

6. Suppose you begin a savings plan in which you deposit $100 per month in an account that pays one half of a percent of interest per month. On the first day of each month the bank adds interest to the account, and you deposit $100. Develop a model for the amount in the account after n deposits. How much will be in the account after 5 years? 10 years? 15 years? How long will it take to accumulate $100,000?

7. In a factory, water is constantly flowing through a large vat, where it is mixed with a solid chemical. A chemist is studying the way the chemical dissolves and develops a mixed growth model. At the start of the experiment 100 pounds of the chemical in a solid form are added to the vat. Each hour, about one tenth of the chemical dissolves, leaving $9/10$ still lying at the bottom of the vat, and another 20 pounds of the solid chemical is added to the vat. Develop a mixed model for this situation. Let s_n be the amount of solid chemical at the bottom of the vat after n hours, starting with $s_0 = 100$. In your answer, include

  a. The first few terms: s_1, s_2, and s_3.
  b. A difference equation for s_n.
  c. A functional equation for s_n.
  d. A discussion of the long range predictions that can be made based on the model.

Group Activities

1. Discuss adapting the drug model to the problem of describing pollution in a lake. In previous work, we looked at a model where no new pollution enters the lake, and the pollution level decreases geometrically. Using the approach of the medicine model, discuss how to include in the lake model the addition of new pollutants on a regular basis. To illustrate your ideas, make up an example.

2. Discuss using the drug model to describe blood alcohol levels that result from regular drinking. You will need to find out (at the library) the rate at which alcohol leaves the blood. Also, decide whether this is a geometric or arithmetic phenomenon. Make some reasonable assumptions about the number of drinks someone might consume per hour, and develop a model for the way the blood alcohol level goes up during drinking and down after drinking stops.

3. Go the library and get some additional data on annual world oil consumption. Does your new data set fit with the mixed model developed in this chapter? Make your best-fit of either an arithmetic, geometric, or mixed model to your data. Use this model to project when the oil will be used up. Compare the results with those in this chapter.

Solutions to Selected Exercises

Mathematical Skills

1. a.   The answer is $\dfrac{1 - .5^{11}}{1 - .5} = 1.999023$.

   d.   This sum is $9(.1^0) + 9(.1^1) + 9(.1^2) + \cdots + 9(.1^{10})$, which equals

$$9(.1^0 + .1^1 + .1^2 + \cdots + .1^{10}).$$

Now inside the parenthesis is the geometric sum of powers of .1. This can be computed using the shortcut as

$$\frac{1 - .1^{11}}{1 - .1} = \frac{1 - .00000000001}{.9} = \frac{.99999999999}{.9}$$

But the final answer is supposed to be 9 times that, which is 9.9999999999. This sum can also be computed directly without using the shortcut. Again look at the geometric sum in the parentheses. We are adding $.1^0 = 1$ and $.1^1 = .1$ and $.1^2 = .01$ and so on out to $.1^{10} = .0000000001$. Just add these decimals up. The result is clearly 1.1111111111, so 9 times that amount is 9.9999999999.

2. The pattern of the first few terms looks like this:

$$a_0 = 5$$
$$a_1 = 5 \cdot 1.3 + 2.4$$
$$a_2 = 5(1.3)^2 + 2.4(1.3) + 2.4$$
$$a_3 = 5(1.3)^3 + 2.4(1.3)^2 + 2.4(1.3) + 2.4$$
$$a_4 = 5(1.3)^4 + 2.4(1.3)^3 + 2.4(1.3)^2 + 2.4(1.3) + 2.4$$

Each of these equations can be rearranging slightly by combining all the terms that include a 2.4:

$$a_0 = 5$$
$$a_1 = 5 \cdot 1.3 + 2.4$$
$$a_2 = 5(1.3)^2 + 2.4(1.3 + 1)$$
$$a_3 = 5(1.3)^3 + 2.4(1.3^2 + 1.3 + 1)$$
$$a_4 = 5(1.3)^4 + 2.4(1.3^3 + 1.3^2 + 1.3 + 1)$$

Now use the shortcut for adding up powers of 1.3.

$$a_0 = 5$$

$$a_1 = 5 \cdot 1.3 + 2.4$$

$$a_2 = 5(1.3)^2 + 2.4(1.3 + 1)$$

$$a_3 = 5(1.3)^3 + 2.4(1.3^3 - 1)/(1.3 - 1)$$

$$a_4 = 5(1.3)^4 + 2.4(1.3^4 - 1)/(1.3 - 1)$$

Based on this pattern, it should be clear that $a_n = 5(1.3)^n + 2.4(1.3^n - 1)/(1.3 - 1)$. That can be simplified as follows. First $1.3 - 1 = .3$. Also, $2.4/.3 = 24/3 = 8$. So the formula becomes $a_n = 5(1.3)^n + 8(1.3^n - 1) = 5(1.3^n) + 8(1.3^n) - 8 = 13(1.3)^n - 8$

3. The b_n values can be found by adding up the a_n values. For example, $b_1 = b_0 + a_0 = a_0$; then $b_2 = b_1 + a_1 = a_0 + a_1$; and $b_3 = b_2 + a_2 = a_0 + a_1 + a_2$. With this same pattern, it is clear that, say, $b_7 = a_0 + a_1 + a_2 + \cdots + a_6$. Using the functional equation for a_n from the preceding problem, we can list out a_0 through a_6, and add them up:

$$a_0 = 13(1.3)^0 - 8$$

$$a_1 = 13(1.3)^1 - 8$$

$$a_2 = 13(1.3)^2 - 8$$

$$a_3 = 13(1.3)^3 - 8$$

$$a_4 = 13(1.3)^4 - 8$$

$$a_5 = 13(1.3)^5 - 8$$

$$a_6 = 13(1.3)^6 - 8$$

$$\overline{b_7 = 13(1.3^0 + \cdots + 1.3^6) - 8 \cdot 7}$$

Now use the shortcut for adding up powers of 1.3, to find

$$b_7 = 13\frac{1.3^7 - 1}{1.3 - 1} - 8 \cdot 7$$

If you try this for b_4, and b_9 and a few other values, you will find that the pattern is this:

$$b_n = 13\frac{1.3^n - 1}{1.3 - 1} - 8 \cdot n$$

As before this can be simplified. Since $13/(1.3 - 1) = 13/.3 = 130/3$ which is approximately 43.3333, we find $b_n = 43.3333(1.3^n - 1) - 8n = 43.3333(1.3^n) - 43.3333 - 8n$.

4. The equation worked out in the example for repeated loan payments was

$$b_{24} = 1.01^{24}b_0 - P \cdot \frac{1.01^{24} - 1}{1.01 - 1}$$

This gives the balance after 24 payments when the interest is 1 percent each payment. Changing b_0 into the variable A, 24 into m, and .01 into i, the equation becomes

$$b_m = (1+i)^m A - P \cdot \frac{(1+i)^m - 1}{1+i-1}$$

The point of this equation is to make the balance equal to 0 after m payments. So replace b_m with 0. At the same time change $1+i-1$ to i:

$$0 = (1+i)^m A - P \cdot \frac{(1+i)^m - 1}{i}$$

Now we need to isolate P. Move the part that has P in it to the other side, obtaining

$$P \cdot \frac{(1+i)^m - 1}{i} = (1+i)^m A$$

and then divide both sides by the fraction next to P. The final result is

$$P = (1+i)^m A \frac{i}{(1+i)^m - 1}$$

Problems in Context

2. If the interest is described as 3 percent per year, that really means one-twelfth of 3 percent, or $3/12 = 1/4$ percent compounded monthly. So using the formula from the last Math Skills problem, the parameter $i = 1/4\% = .0025$. The other parameters are $A = 15,000$, and $m = 60$, because you make 12 payments per year for 5 years. Now substitute those into the formula and obtain

$$P = (1+i)^m A \frac{i}{(1+i)^m - 1}$$

$$= 1.0025^{60}(15,000) \frac{.0025}{1.0025^{60} - 1}$$

$$= 269.53$$

3. $1,206.93$

4. $1,206.93 - 1,100.65 = 106.28$ dollars per month less. This is just about 100 dollars. So it will take 18 months, or a year and a half, for the savings in monthly payments to equal the additional cost of getting the lower interest rate.

5. Let a_n be the amount of drug in the body immediately after the patient is given the nth dose of medicine. The difference equation is $a_{n+1} = .88a_n + d$, where d represents the amount of the repeated dose. One approach to this problem is to try some different amounts of the repeated dose, and see what happens. For example, try $d = 50$. Also set the initial dose $a_0 = 50$. Then the functional equation for the model is $a_n = 50(.88^n) + 50(1 - .88^n)/(1 - .88)$. Now using a numerical or a graphical approach, see where this model levels off. You will find that it levels off at 416.6666. That is too high, so try 25 in place of 50. That gives the equation $a_n = 25(.88^n) + 25(1 - .88^n)/(1 - .88)$, and the leveling-off point is 208.33333 which is too low. By using trial and error, you can figure out the right d to obtain a leveling off at 250.

A second approach is to use what we know about fixed points. We would like the drug level to become steady at 250, and that means 250 will be a fixed point. That is, if a_n is equal to 250, then a_{n+1} will also equal 250. Put this into the difference equation, leaving the repeated dosage, d as an unknown: $250 = .88(250) + d$. This can be solved for $d = 250 - .88(250) = 30$. We can check that this is right by supposing $a_n = 250$ and seeing what the difference equation gives for a_{n+1}: $a_{n+1} = .88(250) + 30 = 250$.

Finally, an even quicker approach is to use fixed point formula from page 271. The model levels off at $d/(1 - r)$, and we want this to equal 250. Now we know $r = .88$. So with that substitution, set $d/(1 - r) = d/.12$ equal to 250. The result is $d/.12 = 250$ so $d = (.12)250 = 30$.

6. Each month, the account gains interest plus a $100 deposit. So at first the balance is 100. After a month you have to add half a percent. This is calculated as $.005 \cdot 100$. Then you have $100 + .005 \cdot 100 = (1 + .005)100 = 1.005 \cdot 100$. As we found in Chapter 9, when you add .5% to the account that is the same as multiplying by a growth factor of 1.005. Then, after you get the interest, add in the next deposit. So from a starting point of 100 we get after one month $(1.005)100 + 100$. This same calculation can be used with any starting amount. If the balance at some point is 456.78, then the next month it will be $(1.005)456.78 + 100$. This shows that the difference equation for the model is $a_{n+1} = 1.005a_n + 100$, and the starting balance is $a_0 = 100$. Now observe that this is a mixed-model difference equation with $r = 1.005$ and $d = 100$, so we can immediately use the functional equation for a mixed model to see that $a_n = 100(1.005^n) + 100(1.005^n - 1)/(1.005 - 1)$. This can be simplified. Note that $1.005 - 1 = .005$, and $100/.005 = 20,000$. So we get $a_n = 100(1.005^n) + 20,000(1.005^n - 1) = 100(1.005^n) + 20,000(1.005^n) - 20,000 = 20,100(1.005^n) - 20,000$. In using this to answer questions about the model, bear in mind that n counts the number of months. After 5 years, that is 60 months, so we compute the balance as $a_{60} = 20,100(1.005^{60}) - 20,000 = 7,111.88$. Similarly, after 10 years the total is $a_{120} = 16,569.87$. When will the balance be 100,000? This time leave n as the unknown, set the balance a_n equal to 100,000, and solve this equation for n: $100,000 = 20,100(1.005^n) - 20,000$. Add 20,000 to both sides: $120,000 = 20,100(1.005^n)$. Divide by 20,100: $120,000/20,100 = 1.005^n$. Now we can find n using logs: $n = \log(120,000/20,100)/\log(1.005) = 358.247$. Remember this is months. Now 360 months would be 30 years, so this is about two months less than 30 years.

13

Logistic Growth

The focus for this chapter, *Logistic Growth*, is a variation on the idea of geometric growth. In the most general formulation, geometric growth refers to a variable that changes by a fixed percentage in any fixed unit of time. A typical example is a population which increases by 10 percent each year. Geometric growth can also refer to something that shrinks rather than grows, such as a radioactive substance that is reduced by half every 28 years. However, the geometric growth models that actually *grow*, and their limitations, are what lead to the development of logistic growth models. For the rest of this discussion, the shrinking models will be excluded: the expression *geometric growth* will be understood to mean models that increase over time.

So let us consider that typical example again: a population which increases by 10 percent each year. For concreteness, we will imagine that this is a population of fish in a lake. At some point, perhaps there are 10,000 fish in the lake. Then, using the ideas developed in Chapter 9, we can compute the population after n years as $p_n = 10,000 \cdot 1.1^n$. After 5 years the population would be $10,000 \cdot 1.1^5 = 16,105$, after 10 years $10,000 \cdot 1.1^{10} = 25,937$, and so on. We can use the model to predict as far into the future as we please, and the further we look ahead, the larger the population will grow. There is no limit to the size of the population in this model. This is unrealistic in the long run, because it ignores forces that act to limit the size of a population—forces such as competition for food and living space. These forces may be safely ignored with little error when the population is small. But as the population grows larger, as competition for resources becomes more pronounced, it becomes unrealistic to expect population to grow by the same percentage that occurred when the population was small.

The limitations of a geometric growth model can also be recognized graphically. We have seen that the graphs of geometric growth models follow a characteristic shape: they curve upward from left to right, becoming increasingly steep as the population grows. Mathematically, these curves rise ever higher with no limits or bounds. However, we know that real populations can not grow that way. There are always limits to growth. It would be much more reasonable to expect a population to eventually level off.

Logistic growth models arise in a natural way from the idea that the population increase must ultimately decline as populations grow larger. The simplest way to model this idea is with a straight-line model. That is what leads to logistic growth. We will see in this chapter that logistic growth models have several attractive features that appeal to our intuitive understanding of how populations grow. We will look at a logistic model for the growth of a population of fish in a lake, and model the consequences of catching or harvesting some of the fish each year. For many many applications, the logistic model predicts an orderly progression of a population toward a steady size that remains constant for the long term. Interestingly, logistic models can also predict chaotic behavior. We will examine this situation in the next chapter.

A Linear Growth Factor Model

The main idea of logistic growth can be developed in the context of a specific example. As usual, we will simplify a complicated situation as a way to get some initial understanding. So, imagine that we are going to introduce a new species of game fish, such as trout, to a large lake without a natural trout population. Initially, it is reasonable to expect the trout population to grow geometrically. Let us examine this idea more closely, with some particular numbers. Say that we stock the lake with 1,000 trout. We know that some fraction of the trout will be fertile females, probably about 50%. Each female will produce some number of offspring, perhaps as many as 100. But not all of the offspring will survive to adulthood. We may suppose that only 10 out of 100 survive. Also, some of the original 1,000 fish will probably die over the year, say one in 10. Now all of these figures can be combined to predict the number of trout in the following year. The exact numbers are not important. What is important is that all of the numbers are reasonably interpreted as fixed percentages. If we start next year with a different population, we can reasonably expect the same percentage of fertile females, the same number of offspring per female, the same fraction of offspring to survive, and the same fraction of the existing population to die. That is why a geometric growth assumption is plausible. Without trying to determine the specific percentages at each stage of the process, we recognize that the overall increase in the population will be a fixed percentage. If the net result is a 10-percent increase one year, it is reasonable to predict a 10-percent increase in the following year, too.

These ideas lead to the geometric growth difference equation:

$$p_{n+1} = rp_n$$

where the parameter r is a kind of net fertility factor. For our population of trout, if the lake can support many thousands of fish, and if there are few natural predators, we expect the population to grow rapidly. For the sake of concreteness, let us say that in one year the population will increase by a factor of 5. That means that after all the eggs are hatched, and taking into account the adults and babies that die during a year, the net result will be that the population will be 5 times larger after a year. In the difference equation, this corresponds to a parameter value of $r = 5$, leading to

$$p_{n+1} = 5p_n$$

Now as the years go by, this rate of growth cannot continue. Eventually, there will be greater competition for food. Fewer of the babies will survive. Indeed, many of the baby fish will be eaten by the adults. So in that case, the increase will be much less than a factor of 5. Maybe the population will only grow by 50%. Then r will equal 1.5 and the difference equation will be

$$p_{n+1} = 1.5p_n$$

As the population continues to grow, it can approach a kind of balance, where the competition for resources will effectively prevent the population from growing at all. At that point, each year's population size will be the same as the previous year's. That means that $r = 1$ and the difference equation is

$$p_{n+1} = p_n$$

What if the population somehow got over the balance point? This could happen if we overstock the lake. Or, if the conditions in the lake change, the balance point might be reduced to a smaller population size. Then, for a while, there might be more fish than the lake could sustain on a continuing basis. In any case, with too many fish in the lake, we expect the population to actually shrink. The r in that case would be less than 1, say for example, .8. This would correspond to a difference equation:

$$p_{n+1} = .8p_n$$

What these ideas suggest is that r should not be expected to be constant. It should change with the population size. If r is as large as 5 when the population size is small, we expect it to be much smaller when the population size is larger.

In Fig. 13.1 a qualitative graph illustrates the kind of relationship we have been discussing. The growth factor decreases as population increases, producing a curve that slopes downward from left to right.

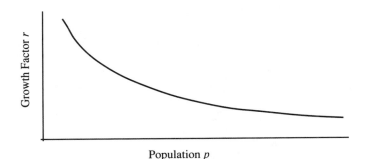

FIGURE 13.1
Growth Factor r Varies with Population

How can we incorporate this into our model? The simplest approach is to use a straight-line model for the variation of r with population p. Then we can easily find an equation for r as a function of p. As illustrated in Fig. 13.1, the true relationship between r and p is probably not actually a straight line. In order to obtain a good approximation with a linear model, we should probably avoid the extreme sides of the graph, where the population is very small or very large. Instead, we want to draw a line that approximates the correct situation for a wide range of populations in the center of the graph.

In a real modeling situation, we would collect data on real fish populations. In this discussion, since we are simply exploring the conceptual ideas of this type of model, we will continue to invent the numbers in the model. We expect that at some point the fish population will be in balance. Suppose that occurs with a population of 50,000 fish. If there are 50,000 fish this year, then there will be 50,000 next year, too. In fact, once the population reaches 50,000, it will remain the same year after year. From that point on, with a size of 50,000, the population is described by the difference equation $p_{n+1} = p_n$, so the growth factor is $r = 1$ when $p = 50,000$. On a graph of p and r, this is the point $(50,000, 1)$.

If the population is lower than 50,000, we expect it to increase over time, and we expect that the smaller the population, the larger the growth factor. A small enough population could even double in a single year. Perhaps this will happen if the population is as small as 10,000. Then, with a population of 10,000, $r = 2$, so that $(10,000, 2)$ is also on the p–r graph. As discussed earlier, with an even lower population size, the rate of growth in a single year could be even higher. However, we will stay away from the extreme cases, such as a growth factor of 5, because we want our straight-line model to be most accurate in the center of the graph. With that in mind, we will use the two data points $(10,000, 2)$ and $(50,000, 1)$ to create the straight-line graph of Fig. 13.2 and derive an equation for the line. We have the points $(10,000, 2)$ and $(50,000, 1)$. This gives us a

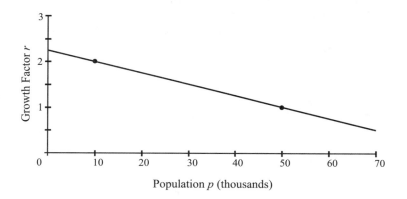

FIGURE 13.2
Growth Factor r Based on Two Points

slope of $-1/40{,}000 = -.000025$. The point-slope equation can now be expressed as

$$r - 2 = -.000025(p - 10{,}000)$$

or more simply,

$$r = 2.25 - .000025p$$

This is a simple linear model for the relationship between the growth factor r and the population size p. It says, for instance, that when the population is 20,000, the growth factor should be $r = 2.25 - .000025 \cdot 20{,}000 = 1.75$. That is, if the population is 20,000 one year, we expect it to grow to $1.75 \cdot 20{,}000$ by the next year. On the other hand, if the population is $p = 70{,}000$, then the growth factor should be $r = 2.25 - .000025 \cdot 70{,}000 = .5$. This means that if the population somehow got to be 70,000, in the following year it is expected to shrink to $.5 \cdot 70{,}000$. Using this linear equation makes the growth factor r depend on the size of the population. For smaller populations there is more rapid growth. For larger populations there is slower growth, or even a decline in the size of the population.

Now we can incorporate this model for the growth factor into our difference equation for the population growth. In place of r in

$$p_{n+1} = rp_n$$

we will use the growth factor that will apply with a population of size p_n, namely $r = 2.25 - .000025p_n$. That gives us the equation

$$p_{n+1} = (2.25 - .000025p_n)p_n \qquad (1)$$

This form of difference equation defines logistic growth. To be more specific, a logistic growth model follows a difference equation of the form

$$p_{n+1} = r \cdot p_n$$

where r is given as a linear function of p_n.

What does Eq. (1) predict for our population of fish? Using a numerical method, we can compute p_n for several years, starting with $p_0 = 1{,}000$. The first few computations are as follows:

$$p_1 = (2.25 - .000025 \cdot 1{,}000)1{,}000 = 2{,}225$$

$$p_2 = (2.25 - .000025 \cdot 2{,}225)2{,}225 = 4{,}882$$

$$p_3 = (2.25 - .000025 \cdot 4{,}882)4{,}882 = 10{,}390$$

Continuing in the same way, the results in Table 13.1 are observed. A graph of the data appears in Fig. 13.3.

The table and the graph show the kind of behavior that we expect to see. Initially, the population growth looks very much like a geometric model. During this stage, the environment is not limiting growth very much. However, as the population continues to grow, the rate of growth tapers off. Each year the percentage of increase is less than for the preceding year. Eventually, the population starts to level off at 50,000. If you

Year	0	1	2	3	4
Population	1,000	2,225	4,882	10,390	20,678

Year	5	6	7	8	9
Population	35,836	48,526	50,314	49,919	50,020

TABLE **13.1**
Data from Logistic Fish Population Model

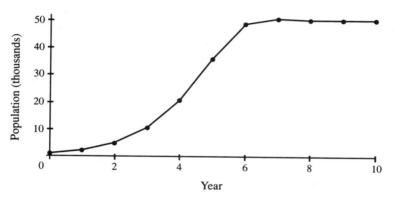

FIGURE **13.3**
Graph of Logistic Fish Population Model

examine the data carefully, you will see that the population actually goes a little above 50,000. Then it jumps back and forth. One year it is a little less than 50,000, the next it is a little more, then a little less, then a little more. The population size continues to fluctuate back and forth, all the while edging closer to the 50,000 figure. Remember that we picked 50,000 as the size at which the population would remain fixed from year to year.

This introductory discussion shows how a simple model can be modified to obtain more realistic behavior. Earlier models with geometric growth assumed that the growth factor would remain constant no matter what the population size. In this presentation we built up a new model that includes variation in the growth factor. Admittedly, using a linear model for this variation imposes its own limitations. But the resulting model is much more realistic than simple geometric growth. In a true application, we would next try to measure the way the growth factor actually varies as the population grows, and fit the best linear model possible to the data. That would give us a logistic growth model that could be tested for accuracy in the real world. As a matter of fact, logistic models are used widely in biology, and in many cases they give results that accurately mimic the way populations grow. In the next section we will look at these logistic growth models a little more generally.

General Features of Logistic Models

In past chapters, we have introduced difference equation models of several different types: arithmetic growth, quadratic growth, geometric growth, and mixed models. In each case, we were able to describe a general difference equation for each type by using parameters. In any actual application the parameters would be replaced by numbers, but the general form with the parameters provided a pattern for all the difference equations for each type. We were also able to find functional equations in each case, again expressed in terms of parameters. And finally, we found general characteristics of graphs of each model type.

Now we would like to proceed in the same way for logistic growth. Unfortunately, these models are inherently more complex than the earlier examples. As a consequence of that increased complexity, we will not be as successful in finding results that apply to all logistic models. In particular, we will not be able to derive a functional equation. However, we can formulate a general difference equation, and make some observations about the graphs. To begin, we state a general description of logistic growth, based on the discussion in the preceding section.

> Logistic growth, like geometric growth, defines a relationship between successive terms of a data sequence a_n. In a logistic model, the change from one a_n to the next involves a growth factor that depends (linearly) on the size of a_n. This contrasts with geometric growth, for which the change from one a_n to the next involves a *constant* growth factor.

Next we will develop general forms for the difference equations that occur in logistic growth models.

Logistic Growth Difference Equations. The difference equation from the example with the fish population took the form

$$p_{n+1} = (2.25 - .000025p_n)p_n \qquad (2)$$

As in previous cases, there are several different ways to express this difference equation, and each one gives us different insights. As it stands, we might introduce the parameters r_0 for 2.25 and m for .000025. This gives the general form

$$p_{n+1} = (r_0 - mp_n)p_n$$

We use r_0 to remind us of an initial growth factor. In the example, this value is 2.25, and for very small populations, the logistic model will grow like a geometric model with growth factor 2.25. However, as the population grows, the growth rate varies according to a straight-line model. The m parameter is the negative of the slope of that straight line.

An equivalent version of Eq. (2) is

$$p_{n+1} = 2.25p_n - .000025(p_n)^2$$

The parameter form of this version is

$$p_{n+1} = r_0p_n - m(p_n)^2$$

This version shows that the logistic model can be thought of as the result of adding one term to a geometric growth model. Where the difference equation for geometric growth had only a constant times p_n, the logistic model has an additional term made up of a constant times p_n *squared*. This is reminiscent of the way polynomials are built up from linear functions. The simplest polynomials are linear, then the next simplest, with squared variables, are the quadratics. Seen in this light, the logistic growth model is a natural extension of geometric growth, and the next simplest form that can be expressed using polynomials. In addition, this form of the difference equation shows how the model starts out like geometric growth. Because the coefficient of $(p_n)^2$ is very small, that term will not have much impact on the final result for p_{n+1} unless p_n is fairly large. With $p_n = 1,000$, for example, we have $p_{n+1} = 2.25(1,000) - .000025(1,000)^2$. That works out to $2,250 - 25$, and subtracting the 25 does not have much effect. We would have gotten pretty much the same result by just computing $p_{n+1} = 2.25p_n$. This illustrates how, for small values of p_n, the formula for p_{n+1} is closely approximated by just $2.25p_n$, which is the kind of expression that produces geometric growth.

Here is one final reformulation of the difference equation from the example. As you can verify using algebra, the original difference equation Eq. (2) is equivalent to

$$p_{n+1} = .000025(90,000 - p_n)p_n$$

In this form, we can easily see some of the limitations of the logistic model. Clearly, if $p_n = 0$, the next value p_{n+1} will also equal 0. That is entirely reasonable. However, if p_n ever reaches 90,000, we can see that the next year's population will be zero. That is not such a reasonable effect. We can imagine that the lake cannot support 90,000 fish, but we would probably not expect the entire population to die off in a single year. Yet that is what the model says: if the population is 90,000 this year, it will be zero the next. Here is an even more questionable aspect of the model. If somehow a population above 90,000 was reached, the difference equation predicts a negative population for the following year. That is just nonsense. These unreasonable aspects are consequences of using a linear model for the way the growth factor varies with the population size. Of course, on a linear model (as illustrated in Fig. 13.2), the line will cross the horizontal axis somewhere, and so predict negative growth factors for some populations. That is not reasonable, and would not occur if we used a graph like Fig. 13.1.

In spite of this questionable behavior for large population sizes, the logistic model is often very useful. We simply have to keep in mind that the model is reasonable for a limited range of populations. In the fish population example, the limited range runs from 0 to 90,000. In the more general setting, we use the parameter L in place of 90,000, to remind us that it is the limiting case of the model. We simply cannot use the model for populations above L. The coefficient .000025 doesn't have such an apparent meaning in this context, so let us simply use the parameter m for that value, as before. Then we have

$$p_{n+1} = m(L - p_n)p_n$$

as the general form for this version of the logistic model. Comparing this equation with the earlier versions, observe that $mL = r_0$.

For easy reference, the three different forms of the logistic growth difference equation are shown below:

> Logistic growth can always be described by a difference equation in each of the following forms:
>
> $$p_{n+1} = (r_0 - mp_n)p_n \qquad (3)$$
>
> $$p_{n+1} = r_0 p_n - m(p_n)^2 \qquad (4)$$
>
> $$p_{n+1} = m(L - p_n)p_n \qquad (5)$$

As in previous chapters, we have formulated these general difference equations using parameters to stand for numerical constants that appear in any specific example. In the previous chapters, we found ways to describe characteristics of a model based on the parameter values. Here is a very simple example from geometric growth. The general geometric growth difference equation is $a_{n+1} = ra_n$. Depending on whether the parameter r is greater than or less than 1, the a_n's will either grow larger or decrease over time. So, by looking at the numerical value of r in a specific example, we can draw conclusions about the behavior of the model.

In a similar way, we can formulate characteristics of logistic growth models in terms of the parameters in their difference equations. That is what we will do in the next few sections.

Keeping Within the Proper Range. Earlier, we saw that a logistic growth model only makes sense when the population stays within an appropriate range. In terms of the parameters in Eq. (5), the model is invalid if p_n gets larger than L, for in that case a negative growth factor occurs. In this section we will see that for some values of the parameters, we can be sure that p_n will always remain in the valid range.

To see how this works, let us consider our numerical example again. The difference equation for this example is

$$p_{n+1} = .000025(90{,}000 - p_n)p_n$$

For this discussion, we will not need to consider a whole sequence of p_n's. Rather, we want to concentrate on just two terms of the sequence. So we will use a different notation. Let x denote the population for some (unspecified) year and write y for the population the following year. That means that in the difference equation above, x will take the place of p_n while y replaces p_{n+1}. Then we have

$$y = .000025(90{,}000 - x)x = 2.25x - .000025x^2$$

In this context we can see that y is a quadratic function of x, and so, using what we know about the graphs of quadratics, we can derive several conclusions. First, the general shape of the graph is a downward opening parabola. The parabola crosses the x axis at the roots 0 and 90,000, and so the highest point must occur just half way between these roots, at $x = 45{,}000$. There, $y = .000025(90{,}000 - 45{,}000)45{,}000 = 50{,}625$. These features are illustrated in Fig. 13.4.

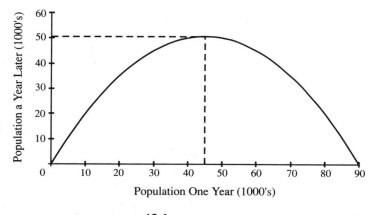

FIGURE 13.4
y stands for p_{n+1}, x stands for p_n

One feature of the graph has special significance: the position of the high point with $y = 50{,}625$. This shows that no matter what we take for x, we will never see a value of y above 50,625. In particular, if x is anywhere between 0 and 90,000, then y will also be between 0 and 90,000. Remembering what x and y represent, this gives us the following conclusion: If the population in any year is between 0 and 90,000, the population in the following year will remain between 0 and 90,000. This shows that the logistic model will never lead to a value at 90,000 or above if we start with an initial population within the acceptable range. In fact we can see a little more. With a starting population anywhere between 0 and 90,000, the population after one year will be between 0 and 50,625, and will remain in that range forever after.

A similar analysis can be carried out using parameters instead of numbers. If the model is

$$p_{n+1} = m(L - p_n)p_n \tag{6}$$

the equation for x and y becomes

$$y = m(L - x)x$$

We can derive Fig. 13.5 following the same steps as before. First, observe that $x = 0$ and $x = L$ both result in $y = 0$. The high point on the graph must occur for x halfway between 0 and L, that is, at $L/2$. But at $x = L/2$ we compute $y = m(L/2)(L/2) = mL^2/4$. We would like this highest value to be less than L, so that the population will never grow too large (meaning so large that our linear model for the growth factor will become negative). In symbols, we should insist that

$$mL^2/4 < L$$

Algebra can be used to simplify this condition to the form

$$mL < 4$$

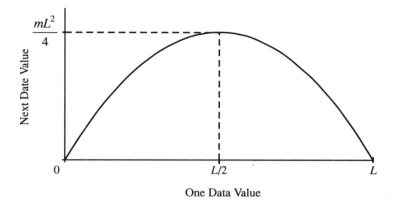

FIGURE 13.5
Using Parameters L and m

So, when we use the logistic model in Eq. (6) we should be sure to check whether $mL < 4$. As long as that is true, we need not worry about the unrealistic behavior that occurs when the population gets too large. On the other hand, if $mL > 4$, we need to be on the alert for unreasonable behavior.

In our original example, $p_{n+1} = .000025(90,000 - p_n)p_n$, so that $mL = 2.25 < 4$, as desired. What would happen if we changed the difference equation to $p_{n+1} = .000085(90,000 - p_n)p_n$? Starting with $p_0 = 1,000$, here is how the population would grow:

$$
\begin{aligned}
p_1 &= .000085(90,000 - 1,000)1,000 &= 7,565 \\
p_2 &= .000085(90,000 - 7,565)7,565 &= 53,008 \\
p_3 &= .000085(90,000 - 53,008)53,008 &= 166,674 \\
p_4 &= .000085(90,000 - 166,674)166,674 &= -1,086,263
\end{aligned}
$$

In this example, the population grows so rapidly that it exceeds the range in which the model is valid. Even though $p_3 = 53,008$ is a valid population size for the model, it leads to $p_4 = 166,674$ which is too large. Our straight-line model for the growth factor is invalid for such a large population, and results in a negative number for p_5. This shows that the model with difference equation $p_{n+1} = .000085(90,000 - p_n)p_n$ can lead to unrealistic results. This can happen any time we use a model with $mL > 4$. To emphasize this fact, it is presented below in a box.

Suppose a logistic growth model has the difference equation

$$p_{n+1} = m(L - p_n)p_n$$

and that p_0 is between 0 and L. Then, if $mL < 4$, every p_n will remain within the range from 0 to L. If $mL > 4$, the model may lead to unrealistic results, including p_n values that are greater than L, or negative.

Where Does the Model Level Off?   One of the attractive features of the logistic model in the fish example is the way the population levels off. If wildlife managers want to introduce a species of fish into a lake, that is just what they would like to see. At first, the new fish population grows, but it eventually levels off and reaches a steady level. That will allow the species to survive in a balance within its environment.

If the population does level off at a steady level, it is easy to predict just what the steady level will be. After all, if the population is steady, it must remain the same year after year. That means that p_{n+1} must equal p_n. Such a population is called a *fixed population*. This is the same concept as the fixed points introduced in the discussion of mixed models. As before, we will find the fixed points or fixed populations by solving the equation $p_{n+1} = p_n$, and once again the answer can be expressed in terms of parameters of the model, although the form of this expression is different for logistic growth than it was for mixed models. To help minimize confusion of the fixed points in the two cases, and to keep in mind that the logistic models are frequently concerned with population growth, in this chapter we will refer to the fixed points as fixed populations.

We can find fixed populations for logistic models using the same method we used for fixed points in the preceding chapter. We seek an unknown population size, x, that is fixed by the difference equation. That is, if p_n ever reaches x, then p_{n+1} will also be x. From the difference equation

$$p_{n+1} = .000025(90{,}000 - p_n)p_n$$

we can therefore obtain

$$x = .000025(90{,}000 - x)x$$

This is a quadratic equation, and can be solved using factoring or the quadratic formula. But recalling the original formulation of the logistic growth model leads to a simpler approach. Logistic growth is geometric growth with a variable growth factor. If the population is fixed, that means the growth factor for that population size must be exactly equal to 1. So one way to find a fixed population is to set the growth factor equal to 1.

On the right side of the difference equation

$$p_{n+1} = .000025(90{,}000 - p_n)p_n$$

the first part, $.000025(90{,}000 - p_n)$ is the equation for the growth factor. That is what we want to equal 1. If the fixed population is x then we must have

$$.000025(90{,}000 - x) = 1$$

This equation is easily solved to find $x = 90{,}000 - 1/.000025 = 50{,}000$. That is a fixed population for this example.

Of course, we already had discovered that earlier. The point of rediscovering it using algebra is that now we can repeat the process using parameters. So starting with the difference equation in the form

$$p_{n+1} = m(L - p_n)p_n$$

we can see that the growth factor is $m(L - p_n)$. Following the same steps we used before, we are looking for a population size x that makes this growth factor 1. So write $m(L - x) = 1$ and solve for x. This leads to the equation

$$x = L - \frac{1}{m}$$

When $p_n = L - 1/m$, we know that $p_{n+1} = 1 \cdot p_n$, which means that the population won't change the next year. This gives one fixed population for the general logistic model.

The original equation for finding fixed populations,

$$x = .000025(90{,}000 - x)x$$

is quadratic. We found that one solution is 50,000. The other solution is 0. You can see this from the equation because, if you set x equal to 0 on each side, the resulting equation is true. But it is also clear conceptually. If your population is 0 this year, it has to be 0 again next year. This reasoning applies in the general logistic model as well. No matter what L and m are, the equation

$$x = m(L - x)x$$

will always have 0 for a solution.

Now we have found two fixed points in the general case, 0 and $L - 1/m$. Which one is the steady population level? The answer depends on the parameters. It is possible for $L - 1/m$ to be negative. Indeed, that will happen exactly when $mL < 1$. In this case, our previous results show that the population must remain between 0 and L, and therefore cannot level off at a negative value. So if the population does level off, it has to be the other fixed population, 0.

On the other hand, if $mL < 1$, then the growth factor will always be less than 1. Look back at Fig. 13.2. Remember that we are using a linear model for the way that the growth factor changes as the population changes. What is more, that straight line has a negative slope. So the highest possible growth factor occurs on the left side of the diagram, that is, at the y intercept. We can find this by using the growth factor formula, $m(L - p)$, and setting $p = 0$. The result is mL, and that is the maximum value for the growth factor. So, the fixed population $L - 1/m$ is negative exactly when $mL < 1$, and in that case the growth factor always stays less than 1. But that means the population always gets smaller from one year to the next because it is being multiplied by a factor less than 1.

Now combine all these facts:

- The population gets smaller each year
- The population stays between 0 and L
- The only possible fixed populations are 0 and something that is negative

There is only one possible conclusion: the population is going to steadily decrease toward 0. So, if $mL \leq 1$, we know what will happen. The population will decrease steadily, leveling off at 0. Eventually, all the fish die.

This is an important possibility, and, if it occurs in a real model, would obviously be cause for concern. At the very least, it would be a signal that further study should be undertaken. But now we want to consider the case where the population does not die out, but levels off at a steady population (greater than 0). This can only happen if $mL > 1$. And then the fixed population would have to be $L - 1/m$. We summarize these results for future reference.

> In a logistic growth model with the difference equation $p_{n+1} = m(L-p_n)p_n$, if $mL \leq 1$, the population will steadily decrease, leveling off at 0. If $mL > 1$, the population may level off to a steady positive value. If so, the steady size of the population will be $L - 1/m$.

When Does the Model Level Off?   Even when $mL > 1$, the logistic model does not always level off to a steady value. This can be illustrated by modifying the example discussed earlier. Originally, we developed a linear model for the way the growth factor for the fish population might change with the size of the population. Recall that we proposed a growth factor of 2 for a population of 10,000, meaning that a population of 10,000 would double in a year. At the same time, we proposed a growth factor of 1 for a population of 50,000, meaning that the population would remain steady year after year, once it reached 50,000. From these assumptions we found the equation

$$r - 2 = -.000025(p - 10,000)$$

Now we will modify this slightly, by changing the growth factor from 2 to 2.8 for the population of size 10,000. That results in a different equation for r and p :

$$r - 2.8 = -.000045(p - 10,000)$$

Notice that changing from 2 to 2.8 produces two changes in the linear equation. The number subtracted from r on the left changed in the expected manner. However, the coefficient $-.000025$ on the left changed to $-.000045$. It is less obvious why this change occurred. To really understand this, you should retrace the steps that lead to the linear model for r in the first place. One of the exercises deals with this issue.

Proceeding as before, we isolate r

$$r = -.000045(p - 10,000) + 2.8 = 3.25 - .000045p$$

and so obtain the difference equation

$$p_{n+1} = (3.25 - .000045p_n)p_n$$

This can also be put into the form

$$p_{n+1} = .000045(72,222 - p_n)p_n$$

If we again start at $p_0 = 1,000$, how will this model work out? Because we simply increased the growth factor in effect with a population of 10,000, you might expect that the population will grow a little faster at first, but eventually level off at the same size as before. Use the equation $p_{n+1} = (3.25 - .000045p_n)p_n$ and $p_0 = 1,000$ to calculate p_1, p_2, and p_3 now, before reading any further.

$$p_1 = \underline{\hspace{1cm}} \qquad p_2 = \underline{\hspace{1cm}} \qquad p_3 = \underline{\hspace{1cm}}$$

Compare the results to the data from the original example shown in Table 13.1. Does the revised model do what you expected?

Although the new model starts out growing like the original model, the long-term behavior is quite different. In Table 13.2, the data are shown for the first 15 years of the model. The data for the first 30 years are shown graphically in Fig. 13.6. For the revised model, instead of leveling off at a steady value, the population jumps back and forth between two different levels, one around 58,000, and one around 36,000. According to this model, in even numbered years, the population will swell to 58,000 fish. But in the odd-numbered years it will be reduced to around 36,000 fish.

This is quite different behavior than what we found earlier. It offers a glimpse of chaos. In the next chapter we will explore this situation in greater detail. Then you will see why chaos has important implications for models in biology (and other subjects). But for this chapter, we want to concentrate on order (not chaos).

To recap the discussion this far, we started out observing that the first model behaved just as we hoped—the population started out resembling geometric growth, but then leveled off. The revised version of the example showed that logistic models cannot be

Year	0	1	2	3	4
Population	1,000	3,205	9,954	27,892	55,641

Year	5	6	7	8	9
Population	41,518	57,365	38,352	58,454	36,215

Year	10	11	12	13	14
Population	58,680	35,759	58,675	35,770	58,675

TABLE 13.2
Data from Revised Logistic Fish Population Model

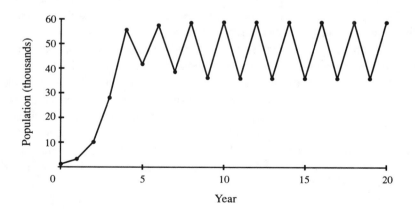

FIGURE 13.6
Graph of Revised Logistic Fish Population Model

counted on to act that way all the time. In fact, with a very minor change in the numerical values of the parameters, we obtained quite a different picture of the future growth of the fish population. As in the preceding section, the parameters of the logistic model can be used to predict which of these cases will occur. The mathematical methods that lead up to this prediction are beyond the scope of this course. They involve applying methods of calculus to study characteristics of polynomial equations. However, the prediction itself is not difficult to understand, and is very similar to the kind of prediction we found earlier. It is stated as follows[1]:

Suppose a logistic growth model has the difference equation

$$p_{n+1} = m(L - p_n)p_n$$

and that p_0 is between 0 and L. Then, if $1 < mL < 3$, the population will eventually level off at a steady value of $L - 1/m$.

We already knew from previous remarks that the steady level, if one occurs, must be either 0 or $L - 1/m$. There are two new items of information given in this box: (1) when $mL < 3$, a leveling off will always occur, avoiding more complicated behavior such as the zigzag pattern of Fig. 13.6; and (2) when mL is both greater than 1 and less than 3, the population will level off at a positive size, and will not die out.

In the original example, we had $m = .000025$ and $L = 90,000$. In that case $mL = 2.25 < 3$, so we predict that the model will level off eventually for any initial value p_0 between 0 and 90,000. The level population is supposed to be $90,000 - 1/.000025 = 50,000$, and that is what was observed. On the other hand, the revised model had $m = .000045$ and $L = 72,222$. For that case we have $mL = 3.25 > 3$, so we cannot be sure if the population will level off. And for the example we tried, with $p_0 = 1,000$ the population did not level off.

When we formulate a logistic growth model, if the parameters m and L satisfy $1 < mL < 3$, we can be sure that the population will stay within the range that makes sense for the model, and, in fact, will level off to a constant population size greater than 0. This leveling-off process can occur in three ways. In some cases the population will simply grow steadily until it reaches the constant level. A second possibility is that we start at a population above the constant level, and then decrease steadily to the constant level. The third possibility is that eventually the population wavers up and down, alternating between a little above and a little below the constant level, but always getting closer to the constant level. That is what we saw in the original example. After an initial period of rapid growth, the population steadied down toward a constant level while wavering up and down. Using the same kinds of methods that lead to the preceding boxed comment, it can be shown how to predict what will happen just by looking at the value of mL, but we will not delve into the details here. We simply observe that this is one more example of how the parameters of the logistic model can convey general information about the behavior of the model. As mentioned earlier, there is no simple functional equation for

---

[1] See *Encounters with Chaos* by Denny Gulick, McGraw Hill, New York, 1992, pages 41–46.

logistic models. At first, this may seem to limit our ability to use these models. In past experiences, the functional equation was one of the most important tools. However, we have seen now that a great deal of information can be determined about the general trends in a logistic model. If $mL < 4$, we can predict confidently that the model will remain within the valid range for the growth factor model. If $mL \leq 1$, we know that the model will decrease steadily to 0, meaning that the entire population will die out. Otherwise, if $1 < mL < 3$, we can conclude that the population will level off. We can even say what it will level off to: $L - 1/m$. Given these results, it is not so critical to have a functional equation. In the next chapter we will explore what happens when the model cannot be relied on to settle down to a steady level. You will see that chaos is lurking in the world of logistic models. We conclude this chapter with a modification of the fish model, to include the effects of harvesting some of the fish each year.

A Logistic Population Model with Harvesting

In the preceding material we have developed the idea of a logistic model and have seen some of the general properties of this kind of model. In particular, we examined an example modeling a population of fish in a lake. With the difference equation

$$p_{n+1} = .000025(90{,}000 - p_n)p_n$$

we found that the population would level off eventually to a steady size of 50,000. Once at this steady level, the fish population will remain unchanged year after year. The steady population level is a balance between the natural tendency of the fish population to grow and the environmental factors that limit growth. The fact that the balance occurs at a level of 50,000 indicates that the lake can support 50,000 fish perpetually.

Now we want to consider an additional factor in the model. If the lake can sustain 50,000 fish year after year, we ought to be able to harvest some of the fish. This might occur as a result of sport fishing, or commercial fishing, or a combination of each. For the purposes of planning, suppose that we will limit the catch to 4,000 fish each year. Then the difference equation becomes

$$p_{n+1} = .000025(90{,}000 - p_n)p_n - 4{,}000 \tag{7}$$

To derive this equation, we imagine that the population grows according to the logistic model for a year, and that, when the new population has been reached, we will remove or harvest 4,000 fish. If we do this year after year, what will be the effect on the fish population? Will the population level off, or will the fish population be eventually wiped out? It seems likely that, if we harvest a small enough number of fish each year, the fish population should be sustainable at some constant level. This would give us a renewable resource of fish for annual consumption. It also seems obvious that, if we harvest too many fish each year, we might drive the fish to extinction. In this section we will study this situation using a combination of graphical and numerical techniques. To conclude the discussion, we will introduce some theoretical results that can be applied to any kind of logistic model with harvesting.

Long Term Effect of Harvesting.   As a first approach, we can simply use Eq. (7) to compute the value of p_n for several years. Originally, without considering any harvesting, we saw that the population would level off at 50,000. So let us assume that that level has been reached before we begin harvesting. Then using Eq. (7), we start with $p_0 = 50,000$ and compute

$$p_1 = .000025(90,000 - 50,000)50,000 - 4,000 = 46,000$$

$$p_2 = .000025(90,000 - 46,000)46,000 - 4,000 = 46,600$$

$$p_3 = .000025(90,000 - 46,600)46,600 - 4,000 = 46,561$$

$$p_4 = .000025(90,000 - 46,561)46,561 - 4,000 = 46,564$$

If more terms of this sequence are computed, we find that the results level off at a population of 46,564. According to the model, then, we can go on harvesting 4,000 fish year after year. The overall effect on the population seems to be a reduction of a little less than 4,000. That is, if the population is steady at 50,000 with no harvesting, then taking out 4,000 fish per year reduces the steady level to about 46,500, a reduction of 3,500.

A resources manager would want to harvest as many fish as possible without endangering the long-term health of the fish population. What will happen if we harvest 8,000 fish annually? 12,000? We can use the model to explore this idea, just as it was used for an annual harvest of 4,000. In Fig. 13.7, the future growth of the fish population model is shown for several different amounts of harvest. In the figure, the variable h represents the number harvested (in thousands). The curve labeled $h = 4$, for example, shows that, if 4,000 fish are harvested each year, the population will level off at about 46,000. This is what we already discovered numerically. With an annual harvest of 12,000 fish, the population seems to level off at about 37,000. On the other hand, if we try to harvest 20,000 or 24,000 fish per year, the population will die out in a few years. The case for a harvest of 16,000 fish is inconclusive. It seems to be leveling off in the figure, but it is hard to be sure.

The figure shows, as expected, that we can maintain a steady population of fish with a fixed annual harvest, as long as we don't harvest too many fish. However, we know

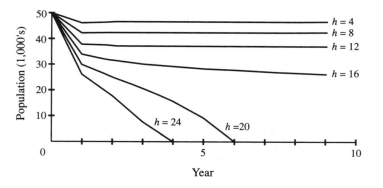

FIGURE 13.7
Future Population with Various Harvest Amounts

that the real situation is more complicated than the model. For this reason, it is a good idea to investigate what happens when the model is modified in various ways. We will take a look at a sample of this kind of investigation in the next section.

Variations in Initial Population and Harvest Amount. Let us consider what will happen for a variety of different starting population sizes. Actually, a detailed analysis of this problem would also consider a variety of different values for other parameters, such as the growth rate (m) and the maximum population (L). That would help us see what kind of future growth might occur if some of the parameters in our model were a little bit off. However, for the purposes of illustration, we will only look at the effects of choosing different starting values, p_0.

In Fig. 13.8, it is assumed that 12,000 fish will be harvested per year. That corresponds to this difference equation:

$$p_{n+1} = .000025(90{,}000 - p_n)p_n - 12{,}000.$$

The figure shows what will happen if the starting population ranges from 38,000 to 65,000. For each of the selected starting population values, the difference equation was used to compute the population for the next 10 years. Accordingly, there is a curve for each different starting population value, showing how the population changed over time starting from that value. For example, the highest curve starts with a population of 65,000 fish. There is a steep drop the first year, followed by a slow increase that levels off to a population of about 37,000 after about 5 years. In fact, all of the curves seem to level off at the same place. No matter what the population is when we start harvesting fish, a steady population of about 37,000 always seems to be reached. These results suggest that harvesting about 12,000 fish per year should be sustainable, even if the population size deviates from the model figure of 50,000.

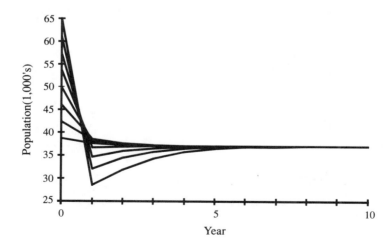

FIGURE 13.8
Harvesting 12,000 Fish per Year

In a similar way, we might see what will happen if we accidentally harvest a few fish too many. After all, it will probably be very difficult to control exactly how many fish are caught. At best, a target figure can be set, and the actual catch monitored to see how close the harvest is to the desired level. But even those results will be approximate. The true number of fish caught will probably be different from the desired level of 12,000. So let us see what will happen if the harvest amount is a little higher than the 12,000 figure used in the preceding calculation.

Using the same approach as before, another figure was created, but this time the difference equation

$$p_{n+1} = .000025(90,000 - p_n)p_n - 16,000$$

was used, modeling an annual harvest of 16,000 fish per year. The result appears as Fig. 13.9. Once again the curves in the figure appear to be leveling off after about the third year. However, this appearance is misleading. If the data are generated for 30 years, rather than for just 10, it is seen that eventually the entire fish population dies off, no matter what the starting population is. This is shown in Fig. 13.10. There are

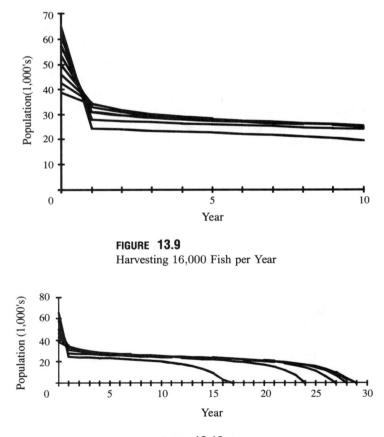

FIGURE 13.9
Harvesting 16,000 Fish per Year

FIGURE 13.10
Data for 30 Years

three important lessons to be learned here. First, the graphical method can suggest trends and patterns, but these suggestions can be misleading. Whenever possible, theoretical methods should be found to substantiate conclusions drawn from graphical appearances. Second, the situation depicted in the graph is just what wildlife managers should be on guard against. For a while the fish population seems to be quite steady, but on each curve, the population ultimately takes a sudden turn, dropping to zero in only 3 or 4 years. Knowing that this kind of phenomenon is possible in the model suggests that similar events could occur in the real world. If so, a sudden drop in actual fish populations might be catastrophic[2]. Third, this example shows that relatively small changes in the parameters for a model can produce dramatic changes in what the model predicts. To emphasize this point even more, Fig. 13.11 shows what the model predicts when the annual harvest is 15,000, rather than 16,000. The figure looks much the same as Fig. 13.8. No matter what the starting population is, the model predicts the population will level off and remain constant. So, the environment will support a harvest of 15,000 fish per year indefinitely, whereas a harvest of 16,000 fish per year will wipe out the population in 10 or 20 years. This shows that the model is extremely sensitive to changes in the harvest level, at least near the level of 15,000 to 16,000 in the examples. Given that we can only get estimates for the actual harvest level, we should be sure to set the target harvest amount well below the point at which the transition occurs from sustainable populations to extinction.

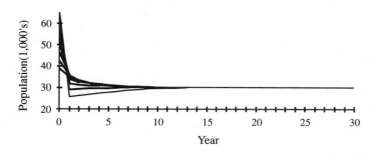

FIGURE 13.11
Harvesting 15,000 Fish per Year

Theoretical Methods. In the preceding examples, the fish populations invariably either leveled off or died out completely. It turns out that the fixed population levels are the key to understanding how these models operate. As usual, to determine possible fixed population sizes, we simply look for a situation where the population one year is equal to the population the preceding year. In symbols, that is the equation $p_{n+1} = p_n$. Proceeding as before, we will use x to represent the unknown fixed population level, and replace

---

[2] The 1994 harvest of Pacific sockeye salmon was 20 percent lower than what was predicted, representing a sudden drop that was completely unexpected. See "Mystery of Vanishing Salmon Puzzles Canadians," *The Washington Post,* 12/31/94, page A23.

both p_{n+1} and p_n with x in the difference equation. Let us see how this works for some of the examples we already considered.

In the first example we assumed that 4,000 fish would be harvested per year. That gave us the difference equation

$$p_{n+1} = .000025(90,000 - p_n)p_n - 4,000$$

Setting $p_{n+1} = p_n = x$ gives this equation:

$$x = .000025(90,000 - x)x - 4,000$$

This is a quadratic equation, and the solutions can be found using the quadratic formula, but first we need to put the equation into descending order. Here is the algebra:

$$x = .000025(90,000 - x)x - 4,000$$

$$x = 2.25x - .000025x^2 - 4,000$$

$$x - 2.25x + .000025x^2 + 4,000 = 0$$

$$.000025x^2 - 1.25x + 4,000 = 0$$

Now we can use the quadratic formula, to find

$$x = \frac{1.25 \pm \sqrt{1.25^2 - 4(.000025)4,000}}{.00005}$$

The results are approximately 3,436 and 46,564. These are the population levels that will remain fixed from year to year according to the model.

Methods of higher mathematics can be used to find out quite a bit about this model, and the two fixed population levels provide key information. One interesting result is this: if the population grows or shrinks so that it levels off over time, it must level off at the larger fixed population, approximately 46,564. To be more precise, because the true population has to be a whole number, what we should predict is that the population will waver a little up and down but settle down very near to 46,564. That is what we observed, too. Another result shows that this leveling off must occur if the population starts out fairly close to the steady level of 46,564. The same is not true for the other fixed population, which is approximately 3,436. Actually, 3,436 is a little less than the theoretical fixed population, and 3,437 is a little more. The population cannot exactly equal the fixed value predicted by the model. In this case, the theoretical fixed value simply serves as a dividing line. Starting above the theoretical fixed value, with a value of 3,437 or higher, the population will grow until it levels off at around 46,564. On the other hand, starting below the fixed value, that is, at 3,436 or less, the population will rapidly decrease until all the fish are gone. There is a symmetric behavior near the upper limit of 90,000, too. If the population starts out at $90,000 - 3,436 = 86,564$, or above, the fish will all die; if the population starts out below 86,564, a steady level of 46,564 will eventually be reached. So, in summary, if the population starts out between 3,437 and 86,564, then it will eventually level off near 46,564. Otherwise, if the population starts smaller than 3,437 or greater than 86,564, the fish are doomed to extinction.

These conclusions can be extended easily to models with modified parameters. Let's look at the case for an annual harvest of 15,000 fish. The difference equation is

$$p_n = .000025(90{,}000 - p_n)p_n - 15{,}000$$

Recall that for this case the population eventually seemed to level off. Following the same steps as before, we can find the fixed populations for this difference equation, and the quadratic equation that we reach,

$$.000025x^2 - 1.25x + 15{,}000 = 0$$

is almost identical to what we had before—only the final constant has changed from 4,000 to 15,000. The solutions this time are

$$x = \frac{1.25 \pm \sqrt{1.25^2 - 4(.000025)15{,}000}}{.00005}$$

This time the roots are whole numbers, namely 20,000 and 30,000. The same conclusions still hold. If the population starts out between 20,000 and $90{,}000 - 20{,}000 = 70{,}000$, then it will eventually level off at 30,000. However, if the starting population is below 20,000 or above 70,000, then the population will decrease down to zero. Notice that the range of initial populations that eventually level off is narrower now than it was for the preceding example, with an annual harvest of 4,000. But it is still quite broad. In fact, all the initial populations in Fig. 13.11 are between 20,000 and 70,000, which is why every curve in the figure levels off at 30,000.

The conclusions of the two preceding examples are typical of a much more general phenomenon. In any logistic model with harvesting, we can attempt to find fixed populations. That will always lead to a quadratic equation. Usually we will find two roots to the quadratic equation, representing a lower fixed population and a higher fixed population. These in turn determine what happens for any starting population size. In Fig. 13.12, the situation is depicted qualitatively. The figure includes a number line, running from 0 to L for the starting population size. Also marked on the number line are the lower fixed population (indicated as *LFP*), the higher fixed population (*HFP*), and the result of subtracting the lower fixed population from L (L - *LFP*). As the figure shows, if the starting population is between *LFP* and L - *LFP*, then the population will level off to a steady population size given by the higher fixed population. In all other cases, the population will die off.

The preceding examples and Fig. 13.12 all predict that for at least some starting population sizes, the population will eventually level off. In contrast, when the harvest

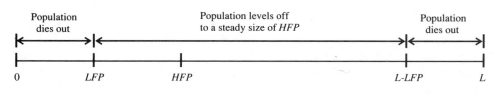

FIGURE 13.12
Starting Population Number Line

was set at 16,000 per year, every curve showed the population dying out. What does our theoretical analysis say about that? Harvesting 16,000 fish per year, the difference equation is

$$p_n = .000025(90,000 - p_n)p_n - 16,000$$

Once again we look for the fixed populations, and once again we arrive at a quadratic equation

$$.000025x^2 - 1.25x + 16,000 = 0$$

But when we try to use the quadratic formula, the expression

$$x = \frac{1.25 \pm \sqrt{1.25^2 - 4(.000025)16,000}}{.00005}$$

does not give us a solution, because the quantity inside the square root is negative. In this case, there are no fixed population sizes, and the conclusions we reached before do not apply. In general, when there are no fixed population sizes, the entire population must die off, no matter what the starting size is.

We can actually push this analysis a tiny bit further. There is a critical change that occurs between 15,000 and 16,000 fish harvested per year. It is simply this: the expression inside the square root in the quadratic formula changes from a positive to a negative quantity. Now that expression is

$$1.25^2 - 4(.000025)\textbf{15,000}$$

for a harvest of **15,000** fish per year, and

$$1.25^2 - 4(.000025)\textbf{16,000}$$

for a harvest of **16,000** per year. Let us introduce the variable h to stand for the number of fish harvested per year. If we look for the fixed population sizes, we will be led to use the quadratic formula, and will have to take the square root of

$$1.25^2 - 4(.000025)h = 1.5625 - .0001h$$

This will be impossible when $1.5625 < .0001h$, for that will give us a negative number. This shows clearly that we cannot make h too large. Indeed, the largest possible choice for h will be the one that makes $1.5625 = .0001h$. That is, $h = 15,625$ is the largest possible annual harvest that will allow the population to level off. Any larger harvest amount will lead to the ultimate extinction of the fish population, according to the model. This reasoning shows why we found such different results for annual harvests of 15,000 and 16,000 fish. It also establishes a theoretical maximum for the amount of fish the lake will provide as a sustainable resource: 15,625 per year. It would be foolish to actually try to harvest this maximum amount. In any year, the true harvest might be more than intended, or there might be errors in the model that caused us to overestimate the maximum safe harvest amount. In either case, the effects on the fish population could be disastrous. With a lower harvest level, say 12,000, the effects of errors would be much less pronounced. So the theoretical maximum of 15,625 should be thought of as a limit rather than as an attainable goal.

This concludes the discussion of the logistic model with harvesting. The point has been to illustrate the surprising richness of this type of model, and some of the power of mathematical models. The future evolution of the fish population according to the model is quite predictable, and the theoretical method leads to a number of interesting conclusions about the model. The development of the logistic model is a very simple variation on geometric growth; the inclusion of a harvesting effect is also a very simple idea. But with these two modifications we find that the unreasonable long-term behavior of geometric growth models is transformed into something that is appealingly plausible, and also analyzable using fairly elementary methods.

Summary

In this chapter we have developed the idea of logistic growth. We started by examining one of the limitations of geometric growth models, namely, the unreasonable assumption that a population can grow by a constant factor no matter how big the population is. We modified the model to reflect factors that limit population growth. Specifically, we assumed a linear relationship between the size of the population and the growth factor. The resulting kind of model is referred to as logistic growth.

Working with a specific example, we found that our logistic model improved on geometric growth as desired. For a small starting population, the model was very similar to geometric growth at first. As the population grew larger, however, the rate of growth slowed, and eventually the population leveled off at a steady size. These conclusions were discovered numerically and graphically.

Following the study of the specific example, a more general discussion was presented, covering properties common to all logistic growth models. Three different versions of the logistic growth difference equation were given, each featuring different combinations of parameters r_0, m, and L. The parameters r_0 and m come from the linear equation for growth factor; r_0 is the initial growth factor, and m is the slope. The parameter L is the limiting size for the population. For a population size larger than L, the growth factor equation becomes invalid, producing negative growth factors. These parameters are connected by an equation: $mL = r_0$.

The values of the parameters in any particular example of a logistic model can be used to determine some of the characteristics of the model. If $mL < 4$ and the population starts out between 0 and L, we can be certain that the population will always remain between 0 and L. If $mL < 3$, we can also be sure that the population will eventually level off to a steady size. In any case, if the population does level off, the steady size will be about $L - 1/m$.

For logistic growth there is no functional equation. That is, there is no simple equation that expresses p_n as a function of n. This is an important difference between logistic models and those we studied earlier. For example, in a geometric growth model, we were able to write equations of the form $p_n = 10{,}000 \cdot 1.3^n$. This equation can be used to compute p_{50} immediately as $10{,}000 \cdot 1.3^{50}$, and we can compute any other p_n in a similar way. There is no such equation for logistic models. For the fish population model

$$p_{n+1} = .000025(90,000 - p_n)p_n; \qquad p_0 = 1,000$$

There is no direct way to compute p_{20} or p_{30}. To find those values we simply have to compute p_1, p_2, p_3, and so on until we reach the 20th or 30th p.

The lack of a functional equation for logistic models is partly offset by general information about the leveling off of logistic models. In the case that $mL < 3$, we know that in the long run the population will approach a steady level of $L - 1/m$. So while we may not be able to compute the exact value of p_{50}, we can be confident that it will be somewhere about $L - 1/m$. The situation is more complicated when $mL > 3$, and logistic models for this case can behave chaotically. That topic will be covered in the next chapter.

As a final topic for the chapter, a modified logistic model with harvesting was considered. For this discussion we modified the original logistic model for a fish population to include the removal of a fixed number of fish per year. That situation is described by the difference equation

$$p_{n+1} = .000025(90,000 - p_n)p_n - h$$

where h is the fixed number of fish to be harvested. Using numerical and graphical methods, we saw that harvesting in this way can lead to two different future developments. If the harvest is not too great, the population will eventually level off. In this case, the harvest is a renewable resource that can be taken year after year. On the other hand, if the harvest is too large, the population will be driven to extinction. This situation can be analyzed theoretically using methods of higher mathematics. The results depend on fixed populations for the model. These are populations which remain the same year after year, and which therefore satisfy the equation

$$p_{n+1} = p_n$$

Combined with the difference equation above, this condition leads to a quadratic equation, and we can use the quadratic formula to find the fixed populations. If h is too large, there are no solutions to the quadratic equation. In this case the population must eventually die off. If h is small enough, there will be two fixed populations, LFP (for the lower fixed population) and HFP (for the higher fixed population.) In this case, as long as the starting population is between LFP and $L - LFP$ the population will eventually level off to a steady size equal to HFP. If the initial population is less than LFP or greater than $L - LFP$, the fish will eventually become extinct.

Exercises

Reading Comprehension

1. Compare and contrast arithmetic growth, geometric growth, and logistic growth. Refer to the concept of growth factor where appropriate.

2. Write a short essay (about a page in length) explaining what logistic growth is and how it is related to geometric growth.

3. What is meant by a fixed population? Why are they important in discussing logistic growth? Why are they important in discussing logistic growth with harvesting?

4. Describe the possible long-term patterns that are observable in logistic growth models. Is it possible to predict for a particular application what the long term pattern will be? Explain.

5. Repeat the preceding question, but this time in the context of logistic growth with harvesting.

6. Write a summary in your own words of the section *Keeping Within the Proper Range*.

7. Explain the example connected with Table 13.2 and Fig. 13.6. What is the significance of this example?

Mathematical Skills

1. For a certain model, when the population is 1,000, the growth factor is 3, and when the population is 25,000, the growth factor is 1. Find a linear equation relating population and growth factor.

2. Repeat the preceding problem if the growth factor is 5 when the population is 1,000 and the growth factor is 2 when the population is 25,000.

3. A logistic growth model has the difference equation

$$p_{n+1} = .00008(17,500 - p_n)p_n$$

What is the fixed population for this model? Can you be sure that the model levels off eventually?

4. A logistic growth model has the difference equation

$$p_{n+1} = .00016(17,500 - p_n)p_n$$

What is the fixed population for this model? Can you be sure that the model levels off eventually?

5. A logistic growth model has the difference equation

$$p_{n+1} = .0002(17,500 - p_n)p_n$$

What is the fixed population for this model? Can you be sure that the model levels off eventually?

6. Solve the equation $x = .00008(17,500 - x)x - 400$.

7. Solve the equation $x = .00008(17,500 - x)x - 600$.

8. A logistic growth model with harvesting has the difference equation

$$p_{n+1} = .00008(17,500 - p_n)p_n - 400$$

Are there fixed populations for this model? If so, find them.

9. Describe the long-term growth pattern for the preceding problem for various starting sizes of the population.

10. A logistic growth model with harvesting has the difference equation

$$p_{n+1} = .00008(17{,}500 - p_n)p_n - 600$$

Are there fixed populations for this model? If so, find them.

11. Describe the long-term growth pattern for the preceding problem for various starting sizes of the population.

Problems in Context

1. On page 300 the equation

$$r - 2.8 = -.000045(p - 10{,}000)$$

is given. Show the steps that lead to this equation.

2. In this problem you will construct a logistic model following steps similar to what was in the reading.

Bacteria are grown in a sealed environment in a laboratory. The size of the population of bacteria is measured daily. Early in the experiment, the size grows very rapidly. For example, one day the population was 500, and the next it had doubled to 1,000. (This means the growth factor for that day was 2.) Later, when the population was much larger, the rate of growth slowed down. When the population reached a size of 10,000 it was only growing by about 10 percent each day. (That is a growth factor of 1.1.)

a.   Find a linear equation for growth rate r as a function of population p. [Hint: The information above gives you two data points: when p is 500 r is 2, and when p is 10,000 r is 1.1.]

b.   Use your linear equation to formulate a logistic difference equation for the bacteria population.

c.   Explore the future growth of the population according to the model. Use numerical and graphical techniques, and assume a starting population of 500.

d.   From your numerical and graphical work, does the population seem to level off? At what level? Can you derive that result theoretically?

e.   Try a variety of different starting populations. Do they all lead to a steady population eventually? Can you derive that result theoretically?

3. A logistic model for the dissemination of information. There are about 4,000 students at Anonymous University. Early one morning the student body president tells the vice president that a famous actor (whose identity is supposed to be kept secret) has agreed to be the commencement speaker. In this exercise you will develop a logistic model for the way that information is spread to the entire student body.

Let p_n stand for the total number of students who have heard the news after n hours. We will assume that each student tells another student every hour. Initially, that would cause p_n to double every hour. However, as the number of students who have heard the news increases, the situation changes. When a student tells the news to another student, he or she might have already heard the news. In fact, once p_n reaches 4,000, it cannot possibly rise any higher, since there are only 4,000 students at the university. So assuming that p_n doubles initially (when $p = 1$) and that p_n remains constant when it reaches 4,000, develop a linear equation for r and p, as in the previous problem. Then formulate a logistic difference equation for this model, and use it to study how long it takes the news to spread over the whole university.

4. Another Fish Model: For this problem, use the following difference equation

$$p_{n+1} = .01(200 - p_n)p_n - 16$$

for a fish population model. Here p_n is the population after n years in units of one thousand. This model includes a harvest of 16,000 fish per year, represented by the figure of 16 subtracted at the end of the difference equation.

   a. Find the fixed populations for this model.

   b. What is *LFP* for this model? What is *HFP*? What is *L*?

   c. Suppose the starting population in the model is between *LFP* and $L - LFP$. What is supposed to happen? Use a numerical method to see whether the prediction is correct?

   d. Suppose the starting population for the model is below *LFP* or is above $L - LFP$. What is supposed to happen? Use a numerical method to see whether the prediction is correct.

   e. Modify the model by replacing the figure of 16 with 20 in the difference equation. That means increasing the harvest to 20,000 fish. Repeat the questions above for this new model.

   f. If we try to raise the harvest to 30,000 fish, the difference equation becomes

$$p_{n+1} = .01(200 - p_n)p_n - 30$$

   Show that for this version of the model there are no fixed populations. What does that indicate about the model? Verify that with a numerical method.

   g. What is the maximum possible theoretical harvest?

   h. If you were the manager for this population of fish, how many fish would you permit to be harvested each year? Why?

Solutions to Selected Exercises

Mathematical Skills

1. Make a graph showing population (p) on the x axis and growth factor (r) on the y axis. We have two data points: $(1,000, 3)$ and $(25,000, 1)$. The slope of the line joining these two points is $(1 - 3)/(25,000 - 1,000) = -2/24,000 = -1/12,000$. Now we can use the point-slope form for the equation of the line:$(y - 3) = (-1/12,000) \times$

$(x - 1,000)$ so $y = (-1/12,000)x + 1,000/12,000 + 3 = (-1/12,000)x + 3\frac{1}{12}$. But really, the variable x should be p, and y should be r, so the equation is $r = (-1/12,000)p + 3\frac{1}{12}$ or in decimal form, $r = -.00008333p + 3.83333$.

3. For this problem, $m = .00008$ and $L = 17,500$, so $mL = 1.4$. Since that is between 1 and 3, we know the model will level off at $L - 1/m = 17,500 - 1/.00008 = 5,000$. To check this, suppose that $p_n = 5,000$. Use the difference equation to find $p_{n+1} = .00008(17,500 - 5,000)5,000 = 5,000$. This shows that 5,000 is a fixed population.

4. For this problem, $mL = 2.8$ so the model does level off, this time to 11,250.

5. This time $mL = 3.5$, so you cannot assume it will level off. You can check with a numerical method and discover that it does not level off; rather it goes up and down repeatedly. There is a fixed population, however: $L - 1/m = 12,500$. If you start your model at that figure, it will remain constant.

6. Put the equation into the standard form for a quadratic:

$$x = .00008(17,500 - x)x - 400$$

$$x = (1.4 - .00008x)x - 400$$

$$x = 1.4x - .00008x^2 - 400$$

$$0 = .4x - .00008x^2 - 400$$

$$0 = -.00008x^2 + .4x - 400$$

Now we can simplify things a little before using the quadratic formula by dividing both sides of this equation by $-.00008$. That leads to

$$x^2 - 5,000x + 5,000,000 = 0$$

Using the quadratic formula, we find

$$x = \frac{5,000 \pm \sqrt{25,000,000 - 20,000,000}}{2} = 2,500 \pm 1,118.03$$

so the two solutions are 3,618.03 and 1,381.97.

8. The fixed populations are the ones we found in the preceding solution: 3,618.03 and 1,381.97.

9. For this model, the parameter L is 17,500. If the starting population is between 0 and 1,381.97, then it will die off. Similarly, if the starting population is between $17,500 - 1,381.97 = 16,118.03$ and 17,500 the population will die out. Otherwise, if the starting population is between 1,382 and 16,118, then it will level off to a stable value of 3,618.03.

Problems in Context

2. a. Using the two points $(500, 2)$ and $(10,000, 1.1)$ the slope is $(1.1 - 2)/(10,000 - 500) = -.9/9,500 = -.000094737$. Using the point-slope formula to get the

equation of a line, $y - 2 = -.000094737(x - 500)$. However, we are using r in place of y and p in place of x, so the equation is $r - 2 = -.000094737p + .0473685$ or, $r = -.000094737p + 2.0473685$.

b.   The logistic growth difference equation is $p_{n+1} = rp_n$ where r, the growth factor, is given by the linear equation we found earlier. That leads to the equation $p_{n+1} = (-.000094737p_n + 2.0473685)p_n$. We can rearrange this to the standard form by reversing the order of terms in the parentheses: $p_{n+1} = (2.0473685 - .000094737p_n)p_n$ and then putting the factor of .000094737 outside the parentheses: $p_{n+1} = .000094737(21,611.076 - p_n)p_n$

d.   Here, $mL = 21611.176$ which is between 1 and 3. So the theoretical results indicate that the model should definitely level off for any starting population between 0 and 21,611. The fixed population level is $L - 1/m = 11,056$, approximately.

4. a.   The equation to find fixed populations is $x = .01(200 - x)x - 16$. This can be put into the standard form for quadratic equations. The result is $-.01x^2 + x - 16 = 0$, and, by dividing both sides by $-.01$, it becomes $x^2 - 100x + 1,600 = 0$. This can be solved using the quadratic formula. The solutions are 20 and 80.

b.   The LFP is 20, the HFP is 80. The value of L is 200.

c.   If the starting population is between $LFP = 20$ and $L - LFP = 200 - 20 = 180$, the population is supposed to level off at the HFP, which is 80.

d.   If the starting population is below 20 or above 180, the population in the model will die off.

e.   For the modified equation, $LFP = 27.639$, $HFP = 72.361$, so the population should level off to 72.361 as long as the starting population is anywhere between 27.639 and $200 - 27.639 = 172.361$.

f.   No fixed populations would indicate that the population will die off, no matter what the initial population is.

g.   The maximum theoretical harvest will occur if the part under the square root in the quadratic formula is equal to 0. Use a variable for the harvest. Call it h. Then the fixed populations will be given by solving the equation $x = .01(200 - x)x - h$. When we put this equation into the same form used earlier, it becomes $x^2 - 100x + 100h = 0$. For this equation, the quadratic formula includes $\sqrt{100^2 - 4(1)(100h)}$, and we want the part under the square root to be 0. That is, we want $100^2 - 400h = 0$, or $10,000 = 400h$. Divide both sides of this equation by 400, and we find $h = 25$. This is the maximum theoretical harvest.

h.   Set the harvest well below the maximum level, say at 20. Then the fish population should level off at about 72. That is, with a harvest of 20,000 fish, the population should level off at about 72,000 fish.

Chaos in Logistic Models

Now we come to the topic of Chaos, a subject that is the focus of research today in important areas of mathematics, engineering, and the physical and social sciences. Chaos in mathematics is closely related to the subject of fractals, widely publicized in the form of stunningly intricate and fascinating color images. A popular book on the subject, James Gleich's *Chaos: The Making of a New Science,* was a best-seller. Chaos was even featured as an important plot element in Michael Crichton's novel *Jurassic Park,* and in the movie based on the novel.

But what exactly is chaos? And why is it of so much interest? These are the questions that will be considered in this chapter. The scope of the discussion will be very limited. Following a brief general account of the main ideas of chaos, we will look in detail at the chaos that arises in logistic models. Although this is only the tip of the chaos iceberg, it should give you an idea of several important aspects of chaos. Gleich's book discusses chaos in many other contexts, and describes the historical development of the subject, without going into much mathematical detail.

An Overview of Chaos

To understand the main ideas of chaos, you need to start with a general awareness of the development and use of mathematical models. That is what you have been studying through the whole course, so it is appropriate here to reflect on some general characteristics of the subject. We will look at two aspects of modeling: the mathematical methods that have been used in this course and the practice of using models in the study of real applications.

Review of Models Studied. In this course, almost all of the models that have been presented were developed along very similar lines. The starting point is a discrete sequence, representing repeated measurements of some phenomenon over time. These measurements can be real data or simply a conceptual image. They can occur once a year, once a month, once a day, or at any other regular frequency. Whatever the frequency is, and whether the data are real or imagined, the common aspect to all of the examples is a progression of numbers—that is, a sequence.

It is a natural idea to examine the way the numbers in that sequence evolve. How is each number related to the one that came before it? One can imagine various ways to use each number to compute the next. For example, in Chapter 2, we considered the sequence of daily world oil consumption levels. As an approximation to the pattern of data, we supposed that each year's consumption level was .3 higher than the preceding year's. That kind of pattern, in which each data value is increased by a fixed amount to find the next data value, is referred to as arithmetic growth. And the mathematical formulation of the pattern is expressed as a difference equation:

$$c_{n+1} = c_n + .3$$

But arithmetic growth is just one possibility for the evolution of a number sequence. We also considered quadratic growth, geometric growth, mixtures of arithmetic and geometric growth, and logistic growth. In Table 14.1 we summarize these different kinds of models.

Implicit in the way we studied these models is the idea of grouping them by type. We studied all of the arithmetic models together, discovering common characteristics of their graphs, and functional equations. In a similar way we studied all the geometric models together. This approach works because arithmetic growth models do share many common features; geometric growth models share a different set of common features; mixed models have still another set of common features, and so on. Here we are focusing on the mathematical form of the difference equation. We are saying that these three equations are alike:

$$a_{n+1} = a_n + .0001$$

$$a_{n+1} = a_n + 102$$

$$a_{n+1} = a_n + 12.435$$

Type	Description	Sample	General Form
Arithmetic	Add a constant	$a_{n+1} = a_n + .3$	$a_{n+1} = a_n + d$
Quadratic	Add an amount that grows linearly with n	$a_{n+1} = a_n + .3 + .02n$	$a_{n+1} = a_n + d + en$
Geometric	Multiply by a constant growth factor	$a_{n+1} = 2a_n$	$a_{n+1} = ra_n$
Mixed	Multiply by a constant growth factor and then add a constant	$a_{n+1} = 2a_n + .3$	$a_{n+1} = ra_n + d$
Logistic	Multiply by a growth factor that depends linearly on the current term of the sequence	$a_{n+1} = .001(1,000 - a_n)a_n$	$a_{n+1} = m(L - a_n)a_n$

TABLE 14.1
Difference Equation Models

whereas these three equations are not:

$$a_{n+1} = a_n + 3$$

$$a_{n+1} = a_n \cdot 3$$

$$a_{n+1} = (a_n)^3$$

The first three equations involve the same mathematical operation—addition—and lead to models with the same characteristics. The equations in the second set involve different mathematical operations—addition, multiplication, exponentiation—and so lead to models with quite different characteristics. In comparing the two sets of equations, we see that the particular number that appears in an equation is less important than the operations that are performed. This leads us to the important concept of a *parameter*.

Parameters are used whenever we study a family of equations which share a similar form, but which involve different numerical constants, as in the first set of three equations. We represent the general form of the equation by introducing a variable in place of each numerical constant. So, the three equations in the first set can all be represented in the form

$$a_{n+1} = a_n + d$$

where we understand that d can be replaced by any fixed constant. This d is a parameter. It represents a constant in any arithmetic growth model, but assumes different values in different particular examples.

In our study of the various types of difference equations included in Table 14.1, we found many properties for each family. These properties were expressed in terms of the parameters appearing in the difference equation for the family. For example, every arithmetic growth model has a graph that is a straight line, and the parameter d gives the slope of this line. Similarly, for every geometric growth model the graph is an exponential curve, and the parameter r gives the base of the exponential. In the case of logistic growth, the parameters m and L indicate whether the model levels off to a steady value: when $1 < mL < 3$ we can be assured that the sequence will approach a steady value of $L - 1/m$.

To summarize the approach to all of these types of models, we began in each case with a simple description of how one number in the sequence can lead to the next. A general difference equation was defined, with appropriate parameters. The common characteristics for the particular type of difference equation were discovered, including in every case except logistic growth, a functional equation for the sequence.

Using the Models. Now let us turn to the practice of actually applying these models. We understand that the model is only an approximation to the actual situation, and that there are simplifications in every model. Often, we try to make the approximation as accurate as possible by choosing the values of the parameters carefully. But we realize that even when we do so, there are measurement errors in our data, and so the model is at best an estimate. Often, we are more interested in qualitative descriptions of whatever we are modeling than we are in highly accurate predictions of the future. Will the amount of the drug in the blood stream level off? Will the population of fish die out? The models give us qualitative answers to these questions.

There is an important feature of the models we have examined that makes this qualitative approach effective: moderately changing the values of the parameters does not significantly change the qualitative behavior. So, for a geometric growth model, if we have an incorrect value of r, we can still be pretty sure that the long-term behavior will be accelerating growth to ever larger values (for $r > 1$) or a decrease to 0 (for $r < 1$). Similarly, for the mixed models, we can be sure that the number sequence levels off, even if we have some errors in our parameters. In a similar way, if we have some uncertainties about the starting point for the sequence, we can still make predictions about the future. This is particularly true for the mixed models and the logistic models where we found a leveling-off occurred. For those models, even if we are unsure of the exact starting point for the model, we can be confident that the long-term behavior will level off at a prescribed value.

An Important Aspect of Chaos.   And now we come to the crux of the chaos concept. A characteristic of chaos is that small errors in the parameters or the starting values for the data lead to vastly different future behaviors. To emphasize this point, here is a comparison of a chaotic model with one that is not chaotic. First, the non-chaotic example.

Imagine that you are a doctor interested in the way medicine is absorbed by the body. Your favorite applied mathematician describes a model based on the assumption that a fixed percentage of the medicine is absorbed in a fixed amount of time. *So, if it is absorbed at a rate of* 10 *percent each hour,* you are told, *and if* 50 *mg are taken every four hours, then over time the amount of drug in the blood will level off at about* 145 *mg.* But you point out that the studies are not conclusive. For some patients the absorption rate is as high as 12 percent an hour. For others it is as low as 8 percent an hour. *In that case, the model still indicates that the amount of drug in the blood will become steady,* you are reassured, *somewhere between* 125 *and* 176 *mg.* That is a reliable model. It allows qualitative predictions even when there are errors in the assumptions.

In contrast, suppose you are trying to forecast the weather. Your favorite mathematician describes a model that takes into account temperature, pressure, humidity, and wind data from weather stations all over the globe. It uses equations that are known to be highly reliable descriptions of how the weather evolves. *And based on this model,* your consultant confidently announces, *it will be sunny and warm in New York two weeks from Wednesday for your campaign speech.*   Then one of the consultant's assistants comes in, and apologizes that the temperature reading from Tokyo was entered in the model incorrectly. It was accidently entered as 67.8 rather than 68.7. So the model was recomputed. *Ah, now we have the correct answer,* purrs your consultant—*expect a blizzard for your speech.* That is typical of a chaotic model. Even very slight changes in the parameters or starting values for the model lead to vastly different qualitative predictions. That is chaos.

The realization that certain kinds of equations lead to chaotic models is a fairly recent discovery. Although Henri Poincaré raised this as a theoretical possibility in the early

1900s, the significance for real models was not appreciated until just a few years ago. It was long suspected that the simple equations that had been used in models would lead to accurate long-range predictions if the parameters could be determined accurately enough. Now we have mathematical analyses that show for specific models that chaos is unavoidable. For these models, even the slightest errors in the parameters and starting values would lead to wildly inaccurate predictions. And since real measurements always include some error, the predictions that come out of the models simply cannot be trusted.

So far, this discussion has been pretty abstract. Shortly we will look at a particular case of chaos, as it appears in the logistic model. There you will see in detail how the ability to make predictions breaks down, and how very slight changes in the starting value can lead to vast differences in the future evolution of the model. However, before proceeding to that topic, one more point should be made about the difference equations in Table 14.1.

Linear and Nonlinear Difference Equations. In all of the difference equations in the table, there is a common pattern. Each equation has a_{n+1} on one side and an expression involving a_n on the other. In fact, we can think of each equation as a kind of function. In every case, the equation allows us to compute a_{n+1} as soon as we know the value of a_n. That means that each equation gives a_{n+1} as a function of a_n, or, in one case, as a function of both a_n and n. Up to this point, when the terms *functional equation* were used in connection with a difference equation, we were interested in giving a_n as a function of n. This was meant to emphasize the distinction between two different approaches to compute a particular term, such as a_{25}. One method is *recursive*: compute a_1, then a_2, then a_3, and continue in this way until a_{25} is reached. The other method is direct: specify that $n = 25$, and compute a_{25} immediately. This is possible if there is an equation that gives a_n as a function of n. But the idea of *function* has a broader meaning than the one just described. Any time one quantity can be computed directly from the value of another quantity, there is a function at work. So, it makes sense in a difference equation to say that a_{n+1} is a function of a_n: as soon as the value of a_n is known, we can immediately compute a_{n+1}. The most general formulation of all the difference equations is

$$a_{n+1} = f(a_n)$$

where the right hand side is meant to indicate a function of a_n. In the case of quadratic growth, we have a slightly different form:

$$a_{n+1} = f(a_n, n)$$

indicating that you need to know both a_n and n before you can compute a_{n+1}.[1]

Having reached this perspective, we can classify the different kinds of difference equations by classifying the functions they involve. For arithmetic growth, the function

---

[1] To make this idea more concrete, suppose I tell you that one of the numbers in a data sequence is 16, and ask you to predict the next one. For every case except quadratic growth, all you need to know is the parameters in the model, and then you can compute the next number after 16. But for quadratic growth, you need to know where 16 shows up in the pattern. If it is the fifth number, you will get one prediction; if it is the tenth number, you will get another. So for quadratic growth, you need to know both a_n and n.

has the form $f(a_n) = a_n + d$. For geometric growth, it is $f(a_n) = ra_n$. And the mixed-model function is $f(a_n) = ra_n + d$. In each of these cases the function f is *linear*. That is, for arithmetic growth, geometric growth, and mixed models, each term in the sequence depends linearly on the preceding term. For quadratic growth, although both the preceding term and n enter into the calculation, the function is again linear. As a result, we refer to the difference equations in all of these cases as linear difference equations. In contrast, for the case of logistic growth, the dependence of each term on the preceding term involves a quadratic function: $f(a_n) = m(L - a_n)a_n$. This is a nonlinear difference equation. It is the nonlinear difference equations that give rise to chaos.

Do not confuse the concept of a linear difference equation with a linear functional equation. We know that in geometric growth, the graph of a_n versus n is exponential, and that for mixed models the graphs are shifted exponentials. When we study how a model evolves over time (or in relation to n), we only observe a linear pattern for arithmetic growth. The idea of a linear difference equation is not concerned with how some term in the sequence depends on n, but rather on how it depends on the preceding term. This distinction is illustrated graphically in Fig. 14.1 and Fig. 14.2. Using the example of the difference equation

$$a_{n+1} = .5a_n + 20 \tag{1}$$

and a starting value of $a_0 = 10$, Fig. 14.1 shows the relationship between a_n and n. In this figure we graph a_1 above the number 1, a_2 above the number 2, and so on. The graph is clearly not linear. In Fig. 14.2 there is a different kind of graph for the difference equation. This graph shows the relationship between any two successive data values. Here, we ignore the value of n, and focus on how a_{n+1} depends on a_n. For example, if some a_n is 10, then using Eq. (1), the next will be $a_{n+1} = .5 \cdot 10 + 20 = 25$. So for this difference equation, a value of 10 is always immediately followed by a value of 25. That gives one point on the graph, $(10, 25)$. Similarly, if $a_n = 16$, the next value will be $8 + 20 = 28$, producing the point $(16, 28)$. Repeat this process for many many different choices of a_n, and the result will be a graph as shown in Fig. 14.2. For a linear

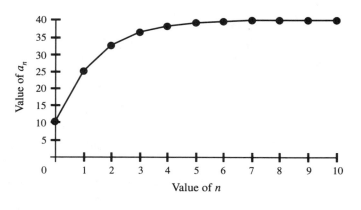

FIGURE 14.1

How a_n depends on n

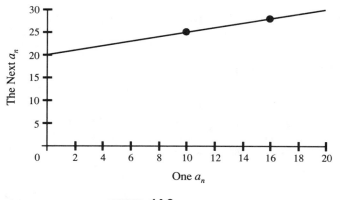

FIGURE 14.2
How a_n depends on a_{n-1}

difference equation, it is this latter kind of graph, showing how each term is related to the preceding term, that is linear. As the two graphs illustrate, even if the difference equation is linear, the graph showing how the data actually evolve over time (like Fig. 14.1) may well be nonlinear.

Whether a model has a linear or nonlinear difference equation has a special significance for a discussion of chaos. Models with linear difference equations are *never* chaotic. Of all the models we have looked at in the course, only the logistic models have nonlinear difference equations. Of all the models, only logistic models sometimes exhibit chaos. How that can occur, and what it means, will be considered next.

Chaos in the Logistic Model

To make the discussion as concrete as possible, we will again consider the fish population model. In the preceding chapter, we developed the equation

$$p_{n+1} = .000025(90{,}000 - p_n)p_n$$

To simplify the discussion here we will modify the equation slightly, by changing the 90,000 to 100,000. That gives the equation

$$p_{n+1} = .000025(100{,}000 - p_n)p_n \tag{2}$$

The numbers in this equation will be easier to work with if we use the population in units of 100,000. To be clear how to do this, we can define the variable a_n to be the population in 100,000's, and recognize that $p_n = 100{,}000a_n$. For instance, if $p_n = 125{,}000$, then in hundred-thousands, $a_n = 1.25$. Similarly, if $p_n = 200{,}000$, $a_n = 2$. These examples illustrate the more general pattern that $p_n = 100{,}000a_n$. Of course, it is also true that $p_{n+1} = 100{,}000a_{n+1}$. In Eq. (2), if each p_n is changed to $100{,}000a_n$, and similarly for p_{n+1}, a difference equation is derived for a_n:

$$100{,}000a_{n+1} = .000025(100{,}000 - 100{,}000a_n)100{,}000a_n$$

Dividing both sides by 100,000 leads to

$$a_{n+1} = .000025(100,000 - 100,000a_n)a_n$$

and that can be simplified with algebra to

$$a_{n+1} = 2.5(1 - a_n)a_n$$

This is the difference equation for our fish population in units of 100,000's.

In the context of the earlier discussion of logistic growth, we can compare this equation with the general form

$$a_{n+1} = m(L - a_n)a_n$$

where $L = 1$ and $m = 2.5$. As described in Chapter 13, by looking at $mL = 2.5$, we can see that this model will always lead to a steady population size of $L - 1/m = .6$.

Now to explore the way chaos occurs, we will look at variations of this model by changing just the parameter m. That is, we will look at examples like these

$$a_{n+1} = 3.0(1 - a_n)a_n$$

$$a_{n+1} = 3.1(1 - a_n)a_n$$

$$a_{n+1} = 3.2(1 - a_n)a_n$$

We will be interested in what happens for m between 3 and 4. From our previous study of logistic models, we already know that mL can give us some information about the way a model operates. In the present example, because $L = 1$, the value of mL is really just m. What do the previous results about mL say? For $m < 3$, the model will eventually level off no matter what we start with for a_0. For $m > 4$, the model can lead to invalid results, such as negative populations and growth factors. But for m between 3 and 4, we should see future population patterns that are valid, but which do not level off.

Example: $m = 3.2$.   Let's start with $m = 3.2$. If the initial population is .7 (remember, that is in units of 100,000's), what will the future hold for our fish population? The answer is revealed in the first part of Fig. 14.3. It shows that starting from an inital population of .7 (actually 70,000) the number of fish fluctuates getting larger and smaller. At first, the fluctuations grow, but after about 15 years they steady down to a regular repeated pattern: .799 in the even-numbered years and .513 in the odd-numbered years.

Although this is different from the pattern we saw earlier, it is still very regular and easily permits future predictions. The situation is similar to the earlier patterns in another way, too. The long-range behavior of the population in the model does not depend on the starting point. In the second part of the figure the same model is used, but this time the starting population size is .4 (or 40,000 fish). Once again the same pattern of fluctuation is observed, and after about 6 years, we see the identical behavior as before: year by year the population switches between .799 and .513.

What the graphs reveal for two starting population sizes can be shown to hold in general, using theoretical methods. We won't go into the details of those methods here, but they are similar to what was used in the last chapter to find the steady population

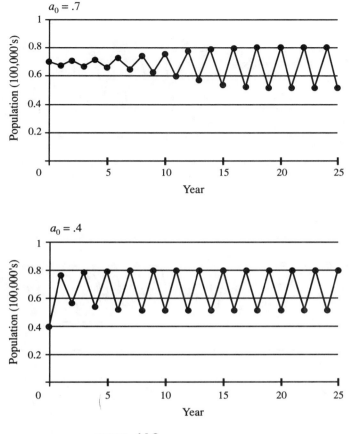

FIGURE 14.3
Population Models with $m = 3.2$

levels. There, for a steady population, we wanted to find $p_{n+1} = p_n$ for any n. This led to an equation which we solved to find the steady population size p_n. Now we are looking for a population that repeats not the next year, but in two years. That is expressed by the condition $p_{n+2} = p_n$. This also leads to an equation for p_n, but it is a little more complicated than the one we found in the previous chapter. Be that as it may, it is possible to find the solutions to the equation (they include .799 and .5134) and to show that in the long run the model will fluctuate between these two figures, no matter what the starting population size is.

Example: $m = 3.4.$ Now we have examined one example, the case for $m = 3.2$. We already knew that, for $m < 3$, the population model would settle down to a steady size. Now we have seen that for one $m > 3$ the population fluctuates between two steady sizes. Will this same situation occur for other values of m? In Fig. 14.4 the situation for $m = 3.4$ is shown. As in the previous figure, there are two examples, one with an initial population size of .7, and one at .4. The long-term results are very similar to what

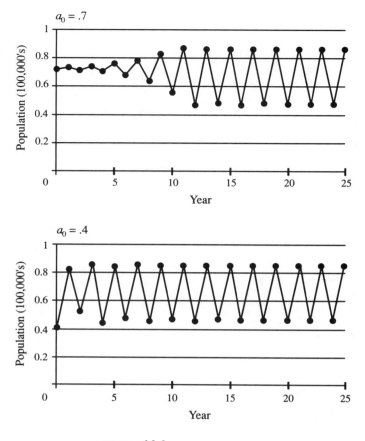

FIGURE 14.4
Population Models with $m = 3.4$

was seen earlier. Once again the population will fluctuate between two steady values, no matter what the starting population is. This time, though, the fluctuations are a little greater in size, from .452 to .842.

As these examples show, even when there is not a single steady population, meaningful predictions can be made about the way the population grows in the model. For both examples, a wildlife manager could expect that a year of high fish population would be followed by a year with a low population, but that the population would spring back to the higher level the following year. If the estimate of the initial population is a little off, that has no effect on the long-range behavior. Even if m changes slightly, that doesn't make a significant difference in the way the model behaves. These conditions are not what you would call chaotic. But as we try ever higher values of m, the model begins to change.

Example: $m = 3.5$. For $m = 3.5$ we observe a new behavior (see Fig. 14.5). This time the population does not settle down to a pattern of fluctuation between two steady

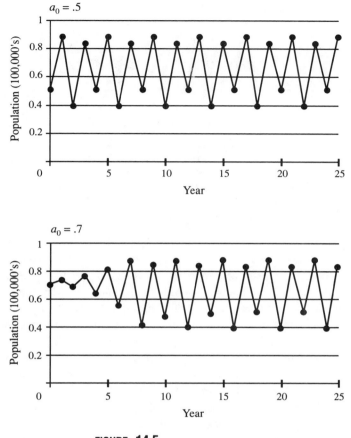

FIGURE 14.5
Population Models with $m = 3.5$

sizes. Instead there are four population sizes that get repeated over and over again. There is still regularity in this pattern, but it is more complicated than the earlier examples, and is not as easy to pick out by eye. It is still true that the starting population size has little effect on the long-term model behavior. In the first part of the figure the population starts out at .5, and in the second it starts at .7. But eventually the same pattern of four population sizes shows up in each case. To illustrate this, Fig. 14.6 shows both of the patterns combined on a single graph, one with black dots and the other with gray dots. The way these dots line up shows that both patterns cycle among the same four values.

This pattern is more complicated than the ones shown earlier, but is still not chaotic. All that has been illustrated up to this point is that as m gets larger and larger, the pattern of the model gets more and more complicated. Interestingly, there is a point at which the behavior is so complicated that all patterns break down. At that point, small differences in m have a tremendous impact on the way the model behaves. To illustrate chaotic behavior, we will consider an example for $m = 3.8295$. The reason for choosing that particular m will be discussed a little later.

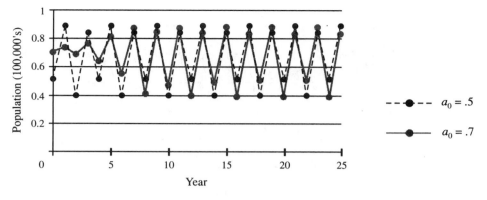

FIGURE 14.6
Combined Pattern for Two Starting Populations

An Example of Chaos.   In Fig. 14.7 there are two graphs for the model with $m =$ 3.8295. Each projects the fish population for 50 years. In the first graph, the starting population is .9000, in units of hundred-thousands, or 90,000 fish. In the second graph, the starting population was taken to be .9001, meaning 90,010 fish. The two graphs show very similar patterns of development for about 10 years. But then they separate, and become quite different. Notice that in the first graph for about 6 years in a row the population holds very steady at about .75. This occurs after about 19 years of fluctuating populations. There is no such pattern in the second graph. There seems to be no order at all in the long-term pattern of either graph. For some periods the population goes into a repeated three-year cycle of low, medium, and high populations. But this pattern does not continue long. At another place on the graph there is fluctuation between high and low population amounts, but that too lasts for only a few years. To make it clearer where the models are similar and where they look completely different, the two graphs have been combined in Fig. 14.8.

This is a chaotic model. There just isn't any pattern that can be used to predict the future of the model. What is more, even the slightest change in the starting population value leads to completely different longe-range patterns of population activity. This is what is referred to as the *butterfly effect*. Remember the example at the start of the chapter about weather prediction? There, a slight error in one temperature leads to a dramatic difference in the weather predicted by a model. With the incorrect temperature the prediction calls for warm sunny weather; with the correct temperature a blizzard is predicted. A variation on this idea involves the beating wings of a butterfly instead of an error in the temperature. One prediction includes the effects on the atmosphere of the butterfly's wings, the other does not. The idea that a single butterfly in Tokyo can alter the predicted weather for New York dramatizes how sensitive the model is to the slightest changes in the parameters. That is what is meant by the butterfly effect.

We can see the butterfly effect at work in the fish population model in the preceding example. In one case the model is based on 10 fish more than the other. After 10 years, the effect of those 10 additional fish is a completely different population projection.

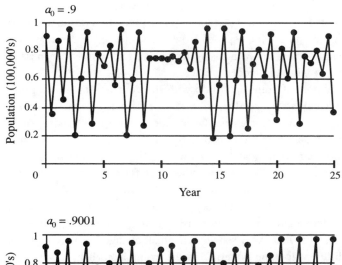

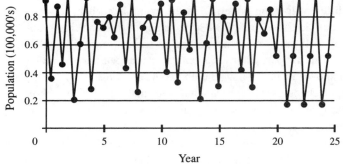

FIGURE 14.7
Population Models with $m = 3.8295$

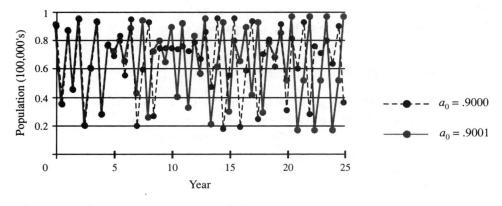

FIGURE 14.8
Combined Graphs for $m = 3.8295$

There are two important consequences of the butterfly effect as it appears in the fish population model. First, it shows that longe-range prediction of the real world (as opposed to the model) is essentially impossible. However accurate we can make the model, there will be some slight errors in our measurements. We know now that these errors will lead to major effects in what the model predicts.

The second important point has to do with the complete lack of a predictable pattern in the model for any starting value. If a chaotic model is accurate, looking at real fish population data for several years really provides no help in making future predictions. The actual fish population could remain essentially unchanged for several years (as in the first graph from year 19 to year 25). How tempting to predict that the fish population is in balance, and will remain at the same level for many years to come. But as the graph shows, with no apparent warning or explanation, the population suddenly starts to fluctuate between very high and very low populations. The wildlife manager cannot count on any sensible pattern for future fish population figures. That is chaos.

The parameter m is itself very sensitive to small changes. Changing it very slightly can lead to very different looking patterns. In the example earlier, m was set to 3.8295. If, instead, a value of 3.829 or one of 3.830 is selected, the resulting graphs are completely different. They will still follow no particular pattern, and will still be subject to the butterfly effect. But the particular arrangement that appears in the graph will be different from the ones shown in Fig. 14.8. The choice of 3.8295 has no special significance for the discussion of chaos. After experiments were made with many choices of m, 3.8295 was chosen because for that value one graph shows nearly constant populations for one period of several years.

Now you have seen that chaos arises in the logistic growth model for fish populations. It arises in many other kinds of models as well. This is one of the reasons that the study of chaos is so important today. The examples you have seen show that there is essentially no order that can be used to predict future fish populations in the chaotic model. But that is not the same thing as saying that there is no order whatever. It turns out that there is a kind of order in chaos. One aspect of this order is the ability to predict when a system will become chaotic, and in some cases, what to do to keep chaos from occurring. We will take a brief look at this idea as the final topic for the chapter.

Order in Chaos: A New Kind of Graph

Take a moment to review what we know about the logistic fish population model. What happens in the long run depends on the parameter m. If m is too small, that is, when $m < 1$, the population will eventually die out completely. For larger values of m, from 1 up to a value of 3, the population will settle down to a steady value and remain the same year after year. Just slightly above 3, we observed a fluctuating pattern between two steady populations. Then for m a bit larger, we found fluctuations among four steady values. Finally, when we made m big enough, chaos appeared. Now we will study the effects of making m larger and larger in a more systematic way. In the process we will see how the logistic population model makes a transition from a non-chaotic to a chaotic system. In this transition to chaos, a new kind of order will emerge.

Chaos theoreticians invented a new kind of graph to display the effects of m and the onset of chaos. The idea is to make a graph showing different values of m on the horizontal axis, and the long-term behavior of the population model on the vertical axis. To understand how this idea works, let us look again at examples we examined earlier.

Consider the case for $m = 3.2$. In the first graph of Fig. 14.3 the population values are plotted year by year. Now we will ignore the year, putting each of the points on a graph with population on the vertical axis and with m on the horizontal axis. Every one of these points has $m = 3.2$, so they all line up directly above 3.2 on the m axis. The result is shown in Fig. 14.9. The points are so close together that they seem to create a solid black line. However, many of these points are not really significant for long-term prediction.

We already saw that for $m = 3.2$, eventually the population always reaches a pattern of fluctuation between two steady sizes. At the beginning, the choice of the starting population does have an effect. Different starting values will produce different sequences of population values for the first several years. But in the long run, these first few data points are not really significant. Ultimately, the population will fluctuate between the same two fixed values, no matter what the first few data values are. For this reason, it makes sense to screen out the short-term effects, and focus on what happens in the long run. To do so, we will ignore the predicted population figures for the first 100 years. Then we can graph the next 20 years' worth of data points. That gives the graph shown in Fig. 14.10. Rather than the appearance of a solid line shown in Fig. 14.9, now there only seem to be two data points. That is because after 100 years, the population will just jump back and forth between the same two steady population sizes. So 10 of the points all land in one place, and 10 all land in the other, giving the appearance of two points. This graph captures the long-term behavior of the model with $m = 3.2$. In the long run, there are only two population sizes that occur, and these are shown by the two points on the graph. What is more, we would get exactly the same graph no matter what starting population size we choose. After 100 years, the graphs will all look the same.

In Fig. 14.11, a similar graph is shown for $m = 3.5$. As for the preceding graph,

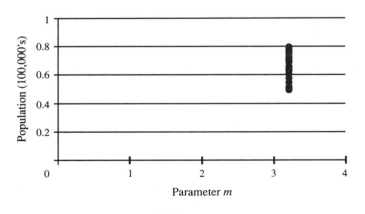

FIGURE 14.9
Population and Parameter m

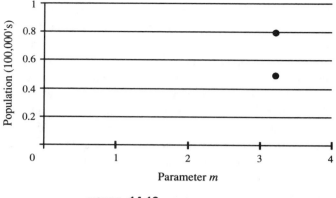

FIGURE 14.10
Long-Term Behavior for $m = 3.2$

we used the model to project 100 years into the future, but did not graph those data points. Then we computed and graphed the data points for the next 20 years. In our earlier discussion for $m = 3.5$, we observed a long-term behavior of fluctuation between four steady values. These four values are clearly revealed in Fig. 14.11. The long-term behavior of the model is quite predictable as it cycles among the four steady values.

For a contrasting view, consider $m = 3.8$. That leads to chaos. We project forward 100 years and then plot the next 20, as before, to make Fig. 14.12. But this time, there aren't just a few points on the graph. The data points spread across most of the range from .2 to 1. The appearance of chaos is revealed in this spread of values. Even after 100 years, the population can be virtually anything.

The last three graphs can all be combined into a single graph. Each has a set of points for a different choice of m, so we can just plot them all on one graph. At the same time, we will add data points for $m = 2.9$ and $m = 3.4$. This gives Fig. 14.13. There is only one data point for $m = 2.9$. That is because, with $m < 3$, the population eventually

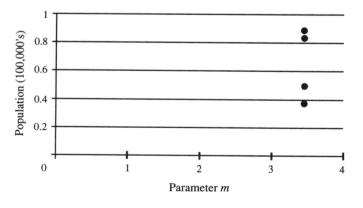

FIGURE 14.11
Long-Term Behavior for $m = 3.5$

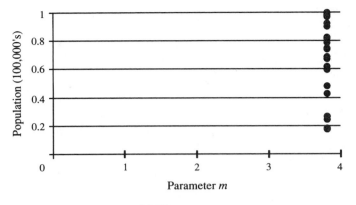

FIGURE 14.12
Long-Term Behavior for $m = 3.8$

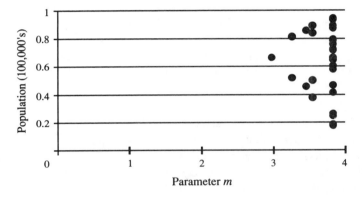

FIGURE 14.13
Combined Graph for Several Values of m

settles down to a single steady value. There are two data points for $m = 3.4$, the two steady values that appear in Fig. 14.4. When you look at the combination chart, you start to see some order in the transition to chaos. In Fig. 14.14 there is a much more detailed version of the graph. Data were generated for this graph using very closely spaced values of m between 1 and 4, and plotted using very small points.

The graph appears to have a smooth curve between $m = 1$ and $m = 3$. This curve shows how parameter m affects the steady population.[2] But at $m = 3$ something interesting happens. There is a transition from models with a single steady population that holds constant year after year, to one in which the population switches back and forth between two different values. At $m = 3$ this appears as a fork where the curve splits into two parts. A little farther on, each of those forks is itself split. That reveals values

---

[2] For instance, with $m = 1.6$, the point on the curve has a y coordinate of about .4. This indicates that, for $m = 1.6$, the model will eventually level off and remain at .4, or 40,000 fish.

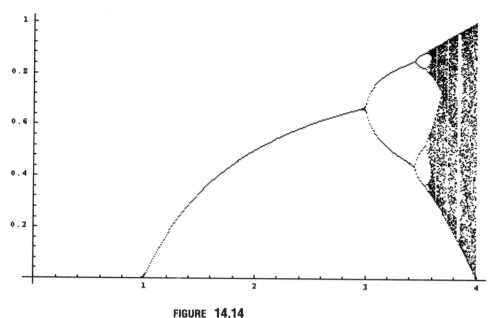

FIGURE 14.14
Detailed Graph for $0 < m < 4$

of m giving rise to four different population values, with the long-term behavior of the model alternating among the four. We saw that kind of behavior earlier for $m = 3.5$. Although it is hard to see in the graph, farther on each of the four branches splits again, and then again, and eventually the curves degenerate into a kind of cloud of dots. That is where chaos arises. Beyond that point, instead of just a few population values that repeat over and over again, there is just a confused list of future populations with no order or pattern. Remember that the diagram shows the behavior of the logistic model for many many different choices for m. On a vertical line directly above each m on the horizontal axis, data values are shown for the logistic model for that m. Within the cloud of points, data values on each vertical line are spread all over the place. They do not settle down to one value, or to just a few values. It is the way this chaotic behavior occurs for many different values of m that produces the appearance of a cloud.

Considered from the viewpoint of longe-range prediction, the diagram reveals chaos for values of m near 4. For m in that range, the long-term behavior of the logistic model seems to follow no order or pattern. And yet, in this diagram of the onset of chaos, there appear to be patterns of a different type. There are regular gaps and spaces, and some darker areas which appear to reveal curious curves. To see these patterns in even more detail, another graph was created (Fig. 14.15). It was produced in exactly the same way as the previous graph, but it concerns only the values of m between 3.5 and 4.0. This has the effect of spreading out the chaotic part of the diagram, and makes the patterns clearer.

These patterns have many interesting aspects. In addition to the obvious visual symmetries of dark areas and gaps, there are other less obvious features. Look at the

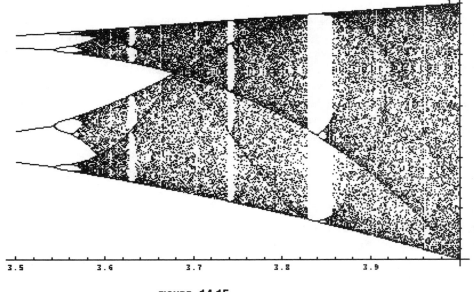

FIGURE 14.15
Closeup Graph for $3.5 < m < 4$

large white gap that is above 3.84 on the m axis. In the center of the gap there is what appears to be a kind of white circle within a triangular patch of black dots. If that part of the figure is enlarged and explored in greater detail, it turns out to have a shape that is remarkably similar to part of Fig. 14.14. That is, there is a miniature chaos diagram within the larger chaos diagram. This is an aspect of chaos that turns up quite frequently, and it is given a special name: self-similarity. When a diagram is self-similar, it is possible to zoom in on one part and find something that looks just like the original figure. Then you can zoom in again on another part and find an even tinier replica of the original figure. In this way, no matter how much the original figure is magnified, it is still possible to recognize similar features to the original graph.

These patterns in the diagrams of chaos for the logistic fish population model can be analyzed in great detail using theoretical methods. Many of the results are typical of a large variety of chaotic models, and are not restricted to logistic growth. Chaos is one of the most active areas for research in applications of mathematics; indeed, chaos has significance in a large variety of fields, including economics, biology, aerodynamics, medicine, and many others. These applications generally involve models that are more subtle and complicated than the logistic growth models presented here. Nevertheless, even in very complicated models, chaos is accompanied by characteristics that are quite similar to the ones you have seen for the logistic growth model. A thorough understanding of this chapter should give you a pretty good idea of what chaos is.

This completes the discussion of chaos, and it is a fitting place to end the book. Throughout the book you have seen ever more sophisticated models, and how these models can be used to predict future conditions in real problems. For very simple models, it should not be surprising that very simple patterns appear. But as more sophisticated models are considered, the behavior can become more complicated. Logistic growth is very important in applications, and under suitable assumptions it behaves very nicely. But it is a complicated enough model to permit chaos to creep in. Chaos should not be viewed as a stumbling block for the modeling process, though. Instead, it provides additional opportunities for applying models. In some applications, it is enough to be aware that chaos can occur, and to know how to recognize the onset of chaos. In other cases, it may be possible to take steps to prevent the start of chaos. There are even applications where chaos is desirable, for it provides a rich range of possibilities at any time in the future. The details of how that all works out is a subject for another course.

Summary

This chapter has introduced the idea of chaos, and has shown how it can arise in logistic growth models. We began with a review of the various kinds of models that have been considered throughout the book. It was pointed out that for all of the applications that we considered, slight errors in values of parameters and data would have little impact on the general pattern of future predictions. Chaotic systems, in contrast, are highly sensitive to even the tiniest variations in parameters and data. This idea was dramatized using the butterfly effect, which refers to the idea that even an effect as seemingly trivial as the beating of a butterfly's wings can dramatically influence the predictions of a weather model. For a system that is not chaotic, whether the butterfly is included in the model or not, the predicted results would be essentially the same. But in the presence of chaos, the inclusion of a butterfly's wings can completely change the character of predictions made by the model.

Chaos only arises in one of the families of models we have studied in this book, the logistic growth models. We saw in this chapter that all of the other models involve difference equations that can be classified as linear. This use of the term linear does not refer to our usual graphs that show how the model evolves over time. Rather, there is a different kind of graph involved, one that shows the relationship between any two successive values determined by the difference equation. If this different kind of graph is linear, then the difference equation is called linear. All of the models we studied, except logistic growth, are characterized by linear difference equations. More generally, it is known that chaos only occurs in models with nonlinear difference equations.

The way chaos arises in logistic growth was explored through a series of examples. All the examples involved a single difference equation for a fish population model: $a_{n+1} = m(1 - a_n)a_n$. We looked at several values for m, observing increasingly complex patterns of long-term population behavior as larger and larger values of m were considered. Ultimately, for large enough m, it was seen that future prediction of populations became impossible. There were simply no observable patterns, and even the slightest changes in the initial population for the model produced significant changes in the long-term behavior of the model. This is characteristic of a chaotic model.

Although a chaotic logistic model does not lead to patterns that allow us to predict the future behavior of the model in any quantitative way, there is another kind of order that does appear. This is revealed by yet another kind of graph, showing the effect of the parameter m. Where all the previous graphs showed results for a single fixed value of m, in this final kind of graph, results were combined for many different values of m. The result illustrates how and when chaos occurs in the logistic model. Moreover, the graph reveals interesting and intriguing patterns within the chaotic behavior of the models. These features in the logistic model are observed in many kinds of chaotic models, which arise in a host of application areas, including economics, biology, aerodynamics, medicine, and many others.

Exercises

Reading Comprehension

1. What is meant by long-term predictions?

2. Explain what it means to say that long-term predictions for arithmetic growth, geometric growth, and mixed models are not very sensitive to small changes in parameter values. Contrast this with what can occur in a chaotic model.

3. Explain the butterfly effect, and relate it to the examples of the logistic fish population model in the chapter.

4. Explain what is meant by a linear difference equation. Give examples of linear difference equations, and of nonlinear difference equations.

5. What is the significance of linear and nonlinear difference equations for the study of chaotic difference equation models?

6. Using Fig. 14.14 and Fig. 14.15, it is possible to predict the long-term behavior of the fish population model $a_{n+1} = m(1 - a_n)a_n$ for various values of m. Do this for each part below.

   a. $m = 1.6$, $a_0 = .5$

   b. $m = 1.6$, $a_0 = .2$

   c. $m = 1.6$, $a_0 = .97$

   d. $m = 3.2$

   e. $m = 3.52$ (Use both figures)

   f. $m = 3.56$ (Use Fig. 14.15)

   g. $m = 3.8$ (Use both figures)

   h. $m = 3.84$ (Use Fig. 14.15)

7. Explain the concept of a self-similarity in a graph.

Solutions to Selected Exercises

Reading Comprehension. 6. In all of these problems, you have to find the value of m on the horizontal axis of the graph, then imagine a vertical line going up to the graph from that m. If the vertical line just hits one point of the graph, that must be a steady population value for that m. If there are two points, the population will go back and forth between them. If there are four points, there will be a cycle repeated among all four. And if there is just a cloud of points, the future behavior will be chaotic.

a. The population will level off at about .36, no matter what a_0 is.

d. With $m = 3.2$ there are two points on the graph, and the population will eventually alternate between them: .5 and .76, approximately.

e. There are 4 values for this m. The population will cycle among them, repeating every four years: .36, .5, .82, and .88, approximately.

f. For $m = 3.56$ there appear to be 8 values.

g. For $m = 3.8$, chaos!

h. It looks as though there might be about 3 points here. So the long-term behavior would likely be to jump back and forth among three different population sizes.

Index

arithmetic growth, 37
    application outline, 43
    compared to geometric growth, 181
    compared to other models, 320
    compared to quadratic growth, 81, 84, 102
    difference equation, 37
    functional equation, 41, 42
    negative growth, 38
    statement in box, 37
    sum, 97, 99
ascending order, 122
asymptote, 157, 158, 215

base of an exponential function, 213
bee ancestor example, 25
bends, 172
best-fit line, 165
    for transformed planet data, 245
    with transformed data, 244
break-even points, 136
Burgi, 223
butterfly effect, 330

carbon dioxide example, 1
center line for parabola, 124
chaos, 7, 319
    butterfly effect, 330
    future prediction, 330
    graph of onset, 332
    nonlinear difference equations, 324, 325
    patterns in, 336
    self-similarity, 337
    slight changes in parameters, 322
coefficients
    polynomials, 150
    quadratics, 122, 124
combinatorics, 105
common logarithm, 222
comparison of different kinds of models, 320
complex numbers, 128
computer network example, 92
    as quadratic function example, 118
continuous
    model, 55, 58
    variable, 12, 58, 195
contraction factor, 189
controversial results of mathematical models, 278
cost, 134

decibel scale, 238
deductive proof, 86

degree of a polynomial, 150
    even degree, 152
    odd degree, 152
demand, 39, 133
descending order
    numerator and denominator of a rational function, 159
    polynomials, 150
    quadratics, 122
difference equation, 6, 16
    and functional equation, 6, 41, 323
    arithmetic growth, 37
    computer model, 92
    different types compared, 102, 320
    direct study, 96
    family, 24
    for drug dosage example, 260, 261
    for Fibonacci numbers, 25
    for oil consumption, 15
    geometric growth, 181, 187
    initial value, 41
    linear and nonlinear, 323, 324
    loan payments, 268
    logistic growth, 291, 293–295
    logistic growth with harvesting, 303, 312
    mixed model, 262
    quadratic growth, 84
    water tank, 27
differences, 80
    second, 80
    third and higher, 156
discontinuity in rational function graph, 157
discrete
    data, 12
    model, 58
drug dosage example, 259, 266
    functional equation, 264

e, 225, 226
effective rate, 196
equations, functions, and expressions, 55
error function, 167
even degree polynomial, 152
exponent rules, 217
    in box, 220
exponential function, 193, 213
    asymptote, 215
    base, 213
    base e, 225
    graphs, 214
    linearized by logarithms, 243

no x–intercept, 215
solving equations, 220
y–intercept, 215
expressions, equations, and functions, 55

factor of a polynomial, 155
factored form, 132, 155
falling body example, 91
Fibonacci numbers, 25
and geometric growth, 186
fixed point, 266, 267
in logistic growth, 298
fixed population, 298
formula, 299
in logistic growth with harvesting, 307
functional equation, 6
and difference equation, 6, 41, 323
arithmetic growth, 41, 42
different types compared, 102, 320
for drug dosage example, 264
for oil consumption, 21, 100
geometric growth, 191, 192
inversion, 22, 103, 193, 221
loan payments, 269
mixed model, 263, 265
quadratic growth, 84, 89
functions, equations, and expressions, 55

Galileo, 91
Gauss, Karl F., 86
geometric decay, 197
geometric growth, 181
applications, 196
compared to arithmetic growth, 181
compared to other models, 320
continuous model, 195
difference equation, 181, 187
first definition in box, 184
functional equation, 191, 192
graphs, 187
growing smaller, 188
second definition in box, 185
statement in a box, 194
geometric progression, 256
geometric sum, 255–257, 271
global warming, 1
graphical method, 2
for carbon dioxide data, 3
for oil consumption data, 19
for quadratic equation, 103
graphs
exponential functions, 214
functions of two variables, 172
geometric growth, 187
linear equations, 61
polynomial functions and equations, 151
quadratic functions and equations, 122
quadratic growth, 101

rational functions, 157
relating two successive data values, 324
transition to chaos, 333
growth factor, 184
related to percentage, 185
straight-line model, 290
variable, 289
Gump scale, 239

half-life, 197
hills, 172
horizontal asymptote, 158, 215
house construction example, 97

initial value, 41
intercept
x–intercept
quadratic graph, 125
x-intercept
exponential function graph, 215
polynomial graph, 153
rational function graph, 158
y-intercept
exponential function graph, 215
linear graph, 62
polynomial graph, 153
quadratic graph, 124
interest
compound, 196
simple, 39
invariant, 17
inverting a function, 59, 131
and roots, 126
inverting a functional equation, 22, 193, 221
irrational number
base e, 225
from a square root, 128
roots of polynomials, 153, 155
isotope, 197

Kepler, Johannes, 247

linear
difference equation, 323, 324
equation, 55
graphs, 61
many solutions, 60
no solution, 60
point–slope form, 66
slope, 62
slope–intercept form, 64
two–intercept form, 67
y–intercept, 62
function, 55
model, 69
statement in box, 56
another statement in box, 57
linear regression, 175
linearization, 234, 243

of exponential functions, 243
of power functions, 243
loan payments, 256, 267
  difference equation, 268
  functional equation, 269
  parameters, 270
logarithm
  base 2, 224
  base e, 225
  common, 222
  common, and decimal digits, 236
  natural, 225
  rules, 234
  rules in box, 236
  transforming data, 243
logarithmic functions, 222, 233
logarithmic scales, 233, 236, 242
  decibel scale, 238
  Gump scale, 239
  number line, 237
  pH scale, 238, 241
  Richter scale, 238
Logistic Growth, 287
logistic growth, 287
  chaos, 319
  compared to other models, 320
  difference equation, 291, 293–295
  fixed population, 298
  fixed population formula, 299
  leveling off, 291, 297, 300, 302
  no functional equation, 293, 303
  not leveling off, 300, 326
  parameters, 293, 294
  statement in box, 293
  with harvesting, 303
    difference equation, 303, 312
    fixed population, 307
    no fixed populations, 310
    starting population, 309

magic carpet, 172
mathematical model, 5, 321
  compared to true data, 22
  computer network, 92
  continuous, 55, 58, 195
  controversial results, 278
  discrete, 58
  drug dosage, 259, 266
  falling bodies, 91
  families compared, 320
  family, 24, 79
  geometric growth, 181
  house contruction, 97
  linear, 69
  loan payments, 267
  logarithmic function, 233
  logistic growth, 287
  mixed, 255

polynomial function, 156
  quadratic growth, 79
  rational function, 159
mixed model, 255, 262
  compared to other models, 320
  difference equation, 262
  fixed point, 266
  fixed point formula, 267
  functional equation, 263, 265
  leveling off, 260, 265
  loan ayments, 267
  not leveling off, 270
  oil consumption, 274
  parameters, 261
  sum, 277
mole, 241

Napier, 223
natural log, 225
Newton, Sir Isaac, 247
nominal rate, 196
nonlinear difference equation, 323, 324
nonlinearity
  quadratic functions, 123
  quadratic growth, 101
numerical method, 2
  for best-fit line, 169
  for carbon dioxide data, 4
  for oil consumption data, 19
  for quadratic equation, 103

odd degree polynomial, 152
oil consumption example, 11
  and arithmetic growth, 37
  and geometric growth, 186
  as quadratic function example, 117
  choosing the best line, 68
  difference equation, 15
  mixed model, 255, 274
  reserves
    from arithmetic model, 46
    from mixed model, 256, 274
    quadratic growth model, 99
    running out of oil, 278
order of magnitude, 237

parabola, 102, 123
paraboloid, 174
parameter, 24
  best value, 24
  drug dosage model, 266
  effect on predictions, 322
  geometric growth, 187
  loan payment example, 270
  logistic growth, 293, 294
  mixed model, 261
  quadratic growth, 84
  school enrollment example, 273
Pascal's triangle, 105

pH scale, 238, 241
planet data example, 244
point-slope equation, 66
polynomial approximation
    for exponential functions, 227
    for square roots, 156
polynomial function, 149
    coefficients, 150
    degree, 150
    descending order, 150
    even degree, 152
    factored form, 155
    graphs, 151
    odd degree, 152
    roots, 154
    solving equations, 153
    terms, 149
    wiggles, 153
    x-intercepts, 153
    y-intercept, 153
power function linearized by logarithms, 243
proof, 86
proportional reasoning, 44, 55, 70
    invalid for quadratic functions, 123
    invalid for quadratic growth models, 94, 101

quadratic equation
    in quadratic growth models, 103
    no solution, 128, 130
    sample applications, 119
    statement in box, 120
quadratic formula, 130
quadratic function, 120
quadratic graphs
    center line, 124
    high and low points, 124
    in three dimensions, 172
    summary in box, 125, 131
    vertex, 124
    x-intercepts, 125
    y-intercept, 124
quadratic growth, 79, 80
    and proportional reasoning, 94
    and sums of arithmetic growth models, 99
    applications, 91
    compared to arithmetic growth, 81, 84, 102
    compared to other models, 320
    difference equation, 84
    falling bodies, 91
    functional equation, 84, 89
    graphs, 101
    house construction example, 97
    network problems, 92
    nonlinearity, 101
    parameters, 84
    statement in box, 80
    another statement in box, 81
    another statement in box, 84

radioactive decay, 197
rational function, 149
    asymptotes in graph, 157
    discontinuity in graph, 157
    graphs, 157
    horizontal asymptote, 158
    vertical asymptote, 158
    x-intercepts, 158
recursion, 6, 19
revenue, 134
Richter scale, 238
root, 125
root of a polynomial, 154
roundoff error, 168
rules of exponents, 217
    in box, 220
rules of logarithms, 234
    in box, 236
running total, 97

sales revenue model, 118
second differences, 80
self-similarity, 337
sequence, 12
    terms, 12
shortcut
    sums of geometric progressions, 257, 271
    sums of whole numbers, 86, 87, 104
shrinkage factor, 189
sigma notation, 87, 106
slope, 62
slope-intercept equation, 64
soda demand example, 40
    costs, revenue, and profit, 133
    functional equation, 42
    rational function model, 160
solving equations, 59
    exponential, 220, 224, 226
    graphical and numerical methods, 126, 153, 155, 220, 277
    in factored form, 132
    linear, 59
    polynomial, 153
    quadratic, 126, 130, 131, 139
square root, 128
squared error, 168
subscript, 12
    variable, 13
sum
    for arithmetic growth model, 97
    for geometric growth model, 255, 256, 271
    for mixed model, 277
    shorthand notation, 86
    sigma notation, 87
    three-dot notation, 87
    whole numbers, 104
sums, 97
surface, 172

symmetry
> in geometric growth, 190
> in quadratic graph, 122

terms in polynomial functions, 149
terms in quadratic functions, 119
terms of a sequence, 12
theoretical method, 2
> for carbon dioxide data, 4
> for oil consumption data, 21

three dot notation, 87
total error, 167, 168
transforming data, 234, 243
two–intercept equation, 67

valleys, 172
vertex of a parabola, 124
vertex of a three-dimensional parabola, 174
vertical asymptote, 158

water tank example, 27, 181, 188
wiggles in polynomial graphs, 153
wrinkles, 172

yield, 196